MOBILITY AND MIGRATION CHOICES

The crossing of national state borders is one of the most-discussed issues of contemporary times and it poses many challenges for individual and collective identities. This concerns both short-distance mobility as well as long-distance migration. Choosing to move – or not – across international borders is a complex decision, involving both cognitive and emotional processes. This book tests the approach that three crucial thresholds need to be crossed before mobility occurs; the individual's mindset about migrating, the choice of destination and perception of crossing borders to that location and the specific routes and spatial trajectories available to get there. Thus both borders and trajectories can act as thresholds to spatial moves.

The threshold approach, with its focus on processes affecting whether, when and where to move, aims to understand the decision-making process in all its dimensions, in the hope that this will lead to a better understanding of the ways migrants conceive, perceive and undertake their transnational journeys. This book examines the three constitutive parts discerned in the cross-border mobility decision-making process: people, borders and trajectories and their interrelationships. Illustrated by a global range of case studies, it demonstrates that the relation between the three is not fixed but flexible and that decision-making contains aspects of belonging, instability, security and volatility affecting their mobility or immobility.

Martin van der Velde is Associate Professor at Radboud University, Nijmegen. He is co-founder of the Nijmegen Centre for Border Research and was co-editor of the Journal of Borderland Studies. *In 2014 he was president of the Association for Borderlands Scholars.*

Ton van Naerssen is retired Associate Professor of Development Geography at Radboud University, Nijmegen and currently works as a freelance researcher and consultant on migration and development issues.

T0331390

Border Regions Series

Series Editor: Doris Wastl-Walter, University of Bern, Switzerland

In recent years, borders have taken on an immense significance. Throughout the world they have shifted, been constructed and dismantled, and become physical barriers between socio-political ideologies. They may separate societies with very different cultures, histories, national identities or economic power, or divide people of the same ethnic or cultural identity.

As manifestations of some of the world's key political, economic, societal and cultural issues, borders and border regions have received much academic attention over the past decade. This valuable series publishes high quality research monographs and edited comparative volumes that deal with all aspects of border regions, both empirically and theoretically. It will appeal to scholars interested in border regions and geopolitical issues across the whole range of social sciences.

Mobility and Migration Choices

Thresholds to Crossing Borders

Edited by

MARTIN VAN DER VELDE

TON VAN NAERSSEN

Routledge
Taylor & Francis Group

LONDON AND NEW YORK

First published 2015 by Ashgate Publishing

2 Park Square, Milton Park, Abingdon, Oxfordshire OX14 4RN
711 Third Avenue, New York, NY 10017, USA

Routledge is an imprint of the Taylor & Francis Group, an informa business

First issued in paperback 2018

British Library Cataloguing in Publication Data
A catalogue record for this book is available from the British Library.

Library of Congress Cataloging-in-Publication Data
Mobility and migration choices : thresholds to crossing borders / by Martin van der Velde and Ton van Naerssen.
        pages cm. -- (Border regions series)
    Includes bibliographical references and index.
    ISBN 978-1-4094-5803-6 (hardback : alk. paper) -- ISBN 978-1-4094-5804-3 (ebook) --
    ISBN 978-1-4724-0762-7 (epub)
        1. Emigration and immigration--Social aspects--Case studies. 2. Emigration and immigration--Psychological aspects--Case studies. 3. Immigrants--Psychology--Case studies. 4. Ethnopsychology--Case studies. 5. Identity (Psychology)--Social aspects--Case studies. 6. Assimilation (Sociology)--Case studies. I. Velde, M. van der (Martin) editor of compilation. II. Naerssen, A. L. van, editor of compilation.

JV6225.M637 2015
304.8--dc23

2015004240

ISBN 13: 978-1-4094-5803-6 (hbk)
ISBN 13: 978-1-138-54696-7 (pbk)

Contents

List of Figures and Tables

Figures

Tables

Notes on Contributors

Maruja M.B. Asis is director of research and publications at the Scalabrini Migration Center, Quezon City, the Philippines. She is a sociologist who has been working on migration and social change in Asia for many years. She is currently involved in several research projects dealing with migration and development, among others on youth, employment and migration in the Philippines, and capacity-building of migrant associations and Philippine government institutions as development partners. She is co-editor of the *Asian and Pacific Migration Journal*.

Fabio Baggio is missionary of the Scalabrinian Congregation. From 1998 to 2002, he worked as researcher in the Center for Latin American Migration Studies (CEMLA), in Buenos Aires. He was director of the Scalabrini Migration Center (SMC) in Manila and co-editor of the *Asian Pacific Migration Journal* and *Asian Migration News* from 2002 to 2010. At present, he is ordinary professor at the Pontificia Universitas Urbanian in Rome, invited professor at the Universidad de Valencia Spain, and director of the Scalabrini International Migration Institute in Rome.

Doaa' Elnakhala is a freelance researcher of the Middle East and North Africa and a contributing analyst at Wikistrat. She has a PhD in political science from the University of Texas at Austin, USA. She studied theories of international relations and their application to the Palestinian–Israeli conflict, and EU policies toward the Middle East and North Africa. Her dissertation is on the security outcomes of barriers on borders. Within the context of this research, she studied the Israeli barriers around the Gaza Strip, Jerusalem and the West Bank. Lately, she has developed an interest in physical barriers on borders between communities.

Keina R. Espiñeira is a PhD student at the Political Science Department, Universidad Complutense de Madrid. She obtained an MA in Documentary Cinema from the Universidad de Alcalá (2014). Her research interests lie in postcolonial theories, mobility, border imaginaries and border art. She is currently working in the Geography Department at the Universitat Autònoma de Barcelona within the EU FP7 research projects EUBORDERREGIONS and EUBORDERSCAPES, both focusing on Spanish–Moroccan border dynamics.

Pol Fàbrega is senior associate at Teach For All, a network of 35 partner organizations from different countries with a shared vision of expanding educational opportunities. Prior to this, he worked with a number of academic and non-profit organizations in both Asia and Europe, including the Institute of Economic Research (Berlin) and the University of Hong Kong. In 2012, he conducted a research for a group of NGOs on migrants' vulnerability and mobility patterns between Cambodia and neighbouring countries. He holds a master's degree in international relations from Warwick University (UK) and bachelor's degree in political science and public administration from the Universitat Autònoma de Barcelona, Spain.

Xavier Ferrer-Gallardo is a postdoctoral researcher at Universitat Autònoma de Barcelona. He works in the EU FP7 projects EUBORDERREGIONS: European Regions, EU External Borders and the Immediate Neighbours; Analysing Regional Development Options through Policies and Practices of Cross-Border Co-operation and EUBORDERSCAPES: Bordering, Political Landscapes and Social Arenas; Potentials and Challenges of Evolving Border Concepts in a post-Cold War World. Earlier, he published on the border between Spain and Morocco in journals like *Cultural Geographies*, the *Journal of European Urban and Regional Studies* and *Political Geography*.

Graeme Hugo was ARC Australian Professorial Fellow, Professor of the Discipline of Geography, Environment and Population and Director of the Australian Population and Migration Research Centre at the University of Adelaide. His research interests included population issues in Australia and South East Asia, especially migration. He authored over three hundred books, articles in scholarly journals and chapters in books, as well as a large number of conference papers and reports. In 2002 he secured an ARC Federation Fellowship over five years for his research project, 'The new paradigm of international migration to and from Australia: dimensions, causes and implications.' Graeme died during the final stages of the book production.

Alexander Izotov is a PhD student at the Karelian Institute in Joensuu, of the University of Eastern Finland. He obtained an MA in history from the University of Petrozavodsk, Russian Karelia, Russia (1980), a Candidate of historical sciences from the St. Petersburg State University, Russia (1990), and an MSc in human geography from the University of Eastern Finland (1998). Since 1995 he works as a researcher in the Karelian Institute. He is interested in regional studies and cross-border interaction. More currently he focuses on the process of identity construction in the Finnish–Russian border area.

Zaheera Jinnah is an anthropologist and researcher at the African Centre for Migration and Society, University of the Witwatersrand, Johannesburg, where she also teaches a course on migration and human rights. Her research interests include labour migration and the Somali Diaspora.

Victor Konrad is director of NorthWing Consulting, Ottawa, and adjunct research professor of Geography and Environmental Studies at Carleton University. During the 1970s and 1980s, he was a professor of geography and anthropology at the University of Maine, and director of the Canadian–American Center. He was recipient of the Donner Medal. Together with Nicol, he published *Beyond Walls: Re-Inventing the Canada-United States Borderlands* (2008). He serves on the Board of Directors of the Association of Borderlands Studies and the International Advisory Board of the *Journal of Borderlands Studies*.

Maggi W.H. Leung is associate professor of International Development Studies at the Department of Human Geography and Planning at Utrecht University, the Netherlands. Before, she worked at various universities in Hong Kong and Germany. Her research interests include international, transnational and translocal mobilities and development, the Chinese diaspora in Europe, and the internationalization of education. She has published in a range of geography and social science journals. Her book *Chinese Migration in Germany: Making Home in Transnational Space* (2004) earned her an award from the Chinese University of Hong Kong.

Helena Lim is a programme manager with the Global Institute For Tomorrow (GIFT), a Hong Kong-based think tank dedicated to executive education. Prior to GIFT, she worked on public policy issues and intercultural dialogue within the Council of Europe in France and the Asia-Europe Foundation in Singapore and conducting fieldwork in Southeast Asia in the area of migration and human trafficking. She is currently engaged in initiatives that re-examine the role of business in society and promote more effective collaboration across sectors. Helena obtained her BA and MA at the University of East Anglia (UK).

Caven Jonathan Napitupulu is a PhD candidate at the Department of Geography, Environment and Population at the University of Adelaide, Australia, and at the final stage of writing his dissertation on transit migration in Indonesia. He obtained a master's degree in public administration from the Flinders University of South Australia and a Diploma from the Indonesian Immigration Academy. His research interests are in refugee issues. He works as an Indonesian immigration officer and is currently posted at the Cross-border and International Cooperation Desk in Jakarta and involved in various international government forums on refugees.

Gery Nijenhuis is assistant professor of International Development Studies, Utrecht University. Her research interests include international migration, local governance and local development, and codevelopment policies in Europe. She currently leads the research project 'Committed Diaspora', on the contribution of migrant organizations to development. Among others, she published in *Environment and Urbanization* and the *International Development Planning Review*.

Ninna Nyberg Sørensen is senior researcher at the Danish Institute of International Studies (DIIS). She has worked on migration–development issues for most of her career and published widely on transnational migration, conflict and gender, for example *Work and Migration* (London 2002), *The Migration–Development Nexus* (Geneva 2003), *Living across Worlds* (Geneva 2007), and *The Migration Industry and the Commercialization of International Migration* (London 2013). Her recent work explores undocumented migration of Central Americans, the insecurities migrants encounter and the consequences of massive deportation on local communities and sending states.

Roos Pijpers is assistant professor at the Department of Geography, Environment and Spatial Planning at Radboud University, Nijmegen. Her primary research interests centre on the geography of work and employment, with an emphasis on labour migration. More generally, she draws on various approaches to economic and social geography to understand the changes in the global economy and society, and the localized consequences for working people. The latter include people's strategies to incorporate and resist these changes.

Marie Sandberg (PhD) is associate professor at the Ethnology Section, Saxo Institute, University of Copenhagen. Her research focuses on labour migration within the EU, ethnographies on border practices and experiences of borders in everyday life Europe. Among her recent publications is *The Border Multiple: The Practicing of Borders between Public Policy and Everyday Life in a Re-scaling Europe* (2012), co-edited with Andersen and Klatt. She is steering board member of the Centre for Advanced Migration Studies (AMIS), University of Copenhagen, and editor of *Ethnologia Europaea: Journal of European Ethnology*.

Joris Schapendonk is assistant professor at the Department of Geography, Environment and Spatial Planning at Radboud University, Nijmegen. His current research focuses on the dynamics of African migration, the border politics of the European Union, and the geography of grassroots welcoming acts towards undocumented migrants. His research has been published in, among others, *Annals of the Association of American Geographers* (2014), *Migration Studies* (2014) and *Population, Space and Place* (2014).

Lothar Smith is assistant professor of Human Geography at the Department of Geography, Environment and Spatial Planning at Radboud University, Nijmegen. He has a special interest in the globalization–development nexus and its multifarious influences on the Global South. His publications include 'Constructing Homes, Building Relationships: Migrant Investments in Houses' (2009, with Mazzucato) and *Women, Gender, Remittances and Development in the Global South* (co-editor, 2015).

Tiina Soininen is a PhD student at the Karelian Institute of the University of Eastern Finland, Joensuu, Finland. She obtained an LSc in social sciences at the University of Eastern Finland (2011) and an MSc from the University of Joensuu (2003). Since 2001 she has worked as a researcher and lecturer at the University of Eastern Finland and the University of Tampere, Finland. Her topics include statistical modelling, functions and structures of labour markets, public administration, transportation and cross-border interaction.

Bianca B. Szytniewski works as a PhD researcher in human geography at Utrecht University and Radboud University, Nijmegen. She holds an MA in international relations and European Union Studies and her research interests lie in 'othering', encounters with differences, cross-border mobility and the political and sociocultural development of Europe, in particular Central and South Eastern Europe. Currently, she is working on the topic of unfamiliarity in the context of cross-border consumers' mobility in European borderlands.

Martin van der Velde is associate professor at Radboud University, Nijmegen. His current research interests centre around border-related issues, especially with regard to labour markets, consumer behaviour, migration and European integration. He is also co-founder of the Nijmegen Centre for Border Research and was co-editor of the *Journal of Borderland Studies*. In 2014 he was president of the Association for Borderlands Scholars.

Henk van Houtum is co-founder and co-director of the Nijmegen Centre for Border Research, associate professor of Political Geography and Geopolitics at Radboud University, Nijmegen and research professor of Geopolitics of Borders at Bergamo University, Italy. His most recent border book is *Borderland: Atlas, Essays and Design; History and Future of the Border Landscape* (2013).

Ton van Naerssen is retired associate professor of Development Geography at Radboud University, Nijmegen and currently works as freelance researcher and consultant on migration and development issues. Among others, he was co-editor of *Asian Migrants and European Labour Markets* (2005), *Global Migration and Development* (2008) and *Women, Gender, Remittances and Development in the Global South* (2015).

List of Foreign Words and Phrases

amana	The whole creation
buufis	To inflate or blow (Somali)
Centros de Internamiento de Extranjeros (CIE)	Internment Centres for Foreigners
chabuk chabuk	'Do it quickly' or 'hurry up' in the Turkish language
kat, or *meera*	Herbal substance with narcotic properties commonly used in Eastern Africa and parts of the Gulf
mekhal	Brokers
milagro económico español	Spanish economic miracle
mojados	Undocumented
Padrón Municipal	Municipal population registers
peidu mama	'Study mamas', women who accompany their children to receive education overseas
purdah	Seclusion
repartamiento	Coerced labour
Ummah Islamia	Islamic world that exists beyond the boundaries of countries
viudas blancas	White widows

List of Abbreviations

ADB	Asian Development Bank
APRODE	Asociación Pro Mejoramiento de los Deportados de Guatemala
BPRI	Border Policy Research Institute
CDRI	Cambodian Development Research Institute
CETI	Centre of Temporary Stay for Immigrants
CIE	Centro de Internamiento de Extranjeros /Internment Centre for Foreigners
CONAMIGUA	National Council for Migrant Attention (original Spanish abbreviation)
DHS	Department of Homeland Security
DIMIA	Department of Immigration and Multicultural and Indigenous Affairs
EC	European Commission
EU	European Union
EUROSUR	European Border Surveillance System
FAFG	Fundacion de Antropologia Forense de Guatemala
FLASCO	Facultad Latinoamericana de Sciencias Sociales
GCIM	Global Commission on International Migration
GFMD	Global Forum on Migration and Development
GMS	Greater Mekong Subregion
IDP	Internal Displaced People
ILO	International Labour Organization
IMA	Irregular Maritime Arrivals
IMTC	International Mobility and Trade Corridor Program
INCEDES	Instituto Centroamericano de Estudios Sociales y Desarollo
INEDIM	Instituto de Estudios y Divulgación sobre Migración
IOM	International Organization for Migration
MENAMIG	Mesa Nacional para las Migraciones en Guatemala
MoU	Memorandum of Understanding
NAFTA	North American Free Trade Agreement
NCBR	Nijmegen Centre for Border Research

NELM	New Economics of Labour Migration
NERG	North East Research Group
NES country	non-English Speaking country
PA	Palestinian Authority
PFLP	Popular Front for the Liberation of Palestine
PIJ	Palestinian Islamic Jihad
PLO	Palestinian Liberation Organization
PRC	Popular Resistance Committees
RABIT	Rapid Border Intervention Teams
SEZ	Special Economic Zones
TPS	Temporal Protection Status
TRANSCODE	Transnational Synergy and Cooperation for Development
UN	United Nations
UNHCR	United Nations High Commissioner for Refugees
WHM	Worker Holiday Maker

Preface

This book focuses on the relationship between the decision-making process of people who want to move across geographical borders and their connotations of the borders and the trajectories to their destinations. There are various theories that concern the decision-making of people who move to other places, regions and countries but there is less theorizing on how they perceive and interpret borders and/or geographical routes. Moreover, the theories usually deal with the factors that impact on the decision to move and not with the process that leads to a decision as such. The latter was what we did in 2011 when we published an article in the geographical journal *Area* about (im)mobility in the expanded European Union and launched 'the threshold model' as a tool to analyse cross-border mobility decision-making.

When Ashgate approached us to publish a book in the Border Region Series, it was agreed that we would strive for an edited volume focusing on the model with the objective to clarify its strong and weak points by way of empirical studies from all parts of the world. The subsequent responses from colleagues were encouraging and we decided to hold a workshop to discuss the proposed contributions. For this reason, in October 2012 book contributors and invited resource persons came together in Soeterbeeck near Nijmegen in the Netherlands. During the first day presentations and abstracts of the book chapters were discussed. On the second day in Utrecht discussions of migration and development issues took place among a broader audience of both academicians and practitioners in the field of international migration.

The central theme of the workshop was the threshold model for mobility and migration. During a lively discussion critical voices were raised regarding the focus of the model on labour migration, the lack of inclusion of social and institutional elements and the definition of terms. It became clear that 'model' was a too static notion and therefore we replaced the term with 'approach' that also allowed contributors to the book to associate more freely and move with more flexibility around our ideas. That is what they did and the result is the present book with a great diversity of empirical studies and reflections on the threshold approach. For us, as we explain in the last chapter of the book, the contributions gave reason to a further clarification, expansion and refinement of the approach.

We would like to express our appreciation to all contributors to this book. Whether they received the threshold approach positively or were rather sceptical about its explanatory value, they all patiently contributed to the book. We wish to thank the organizers of the October 2012 events: the department of Geography, Planning and Environment of the Nijmegen School of Management at Radboud University, Nijmegen and the programme Transnational Synergy and Cooperation for Development (TRANSCODE). We also wish to thank Ashgate, in particular Katy Crossan and her colleagues for their trust and confidence in this book and all our colleagues for the comments and suggestions for improvement of which we have benefited.

Our thanks also go to Agnes Khoo who did the proof and language editing of several chapters and to Marisha Maas, the technical editor, who went accurately and meticulously through all chapters. They both could not resist the urge to communicate their own critical remarks on the chapters they edited and often we and the authors accepted their suggestions. Our appreciation also goes to all others who indirectly contributed to the chapters in this book. Finally, we hope and expect that the readers will recognize their own experiences in the different stories of this book.

Martin van der Velde and Ton van Naerssen
Nijmegen, May 2015

PROLOGUE

Chapter 1

The Thresholds to Mobility Disentangled

Ton van Naerssen and Martin van der Velde

The crossing of national state borders is one of the most-discussed issues of contemporary times and it poses many challenges for individual and collective identities. In the article 'People, Borders, Trajectories: An Approach to Cross-Border Mobility and Immobility in and to the European Union', published in the geographical journal *Area* in 2011, we focused on the decision-making process of potential and real labour migrants (Van der Velde and Van Naerssen 2011). Why, how and where do people cross borders in search of employment away from their home country? Structural approaches at the macro level, such as the various push and pull models, are very economics-oriented, even if social factors are taken into account, such as the availability of educational and health facilities and intervening factors, most notably state policies. Various theories introducing the notion of 'bounded rationality' or the idea of 'transaction costs' still stay within this framework that considers migrant behaviour to be guided by rational decision-making processes based on structural economic differences between areas. Other approaches, however, take the migrant as an actor within a social environment as the point of departure and argue that the existence of social networks across borders (Boyd 1987; Faist 2000) or the family and kin (the New Economics of Labour Migration) impact on migrants' decision-making.[1]

All of these approaches[2] explain the major accommodating factors that impact on an individual's decision to migrate but they tend to focus on why people move and not why they do not. To us, the latter question is as relevant as the former, since it is important to note that the majority of people do not cross borders even if their living circumstances are unfavourable. It is usually a minority that decides to move and work abroad for six months or more.[3] Moreover, regardless of the outcome of such a decision-making process – to move or not to move – the prime interest was with how the *process* itself takes place. Finally, our mobility and migration approach includes the dimensions of the *perception* of spaces, borders and routes, and the personal *meanings* people ascribe to these geographical notions; we want to analyse how these impact on cross-border mobility decisions.

Our interest in perception and meaning has been constituted by earlier discussions with and (co-) publications of members of the Nijmegen Centre for Border Research (NCBR). More specifically, these concern the social production of 'spaces of differences and indifferences', meaning that people move easily in and to spaces that are familiar to them and are not as interested or are

---

1 Also migrants are sometimes motivated to leave the country by less tangible beliefs, such as trust in a higher being. It was not only unemployment that pushed young Senegalese in 2006–07 to take the boat to the Canary Islands but also their faith, despite knowing the risks involved. As they said, 'Tous depend de Dieu' (Adjamah 2012).

2 For an overview of the different approaches, see Hoerder (2002, Chapter 1), Castles and Miller (2009, 20–33) and Van der Velde and Van Naerssen (2011).

3 At the global level, only 3 per cent of the population (including dependents of labour migrants) is born outside the country of their residence while a recent Eurobarometer confirms that only 17 per cent of EU citizens envisage working abroad (see UN 2011; EC 2010).

simply indifferent to other types of spaces: 'That what is beyond the self-defined differentiating border of comfort (*difference*) is socially legitimised to be neglected (*indifference*)' (Van Houtum and Van Naerssen 2002, 129; italics by the authors). Related is the concept of a bordered *space of belonging*: the importance for people to belong somewhere or to be at home in a specific locality or region. The notion of a space of difference or belonging is the consequence of the mental distance created between places and spaces on both sides of the border. On one side of the border, the space of belonging refers to the mental nearness to the other inhabitants – the 'we' in the 'here' and a space of comfort in which one easily moves into, whilst on the other side of the border there is another 'world' – 'they' in the 'there' (Van Houtum and Van der Velde 2004).

It is not surprising that these ideas have been developed in a research centre (among others) that focuses on border regions. In border regions, the distance to the other side of the border is relatively easy to bridge. Moreover, on both sides of the border, differences, as well as cultural and historical similarities, coexist. The other side of the border is familiar and unfamiliar at the same time and by consequence, both barriers and opportunities exist for mobility, such as tourism, trading and shopping. The challenge is to bring concepts associated with borders and border regions into the realm of (long-distance) migration studies. The notions of difference (belonging) and indifference (concerning 'the others') are indeed transferable since the existence and growth of migrant or transnational communities increasingly create a 'we' in 'there' and spaces of belonging that allow, facilitate and initiate cross-border migration (Madsen and Van Naerssen 2003; Collier 2013, 27–53). If one accepts the idea that the spatial is integral to the social, which implies that the former is not merely the resultant of the social but also its producer (Massey 1994, 2005), one understands that borders and spaces of difference and indifference also engender social processes and that both the social and the spatial are continuously susceptible to change.

Our article in *Area* was one of the steps in an ongoing endeavour to outline a geographical approach to cross-border mobility and migration decision-making by individuals. Our approach certainly does not exclude family members, friends and/or transnational social networks that influence the decision-making process. However, we are of the opinion that ultimately, it is the individual who decides. Considering this, our approach to cross-border mobility consists of three constitutive parts: people, borders and trajectories. These match the three crucial thresholds, which are to be crossed in the decision-making process. The individual's state of mind or disposition towards mobility has to be taken into account as well as the intended goal and destination of the mobility. Moreover, mobility involves a route or trajectory.

First of all, it should be in the mindset of *people* to become a migrant. Based on the observation that even in times of crisis, a large proportion of the population remains immobile (see, for instance, Van der Velde 2014), already in the late 1980s Straubhaar (1988) developed the so-called 'insiders advantage approach' that proposes that an economic value be attached to remaining immobile, for example, the acknowledgement and acceptance of one's diplomas, skills and competences. If the employer on the other side of the border is not interested in these credentials, the potential migrant has to weigh the profits that can be made in moving against the loss of competences. This is known as a 'keep factor'. Likewise there are 'repel factors', connected to the potential new place of work that may prevent a move. In the latter's case the individual will decide against moving. However, keep and repel factors are not always 'objectively' measurable, since they are enmeshed with feelings and senses of belonging. Hence, we call this the *mental threshold*.[4]

---

4   Strictly speaking, the name is a misnomer since the entire decision-making process is a mental one. However, we have retained the name, given that this is the one the contributors to this book know from the *Area* article.

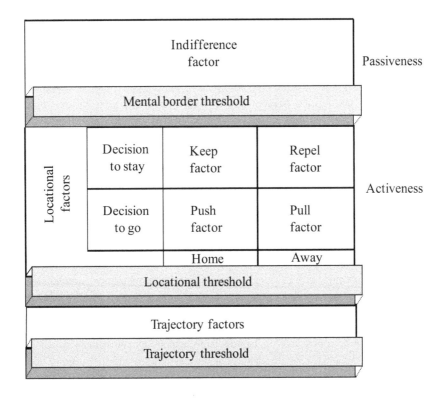

Figure 1.1   Factors and thresholds for spatial migratory behaviour

Source: partly based on Van Houtum and Van der Velde 2004.

Second, people can be mobile or immobile, depending on the destination they are considering during the decision-making process. The destination of migration is not chosen at random. There tends to be a certain familiarity with the destination in question. For instance, the availability of supportive social networks and/or similarities in language or religion are contributing factors to migration. *Borders* also play a role. If the border control is very strict, to the extent of physical barriers like fences and walls, people may give up the idea of leaving or they may choose an alternative destination. Once the destination is decided, however, the migrant has passed a second threshold, known as the *locational threshold*.

Another stage in migrant decision-making concerns the *route* to the destination. This tends to be overlooked in migrant decision-making models even though in reality, it is of critical importance. A potential migrant should think carefully how he or she will reach a chosen destination. This includes considering the safety of the route, by connecting (or not) to intermediaries who will arrange the travel and by gathering resources to pay for the journey, which often leads to incurring high debts. Thus, a potential mover can change the route out of safety consideration or concern about the high cost involved. This will induce a feedback mechanism whereby the decision to stay or to deploy a new destination (or none) can occur. For instance, many Filipino labour migrants would prefer to leave for EU countries but due to the high expenditures required, they end up in neighbouring countries, such as Malaysia. Of course, this mechanism also works the other way round as 'well-serviced' routes might favour certain locations over others. Once a decision is made on the trajectory to follow, the *trajectory threshold* is passed. In that sense, mobility and immobility are relative notions since they are related to specific destinations. For a potential migrant, there are options. As for the spatial trajectories, these also comprise certain aspects of instability, insecurity and volatility, which can affect their mobility. Weighing the pros and cons of alternatives during the decision-making process includes both 'hardware' (knowledge, finance, social networks) and 'software' (feelings of belongings, perceptions of opportunities and risks and self-confidence).

In our opinion, the decision-making process that concerns mobility and migration movements needs to go beyond reductionist explanations that consider motivations to migrate as a matter of rational and measurable choice. That means, decision-making involves both cognitive and emotional processes (Edwards 1954; Pfister and Böhm 2008). Hoerder (2002) therefore, speaks about the necessity of a 'holistic material-emotional approach'. The threshold approach, with its focus on processes affecting whether, when and where to migrate, aims to understand the decision-making process in all its dimensions. In our opinion, this will lead to a better understanding of the ways migrants conceive, perceive and undertake their transnational journeys.

Conceiving decision-making as a mental process that involves crossing the three stages (thresholds), does not imply that mobility is a linear movement from the origin to the destination. As a matter of fact, many migrants have a clear plan where and how to move, such as travelling over land, by sea or air to their intended destination. Labour migrants from Latin America pretending to be tourists will fly to Spain and many seasonal workers from Central and Eastern Europe will take the bus that brings them to one of the Western EU states within a day. But there are also situations in which a migrant has to reconsider his/her decision, especially when the journey is long or the migrants concerned have only a vague idea about the destination and the route to take. A migrant may meet employment opportunities along the way or the borders may act as unexpected obstacles preventing him or her from reaching the desired destination. It can also happen that smugglers are involved who unscrupulously abandon their clients before arriving at the said destination (Pang 2007; Schapendonk 2009).

The publication of our article and discussions with our colleagues and students have compelled us to test our approach empirically. Moreover, first, our approach is concerned with labour migration within the EU particularly, in view of the addition of new EU member countries, as well as increased migrant numbers from the Middle East and Africa to Europe. This raises the question whether our approach can be expanded to (1) other types of mobility and migration and (2) other regions in the world. Henceforth, we invited scholars to test our approach on the types of mobility and regions that are of their interest. This has resulted in 16 case studies, which we have organized into three parts, according to what we consider as the core interest of the contributor: people's options, borders or spatial trajectories. However, it should be noted that it was not always possible to classify a contribution neatly under one of the three labels. In fact, the parts are overlapping fields within one approach.

People

The first part of the book is entitled 'Mobility as an Option' and comprises five chapters. They all focus on people, in particular the reasons for their indifference or motivation to cross borders. Some of the reasons are one's familiarity with the other side of the border, the existence of social networks, the social and institutional environment and gender.

In the first contribution (Chapter 2), Szytniewski uses the concepts of familiarity and 'otherness' to explain cross-border mobility in a Polish bazaar city near the German–Polish border. The current border originated in 1945 and many Germans were expelled while the Poles settled down in the current Polish border region. New communities emerged on both sides of the border and people were largely unfamiliar with their neighbours across the border. After the fall of the Berlin Wall (1989) and the opening of the borders, people who live in the border region became mentally less distant and unfamiliar with the places on the other side of the border. This is further accentuated by circular mobility in the border region. At present, people know what to expect on the other side of the border and use this knowledge to decide if they will cross the border. On the other hand, border crossings no longer cause as much unease and avoidance like before. For instance, before the opening of the East-West German border, Germany in the eyes of the Poles and Poland for the Germans remained 'different'. Having said that, however, some differences and similarities are perceived to be attractive and comfortable, which can contribute to mobility, whilst others are regarded as unattractive and uncomfortable, which can, in turn, prevent people from crossing the border.

The next two chapters focus on the issues of change and concern respectively, that is, changes in the perception and migration idea of potential migrants and changes in the external socio-economic contexts that impact on these ideas. Smith (Chapter 3) explores the decision-making of young adult Ghanaians who aspire to migrate and wonders about their rationality for remaining immobile versus becoming mobile. Migration as an option and a potential step to make is presented through four cases of young people whom the researcher followed over several years. The cases illustrate how aspirations for migration and non-migration develop. In the first case, even though the respondent claimed that she was ready to leave Ghana, in reality she has remained in Ghana for the past ten years. She got married and gave birth to their first child instead. In the second case, the respondent made several visits as a student and/or temporary worker abroad and decided to stay in Ghana when he obtained a well-paid job. The third respondent who talked about going abroad for some years, finally migrated to the Netherlands, most likely due to family pressure. The fourth respondent had a fairly well-established family network abroad and often considered migration as an option but for various reasons this did not materialize. This change in perception is referred to as 'the dynamism of immobility and mobility'.

In Chapter 4, Jinnah applies the threshold approach as her theoretical framework to understand the role that personal and structural factors, as well as processes, play in mobility and immobility. How do potential migrants understand and engage with hurdles to international mobility? Are there personal limitations or state-imposed restrictions? How are these perceived differently, if at all? She argues that the approach has yet to fully grasp the interaction between structural factors and personal ones. Central to her argument is that people exercise what she calls 'strategic and subjective rationality in the migration decision-making process'. Three narratives are presented to illustrate how the threshold approach helps in exploring the subjective meanings of migrants before, during and after migration and how these develop in the interaction with structural ones. At the personal level, social networks play a role but also the level of available information. At the structural level, indifference to crossing the border can change when the socio-economic and/ or political situation in the country of origin or host country changes – the political situation in Somalia is a case in point. Nevertheless, personal, subjective realities are ultimately the real or final driving force in considering migration.

While the threshold approach was originally devised to explain the direction of labour migrants, Leung (Chapter 5) tests the value of the threshold approach on a rather different social group: Chinese academics whose mobility projects tend to be temporary. She argues that a mix of social, cultural, economic and political forces have worked together to lower the indifference (mental) threshold for young people and their parents to pursue educational mobility projects. However, rather than moving about the globe freely, the mobility of the students and scholars concerned is influenced by a range of factors that anchor them in diverse places (home, work, China, overseas and the virtual space). Also, one can observe the need to rescale academic staff mobility as a strategic act performed within a large social unit, namely the mobile academic's research group, department and university. Thus, the locational threshold is shaped by multiple actors, institutions and macrostructures, as well as coincidence. One or a series of coincidental or unplanned (small) events could shape one's mobility decision or trajectory.

Overall, women represent half of the migrant population in the world and contrary to the past, many of them migrate independently and cross borders with the aim of supporting themselves and their families. This seems to suggest that there is gender equality in crossing borders but the reality is different. In Chapter 6, Van Naerssen and Asis show with examples from the Asian continent that thresholds for migration are gendered since there can be different meanings and impacts of migration decision-making for women and men. Three sets of factors are distinguished, the global labour markets, sociocultural traditions and migration policies. The wage labour market is gender segmented, since women are usually employed in different sectors from men. Second, sociocultural norms and practices play a role in nearly all Asian countries. The social barriers for women to migrate are greater than for men, although the degree may vary by country and region. Third, while many low-income countries in Asia actively pursue overseas jobs for their nationals, only Indonesia, the Philippines and Sri Lanka allow the deployment of women migrants without much restriction. In short, the meaning of thresholds, in particular the mental threshold, is gendered.

Borders and Bordering

The second part of the book focuses on borders and consists of six chapters. In Chapter 7, Izotov and Soininen discuss mobility and immobility in the Finnish–Russian Karelia region. This was previously one Finnish region but a part became a Russian republic after the Second World War. The border was closed during the Cold War. It opened at the beginning of the 1990s and since then, there is increasing interest in cross-border 'integration' and in encouraging cross-border social and economic interaction. This is now a public debate at the state level in Russia and at the supranational level of the EU. The authors argue that residents on both sides of the border are involved in what they call micro processes of 'self-initiated and self-motivated integration and separation', but that the actual levels of mobility are low. In a survey assessing the interest for a regular passenger train connection between two cities on both sides of the border, the authors describe the differences in opinions based on age, occupation, gender and income, which indicate that thresholds and border crossings differ between different social groups.

In Chapter 8, Fàbrega and Lim highlight the importance of border dynamics, such as the opening of the border to the market economy and how changes in government regulations shape migrants' decisions. Factors that impact on the mental threshold by respectively lowering difference and indifference on both sides of the border are simultaneously present and are constantly being renegotiated on the ground. Their case is situated in Southeast Asia, more specifically the Cambodian-Thai border region. In the late 1980s, the border became more open and space was given for market economies. As the average income in Thailand is substantially higher than in Cambodia, the border area attracts many internal migrants from all over Cambodia. It also serves as an important springboard for people who wish to move into Thailand. They are driven by a combination of push factors from the place of origin, such as chronic poverty, lack of employment, debt, lack of access to markets and health issues. In turn, the pull of higher earnings and the rapid development of infrastructure and transportation encourage greater movement to places previously considered remote or distant. Finally, depending on the precise destination (the market, the Thai agricultural lands near the border or provinces deeper into Thailand), different trajectories are preferred and each trajectory has its own set of challenges and risks.

Contrary to the foregoing chapters that concern changes accompanying a greater openness of the border, the two following ones discuss the consequences of a stricter border control. Konrad (Chapter 9) assesses the border dynamics between Canada and the United States. During the twentieth century, the border was referred to as the 'longest undefended border in the world' and symbolized integration and coordination between the two countries. Thresholds for commuting and migration barely existed. However, after 9/11, this 'thin border' was transformed into a 'thick border' and cross-border mobility changed considerably and became restricted and managed through security measures. The spontaneous travel and relaxed cross-border migration for business, recreation and family visits is increasingly replaced by planned and rule-bound mobilities. The border is now perceived more of as a barrier and a constraint than before 9/11. The mental border threshold has emerged from relative insignificant in the migrant decision-making process to being a primary consideration and a substantial initial barrier to short-term crossing. Moreover, more thoughts and planning are now required in deciding where to go and where to cross, as well as the trajectory of crossing.

Related to the security of states but even 'thicker' than the US-Canada border is the fortified one between Israel and Palestine, the subject of Elnakhala in Chapter 10. She explains that concepts of difference/indifference or familiarity/unfamiliarity are pushed aside by issues of hostility and violence. The threshold approach can offer a framework for understanding cross-border violence between Israel and Palestine, provided it incorporates the broader concept of border-crossing tactics. The wall built by Israel is meant to keep away Palestinian militants. However, as the latter continue to try to cross the border, they are still involved in decision-making regarding the destination of their attacks, the trajectories to follow and the tactics to employ. Similar to migrants who change their destinations and routes, militants change border-crossing tactics and destinations too.

Violence is also the centre of interest in the contribution by Nyberg Sørensen (Chapter 11) that deals with security issues, in particular the effects of current US border enforcement on forced return migrants. In 2012, nearly 410,000 undocumented migrants were repatriated to their home countries, among them many Guatemalans. Driven by poverty in the home country, labour migration to the United States is the result of more or less conscious decision-making that usually involve negotiations with transfer agents, taking into account the uncertainties and dangers along the route. Many succeed in overcoming locational and trajectory thresholds before border enforcement imposes deportation. For migrants, deportation is a catastrophe. They return to their country of origin without an effective state programme in place for proper reception and integration. Repatriation can change the destination of returnees from a poor but secure home community into a poorer and more insecure location and therefore raise the locational threshold. In other words, the familiar could become unfamiliar. While some among them may eventually return to the United States, those without the financial resources will have to stay and cope with poverty, debts and threats from loan sharks and racketeers. Migrants risk their lives travelling on increasingly dangerous routes and also while travelling back to their home communities.

The relationship between human mobility and national borders poses ethical questions, which is the subject of Chapter 12 by Baggio. He reflects on the EU border policies but the same questions apply to other countries and regions in the world with strong border controls. Three kinds of borders are distinguished: national-territorial borders, external-territorial borders and internalized borders.

The first one concerns the fortification of borders and stricter national immigration regulations. The second refers to the tendency of countries to externalize their borders by influencing migration legislation, policies and practices in the countries of transit and departure, through diplomatic and/or financial measures. The third border concerns feelings, behaviour and attitudes of locals towards foreigners, asylum seekers and migrants, which can lead to exaggeration of immigration figures, criminalization of irregular migration and xenophobia. The three kinds of borders call for ethical reflection and in connection to this, Baggio identifies ethical principles that can help address questions of border and human mobility. These are, just to name a few: respect for equal and inalienable human rights based on the Universal Declaration of Human Rights, approved and adopted by the UN General Assembly on 10 December 1948; the principle of the 'common good', including the foundation of 'distributive justice', which regulates the relationship between the individual and society as a whole; and global citizenship, the ethical principle that challenges contemporary immigration policies most.

Places of Transfer and Trajectories

In 'Doing Borderwork in Workplaces', the first chapter of Part III (Chapter 13), Sandberg and Pijpers compare Polish circular migrants in Denmark and the Netherlands and expand beyond the threshold approach to what they consider as a frame of 'explanation that solely focuses on the various acts of choice' of individuals. They approach the decision-making of circular migrants with central concepts, such as 'assemblage' and 'borderwork'. The former refers to the fact that it takes a whole range of heterogeneous entities, settings and devices to choose mobility and make migration possible, while the latter as explained by Rumford, concerns spatial displacement and dispersion of borders within societies. Processes of bordering are no longer exclusively reserved for states and are negotiated in everyday life practices. Drawing on the two case studies, both authors propose that the migration decision-making process, especially for circular migration, involves a rhythm, a going back and forth which can partly be traced to migrants' own preferences and more importantly, to the available employment terms and conditions, sector-specific agreements, transitional restrictions and so on.

In Chapter 14, Nijenhuis pays special attention to the changing meaning of the locational threshold for Bolivian migrants to Spain, caused by what she calls the 'changing contexts of exit and reception of the migration corridor'. She argues that the border between Spain and Bolivia as a territorial demarcation line has changed from a relatively low barrier with hardly any restriction on entry into a high barrier, following the introduction of visa requirement for Bolivians in April 2007, in line with EU policies. The economic crisis in Spain, which caused unemployment to rise dramatically, is a high barrier in itself. In other words, not only the trajectory threshold but also the locational threshold have changed for Bolivian migrants. The contexts of exit and reception explicitly form part of locational thresholds. The origin and destination of migrants are given but a time dimension is added to explain how changes at both ends of the corridor alter the working of the locational threshold. The context of exit concerns the motivations and aspirations of the migrant, the economic conditions in the area of origin and the attitude of the sending country towards migrants. Factors of the context of reception are the societal response, the labour market and the opportunities for integration in the receiving country.

Although the majority of migrants enter Australia legally (they may become irregular migrants due to overstaying), asylum seekers trying to enter by sea have attracted most of the attention. In Chapter 15, Hugo and Napitupulu apply the threshold approach to these seashore asylum seekers. Many of them come from Afghanistan, Iran, Iraq and Sudan. The passage of the threshold of indifference is driven by the push factors of civil unrest, violence and war in these countries. Pull is exerted by the social networks in Australia, which also determine the passing of the locational threshold. Many asylum seekers have family and friends in Australia who are not only sources of information but also financial assistance, paying for their travel and assisting them after arrival. Thus, the uncertainty associated with migration is relatively low, given that the destination is clear. However, the travel can take a long time and almost all asylum seekers have to rely on people smugglers. The journey is not without risks, but as the authors argue, the biggest barrier faced by asylum seekers is Australia's border policy, in particular the so-called Border Protection Bill 2001, a policy of externalized borders that concerns interception and detention in asylum centres in certain Pacific countries even before the asylum seekers can enter Australia's migration zone.

Schapendonk's study in Chapter 16 concerns migrants from West and Central Africa who use Turkey as their 'springboard to Europe'. By using a biographical approach in combination with a longitudinal research methodology (that is, the researcher studies the respondents over a long period of time), information was gathered about the migrants' journeys to Istanbul, the role of the metropolis as a migration hub and the continuation of migrants' trajectories after a transit stay there. Moreover, the longitudinal research approach allowed the researcher to elaborate on the different ways migration trajectories evolve through the passage of time. The Turkish migration policies keep migrants from moving onwards to Europe, which results in experiences of involuntary immobility among some migrants, while others manage to find opportunities in the place they had only intended to pass through. However, the precarious living conditions tend to contribute to the feelings of being in transit among migrants, to the extent that some migrants who have intended to live in Istanbul may, as a result, decide to move to Europe. Thus, migration aspirations can change, leading to shifts in geographical orientations and the locational threshold.

While Istanbul still offers some employment opportunities for 'transit' migrants, the situation is different in Ceuta. In the final chapter by Ferrer-Gallardo and Espiñeira (Chapter 17), this is characterized as a 'limboscape' that immobilizes migrants. The city, part of the EU but located in northern Africa, is a symbol of 'Fortress Europe'. Sub-Saharan migrants who succeed to cross the EU-North African border fence by irregular means may find themselves stranded in Ceuta. Under these circumstances, the city becomes a limbo-like landscape, a zone where the migrants' trajectories towards European-EU are spatially and temporally suspended, a transitional zone or midway territory between two borders, the one of repatriation and expulsion and the other of regularization. Those who manage to cross the border fence are forced to wait in the city for a long time. This gives Ceuta a new territorial idiosyncrasy that illustrates the proliferation of confinement and encampment practices across the EU, as well as its related forced migrant immobility dynamics.

Close

The contributions in this book cover a diversity of mobility and migration categories and also introduce new themes, such as academic mobility, thresholds in forced return migration and militants' border-crossing tactics. It is interesting to note that only one contribution explicitly deals with the question of decision-making and immobility. This concerns the Ceuta case. In the case of Ghana, immobility is more implicitly seen in the perspective of time and changing personal situations, which is the reason why young Ghanaians seem to be able to switch easily from immobility to mobility and vice versa. Five chapters in this book deal with mobility specifically in border regions; four of them are about commuters, tourists and shoppers, three about changing situations since the opening of borders and onecussions about social class and ethnicity.

Special cases concern the cross-border activities of militants from Palestine into Israel and the cross-border return mobility of undocumented migrants from the US to Guatemala. These and others, like the seashore refugees in Australia and the sub-Saharan travellers trying to cross the barrier with Ceuta, show the systematic violence that accompanies mobility and migration.[5] While the use of violence may deter potential migrants, the inverse can also occur as demonstrated by the Cambodia-Thailand case whereby the porosity of the border offers easy passage for irregular migrants. However, the lack of protection and safeguard can also make the latter vulnerable to corrupt practices and abuse. This calls for further reflection on the ethics of border polices – the subject of one of the chapters. Thus, in opening up windows for new perspectives on the threshold approach, the reader may be confronted with seemingly contradicting conclusions. The challenge this book presents for the reader is to reconcile with them.

References

Adjamah, U. 2012. "Les motivations socioculturelles des départs en pirogue artisanale du Sénégal vers les îles Canaries (Espagne)." In *Les migrations africaines vers l'Europe: Entre mutations et adaptation des acteurs sénégalais*, edited by P. Demba Fall and J. Garreta i Bochaca, 103–18. Lleida: Remigraf-ifan/GR-Ase.

Boyd, M. 1989. "Family and personal networks in migration." *International Migration Review* 23 (3): 638–70.

Castles, S., and M.J. Miller. 2009. *The Age of Migration: International Population Movements in the Modern World*. 4th ed. Basingstoke: Palgrave Macmillan.

Collier, P. 2013. *Exodus: Immigration and Multiculturalism in the 21st Century.* London: Allen Lane.

EC (European Commission). 2010. "Geographical and Labour Market Mobility." Special Eurobarometer 337. Brussels: EC.

Edwards, W. 1954. "The Theory of Decision Making." *Psychological Bulletin* 51 (4): 380–417.

Faist, T. 2000. *The Volume and Dynamics of International Migration and Transnational Social Spaces.* Oxford: Oxford University Press.

Hoerder, D. 2002. *Cultures in Contact: World Migrations in the Second Millennium.* Durham, NC: Duke University Press.

Madsen, K.D., and T. van Naerssen. 2003. "Migration, Identity and Belonging." *Journal of Borderland Studies* 18 (1): 61–75.

Massey, D. 1994. *Space, Place and Gender*. Cambridge: Polity Press.

Massey, D. 2005. *For Space*. London: Sage.

Pang, C.L. 2007. "Chinese Migration and the Case of Belgium." In *Migration in a New Europe: People, Borders and Trajectories*, edited by T. van Naerssen and M. van der Velde, 87–110. Rome: IGU- Home of Geography.

Pfister, H.-R., and G. Böhm. 2008. "The Multiplicity of Emotions: A Framework of Emotional Functions in Decision Making." *Judgment and Decision Making* 3 (1): 5–17.

---

5 Violence not only as 'subjective', physical and visible violence but also 'objective' violence, systematic and anonymous (see Žižek 2008, 9–15).

Schapendonk, J. 2009. "The Dynamics of Transit Migration: Insights into the Migration Process of Sub-Saharan African Migrants Heading for Europe." *Journal for Development Alternatives and Area Studies* 28: 171–203.

Straubhaar, T. 1988. *On the Economics of International Labor Migration*. Bern: Haupt.

UN (United Nations). 2011. *The Age and Sex of Migrants Wallchart*. New York: UN.

Van der Velde, M. 2014. "Cross-Border (Im)mobility in Times of Crisis." In *Cross-Border Cooperation Structures in Europe: Learning from the Past to the Future*, edited by L. Dominguez and I. Pires. Brussels: Peter Lang.

Van der Velde, M., and T. van Naerssen. 2011. "People, Borders, Trajectories: An Approach to Cross-Border Mobility and Immobility in and to the European Union." *Area* 43 (2): 218–24.

Van Houtum, H., and M. van der Velde. 2004. "The Power of Cross-Border Labour Market Immobility." *Tijdschrift voor economische en sociale geografie* 95 (1): 100–107.

Van Houtum, H., and T. van Naerssen. 2002. "Bordering, Ordering and Othering." *Tijdschrift voor economische en sociale geografie* 93 (2): 125–36.

Van Naerssen, T., and M. van der Velde, eds. 2007. *Migration in a New Europe: People, Borders and Trajectories*. Rome: IGU- Home of Geography.

Žižek, S. 2008. *Violence*. New York: Picador.

PART I
MOBILITY AS AN OPTION

Chapter 2

Shopping for Differences: Mental and Physical Borders in the German–Polish Borderlands

Bianca B. Szytniewski

Despite long periods of controlled border movement during Soviet times, the German–Polish borderlands show a lively border-crossing tradition. Since 1990, cross-border trade and shopping practices have contributed to the so-called bazaar phenomenon on the Polish side of the state border. These bazaars, or open markets with covered structures (see also Sik and Wallace 1999; Marcińczak and Van der Velde 2008), have become places where Polish market vendors and German customers come together on a daily basis. However, frequent cross-border interactions between Poles and Germans already took place before the fall of the Berlin Wall in 1989 and were largely shaped and influenced by border policies. At the end of the 1980s, border restrictions changed for Poles from a period of strict border control under martial law to partially open state borders, not with East Germany, but with West Berlin. People were permitted to travel through East Germany to West Berlin and subsequently started to visit the city in large numbers as trade tourists, arriving with suitcases packed full of goods for trade in the streets and parks of the city. The fall of the Berlin Wall brought an end to Polish trade tourism in West Berlin, but in response to visa-free travel possibilities to Poland for East Germans, trade flourished on the streets and later on the bazaars in the Polish borderlands.

Cross-border trade and shopping mobility thus seems to be an ongoing practice in the German–Polish borderlands. Although these practices are for a large part influenced by changing border policies, borders could also be regarded as continuously evolving social constructs – constructs that go beyond state borders by addressing the meaning people attach to different places, people and practices (Anderson and O'Dowd 1999; Van Houtum and Strüver 2002; Anderson et al. 2003; Newman 2006; Agnew 2008). Moreover, these mental borders can contribute to significant levels of perceived otherness, encouraging or discouraging cross-border mobility (Van Houtum 2002; Van Houtum and Van der Velde 2004). Following this line of thought, Van der Velde and Van Naerssen (2011) argue that before cross-border mobility takes place, a mental border threshold needs to be overcome. People then enter a space of difference – a space where otherness across the state border is noticed and a differentiation between 'self' and 'other', 'us' and 'them', and, in a spatial sense, 'here' and' there' becomes part of daily life (Kristeva 1991; Duncan 1993; Riggins 1997). Although the 'there' needs to become part of the decision-making process to overcome the mental border threshold, Van der Velde and Van Naerssen (2011) do not extensively elaborate on how this takes place. The present form of the mental border threshold seems to be regarded as rather static, whereas images of otherness are subject to continuous interpretation processes through different perspectives and practices, direct and indirect experiences and changes in obtained and assessed knowledge (Massey 2005; Szytniewski and Spierings 2014).

This contribution will introduce the theoretical concept of (un)familiarity as an additional layer towards understanding the above-mentioned dynamic features of the mental border threshold.

Feelings of (un)familiarity could offer interesting insights to cross-border practices, in particular with regard to the choice of destination for trade and/or shopping purposes. According to Spierings and Van der Velde (2008, 2013), there is a certain degree of familiarity and/or unfamiliarity which people are prepared to accept when becoming mobile. Some differences and similarities may be perceived as attractive and comfortable and contribute to mobility, whereas others are regarded unattractive and uncomfortable, holding people back from visiting new destinations. As borders are continuously constructed, deconstructed and reconstructed, the degree of (un)familiarity may vary in different spatio-temporal circumstances (Massey 2005), subsequently influencing cross-border mobility.

In order to reach a better understanding of this multifaceted relationship between feelings of (un)familiarity and mental and physical borders, this chapter will study cross-border trade and shopping practices in the German–Polish borderlands. Since the Second World War, cross-border mobility between Germany and Poland occurred for a large part in times in which knowledge about and experiences with people and places across the state border did not develop freely, but were influenced by controlled borders and the political state of affairs. This in turn not only affected the degree of (un)familiarity, but also the decision to take the other side of the state border into consideration for cross-border practices. As a result, the above-mentioned mobility changes in the German–Polish borderlands reveal some interesting insights on the role of (un)familiarity in overcoming the mental border threshold.

This chapter begins with a theoretical part, followed by an empirical section on cross-border trade and shopping mobility in the German–Polish borderlands, employing observations, newspaper articles from regional newspapers, the *Märkische Oder Zeitung*, *Gazeta Lubuska* and the monthly *Gazeta Słubicka*, and personal conversations with German and Polish nationals in Frankfurt (Oder) and Słubice in the summer of 2012. The border-crossing town Frankfurt (Oder) and Słubice has been selected as case study in this chapter. Located at the German–Polish state border and consisting of a German and a Polish part which are connected by a bridge over the Oder river, the town is particularly interesting as it is situated approximately 100 km from Berlin, offering a connection point between Berlin and the rest of Poland. Moreover, after the fall of the Berlin Wall, one of the bigger bazaars in the German–Polish borderlands emerged in Słubice. Even now this bazaar could be regarded as a place of daily cross-cultural encounters between Polish market vendors and German customers.

(Un)familiarity and Cross-Border Mobility

Borderlands may reflect both physical and mental borders. Some differences and similarities in for instance political and economic systems, cultures and social interactions may be noticed and acted upon, others remain unnoticed or are forgotten. People may also be aware of different people and places, but indifferent to taking them into account in their daily lives, creating – or maintaining – a mental distance between the 'here' and 'there' (Van der Velde and Van Naerssen 2011). Depending on personal and spatio-temporal circumstances, people select and activate – consciously or unconsciously – different parts of their knowledge when perceiving different people and places (Massey 2005). Attitudes can change when former experiences, memories, beliefs or stereotypes are reassessed and reinterpreted (Schütz 1962; Moscovici 1988; Bauman 1995). As a result, some places, situations or encounters may be regarded familiar at one point in time, but when considering them from a different perspective, the formerly familiar attributes can become rather unfamiliar (Szytniewski and Spierings 2014). Feelings of (un) familiarity are therefore regarded as complex and dynamic processes, continuously changing and depending on various factors such as personal attitudes, previous experiences, available and assessed information, and physical and also mental proximity (Baloglu 2001; Prentice 2004).

Both mental borders and feelings of (un)familiarity may develop in a state of physical mobility as well as immobility. As put forward by Cresswell (2010), mobility involves an interconnected relationship between movement, representations and practices. Not only spatio-temporal circumstances but also personal characteristics may lead to different ways of perceiving and evaluating features of otherness. When crossing the state border to trade or shop, some people may only pay attention to the physical features of a place they visit rather than to social differences or similarities, while others are engaged in social encounters to the extent that they do not even notice differences in the physical surroundings (Szytniewski and Spierings 2014). Moreover, people may not necessarily be interested in familiarizing themselves with a place, but may only recognize the places they visit or the people they sell to or buy from. The latter may create a feeling of 'familiar strangeness' (Paulos and Goodman 2004; Pearce 2005).

People that are mobile encounter different people and places when travelling and crossing state borders, than people who are immobile, who may meet different (un)familiar others in their hometown. As a result of, for example, cross-border trade, shopping or labour mobility, attitudes of indifference may change and subsequently lead to the consideration or reconsideration of places across the state border (Ernste 2010). Moreover, social networks can play an important role in overcoming the mental border threshold. Not only networks with people across the state border, which are often emphasized in migration studies (Massey et al. 1998; Faist 1998; Conradson and Latham 2005; Van der Velde and Van Naerssen 2011) but also networks on the 'home' side with people that have already taken up cross-border practices could be important triggers for cross-border mobility. Consequently, a state of indifference does not necessarily mean that people are not aware of places, practices and people elsewhere, it only means that they themselves do not consider crossing these borders. As a result of increasing information and communication networks, knowledge about different others and different places has become more common than before (Castells 2005). Not only does this affect people's degree of (un)familiarity and their mental borders, but it can also motivate or discourage cross-border mobility.

Practices of cross-border petty trade and shopping could be seen as a form of circular mobility, which are often repetitive, of variable duration, and with large seasonal variation. For that reason, contacts and visits across the state border are mostly swift and in passing. In contrast to circular migration, they may last for an hour or a day, but more often than not they end in a return home (Bell and Ward 2000). In addition to border stories and experiences of border crossers which can encourage or discourage cross-border mobility, border restrictions and policies may also influence the degree of (un)familiarity and cross-border practices in a borderland (Anderson and O'Dowd 1999). Strict border controls for instance may reduce cross-border interactions and formal and informal information flows as 'the other side of the border becomes partially invisible and unknown' (Newman 2006, 152). Border liberalization on the other hand may contribute to encounters with the unknown, mysterious or previously known otherness (see Newman 2006; Baláž and Williams 2005). Subsequently, the degree of (un)familiarity may be interpreted and dealt with in different ways in different spatio-temporal circumstances. When someone, or for that matter something, a situation, a place or an activity is familiar, it may evoke feelings of recognition and comfort and contribute to mobility. However, as Gabriel and Lang (1995, 107) argue, 'Experiences fade with repetition,' resulting in practices where people seek 'novelty, uniqueness and adventure'. Over time, people may decide to seek new experiences and opportunities. Moreover, as mental borders are reconsidered, people may look for the unfamiliar instead of the familiar. The unfamiliar may then be associated with opportunities, resulting in cross-border attention and interaction, but it could also have the opposite effect, generating feelings of unease and avoidance (Bauman 1995; Spierings and Van der Velde 2008, 2013; Jagetić Andersen 2013).

The following section will elaborate further on the complexities of (un)familiarity by examining cross-border mobility patterns in the German–Polish borderlands, focusing on cross-border traders and shoppers, people that have overcome the mental border threshold.

Border Practices

The German–Polish borderlands, and in particular the border-crossing town Frankfurt (Oder) and Słubice, will be at the centre of the following empirical section. The German–Polish state border is particularly interesting, as it was constructed in 1945 as a result of an agreement between the allies of the Second World War. Under pressure from the Soviet Union, the Polish territory was moved westwards. The Oder-Neisse border became a reality, but remained a sensitive issue in German–Polish bilateral relations. As part of the communist and nationalist propaganda, many Germans were expelled and Poles throughout Poland, in particular from former Polish territories in the east, settled in the borderland. As a result of these migration flows, whole new communities emerged on both sides of the German–Polish state border (Stokłosa 2012). In both the newly established *Deutsche Demokratische Republik* – East Germany – and the western regions of the People's Republic of Poland, people were largely unfamiliar with their 'new' neighbours just across the state border.

For most of the period between 1945 and 1989, until the fall of the Berlin Wall, people living in the German–Polish borderland remained unfamiliar with the other side of the state border. However, as a result of attempts at rapprochement at national level in the early 1970s, border restrictions and policies loosened and a new border-crossing tradition emerged. Many Germans who had been expelled after the Second World War visited their former homes, cultural and educational initiatives were undertaken, friendships were formed and cross-border tourism, consumption and labour mobility increased significantly (Jajeśniak-Quast and Stokłosa 2000; Chessa 2004; Stokłosa 2012). Initially, curiosity prompted many people to engage in cross-border practices, seizing the opportunity to get to know and experience the other side of the state border for themselves. In the years that followed, however, cross-border mobility declined and the novelty of the new border situation between East Germany and Poland appeared to wear off (Jajeśniak-Quast and Stokłosa 2000; Stokłosa 2003).

Following the rise of the solidarity movement in Poland, the East German government sought to prevent antisocialist ideas from crossing the state border. Consequently, a year before Poland's communist leader Wojciech Jaruzelski declared martial law, which lasted from December 1981 until July 1983, East Germany decided on putting border restrictions back in place. In that period, the state border between East Germany and Poland was heavily controlled, practically closed, except for some cross-border labour mobility, since Poles continued to be needed in East German manufacturing firms. Before the end of the 1980s, however, before the Polish bazaars emerged, renewed cross-border trade and shopping flourished.

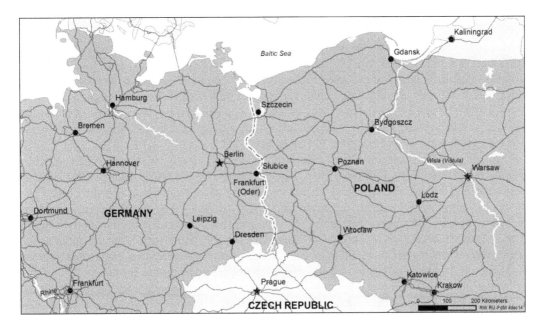

Figure 2.1   Map of the post-war German–Polish state border

Circular Mobilities: Suitcase Tourism in West Berlin

The bazaar in Słubice, and for that matter other bazaars in Poland in general, is not the only place in recent history where customers and traders from different regions and countries come together on a regular basis. The phenomenon can be traced back to older practices of trade and shopping across borders during Soviet times, where open-air markets and street trading in socialist economies played an important role not only in supplementing people's incomes but also in the (re)distribution of goods (Sik and Wallace 1999; Czakó and Sik 1999; Egbert 2006). In addition to cross-border petty trade, shopping tourism was a common practice, as people also bought goods which were not available or difficult to obtain in the home country and which were used for personal consumption or for reselling back home (Wessely 2002). Changes in travel regulations for individuals and small vehicles in the 1980s enabled Poles to travel abroad more easily. Since many Poles took up cross-border trade and shopping activities in this period, local markets across Central and Eastern Europe became known as Polish markets (Sik and Wallace 1999; Czakó and Sik 1999; Wessely 2002). Subsequently, by the end of 1988, Polish traders were not only looking eastward, but also towards the west, finding entrepreneurial opportunities in Sweden, Austria and West Berlin (*Gazeta Lubuska*, 01 April 1989). Although the Berlin Wall had not yet fallen and the borders to East Germany remained closed, Poles were already expanding their trading practices to different, relatively unknown places, developing new cross-border mobility flows from Poland to West Berlin.

   Many Polish suitcase traders or trade tourists, as they were called, engaged in small trade activities, selling cigarettes, spirits, foods, porcelain and clothes in the streets and parks of West Berlin. By May 1989, between 25,000 and 30,000 Polish trade tourists from different parts of Poland visited West Berlin on a weekly basis (*Gazeta Lubuska*, 26 May 1989). In July of that year the authorities in West Berlin took active measures to put an end to street trading in the city (*Gazeta*

Lubuska, 26 July 1989). As large numbers of Polish trade tourists were taking the train into West Berlin, restrictions began with the deployment of border guards on trains in Poland. Subsequently, trade tourists started to recognize the ticket controllers and customs officers on the train as well as East German border guards at the train terminal between East and West Berlin, and began to resort to various measures to avoid being caught. For instance traders tried to travel as light as possible, travelling with handbags and plastic bags instead of suitcases and backpacks. Also they did not always use seat reservations and moved around the train, trying to cause confusion and avoid ticket and luggage checks (*Gazeta Lubuska*, 17 September 1989).

Not only did the trajectories and places of trade, such as the streets around the *Berliner Philharmonie* next to the *Tiergarten Park*, become familiar to the Polish traders, West Berliner customers also became familiar strangers to them as a result of regular cross-border visits. Over time, the Polish traders gradually organized their visits, delivering their merchandise at particular addresses in West Berlin and then taking the first train back (mentioned by female respondent [1961] from Słubice; *Gazeta Lubuska*, 17 September 1989). However, despite familiarization with practices, people and places, the prospect of unexpected encounters and events lurked in the background, causing a feeling of unease on the one hand and adventure on the other.

Interestingly, communication by word of mouth was the most important reason for the sudden increase in the number of Poles travelling to West Berlin. While Poles started travelling to West Berlin from late 1988, it took a few months for the Polish newspapers to pick up on this trend (or at least before they found an opportunity to write about it). Friends and family members who had been in Berlin recounted their stories and encouraged others to try their luck too. Although cross-border trade and shopping tourism involved the risk of being caught, not only when crossing the state border, but also when selling merchandise on the streets, the thrill of crossing the border and getting to know these places in West Berlin, in addition to making a (small) profit, triggered people's curiosity (mentioned by various market vendors at the bazaar). Consequently, unfamiliar places across the border were suddenly not as distant in people's minds anymore and became part of circular mobilities in the German–Polish borderlands.

Mobilizing the Immobile: Polenmarkt in Słubice

While trade and shopping tourism in West Berlin came to an end, new opportunities in cross-border trade and shopping arose as a result of the fall of the Berlin Wall and the ensuing political changes in both East Germany and Poland. Poland started to produce less for export, causing a decline of export volumes for goods by 40 per cent in 1989 (Economy Watch 2013). As a result, many people seized the opportunity to take up new trade practices or to set up their own business. According to Sik and Wallace (1999, 701), 'People used to behaving resourcefully by combining different sources of economic activity continued to use these skills in a new environment, and the skills used for surviving in a Communist society turned out to be very useful ones for surviving in a post-Communist one as well.'

Change in border policies and restrictions also led to a change in mobility. While Poles had to wait until April 1990 for visa-free travel to East Germany, East Germans could travel to Poland without visas from the first of January 1990. This partial (re)opening of the East German–Polish state border resulted in vast numbers of border crossers from the German side of the border. Poles took advantage of these cross-border movements by setting up stands on the streets of Słubice, selling everything that they thought would sell, from crystals to jeans. As the street vendors were positioned on the pavements and German customers were blocking the streets, the municipality of Słubice soon decided to move the traders out of the city centre to an open field two km from the bridge between Frankfurt (Oder) and Słubice. In a few years time, the bazaar changed into a permanent market place, housing many different stalls with clothing, shoe ware, fruits, vegetables and other fresh produce, cigarettes and alcohol, but also different food and beverage outlets and even a few hairdressers. At first mostly locals engaged in trading. However, before long Poles from Łódź, Warsaw, Kraków and further also took up trading at the bazaars along the German–Polish border (*Gazeta Słubicka*, December 1993).

German customers continued their shopping activities, especially since the German reunification on 3 October 1990 had brought favourable exchange rates for the East German mark. Interestingly, the change in cross-border movement attracted customers not from West Berlin, but from former East Germany. It began with people living in Frankfurt (Oder) and close to the state border. Not only word of mouth but also German newspapers spread the word. Although it took a while before the regional newspaper, *Märkische Oder Zeitung*, printed something about German shopping activities in Słubice, one of the first German news items on shopping in Słubice was headlined 'Chaos in Słubice' (*Märkische Oder Zeitung*, 26 September 1990). The article not only informed people of what was happening directly across the state border, but also generated curiosity among East Germans living outside Frankfurt (Oder). People reading the newspaper might have wondered whether there was anything to see or worth visiting. According to one market vendor, German media unintentionally advertised Polish bazaars. Spread of information and cross-border experiences contributed to awareness of cross-border practices by fellow nationals and influenced people's mental distance between 'here' and 'there'. For many, this meant the beginning of regular cross-cultural trade and shopping interactions.

By the end of the 1990s, differences in price had shrunk and goods were becoming more available, not only in former East Germany but also in Poland. Although some novelty and a glint of curiosity recurred when the new bazaar reopened in 2011 after a fire had destroyed the old market in 2007, the novelty and curiosity that accompanied border liberalization in the light of Poland's accession to the European Union (EU) and the Schengen zone mostly faded (Baláž and Williams 2005). When crossing the state border and visiting the bazaar, people are not necessarily confronted with large differences. As a result of similar pre-war architecture, EU-funded roads and pavements, use of German language in town and at the bazaar and the possibility to pay with the euro, the physical border between Germany and Poland has become less visible. Despite these developments, some differences remain, contributing to the bazaar in Słubice as a meeting place between German customers and Polish traders.

People visit the bazaar from different parts of Germany, not only from Frankfurt (Oder) and the surrounding villages but also from further away, for example Berlin, Hanover and even Bremen. Reasons for visiting range from a practical stop at the bazaar, mostly for cigarettes which are sold close to the entrances of the market, to a day out with friends or family, amid the amiable atmosphere of the market. Most of the time, a visit to the market is a combination visit – both functional and for leisure. Many people seem to be familiar with the bazaar, visiting the market regularly and for many years. However, some first-time visitors may still be found, often for reasons of curiosity.

The merchandise and market-related advertisement at the bazaar in Słubice are not directed at locals, but at German speaking customers. Polish market vendors adjust their merchandise to suit the taste of their German customers, they allow payments with euros and speak the German language. Their main purpose is not to familiarize the German customers with Poland, but rather to create a familiar, comfortable place for their German customers – familiarity in an unfamiliar place or a positive feeling of familiar strangeness. Most German customers seem to experience the Polish market as such, but at the same time they also appreciate the differences in merchandise, people and atmosphere. The market mentality and the feeling that one can still negotiate prices are often mentioned by German customers as a positive difference between the Polish bazaar and the local markets at home. Furthermore, most of the time German cross-border shoppers only visit the bazaar in Słubice. They are not necessarily interested in visiting other parts of the town nor do they stop in Frankfurt (Oder). Differences in products, atmosphere and cross-cultural interactions matter as they contribute to the cross-border shopping experience. When engaging in cross-border mobility, people seem to look for a balance between comfortable familiarity and attractive unfamiliarity (Spierings and Van der Velde 2013).

Changing Mobilities

Changes in border policies and economic and political developments in East Germany and Poland have not only contributed to opportunities for creative entrepreneurship among Poles and cross-border shopping practices for Germans, but they have also influenced the direction of mobility of both traders and shoppers, from west to east. Whereas many Polish traders decided on a permanent stand at the bazaar in Słubice, or at one of the other bazaars along the German–Polish state border, German shoppers became active border crossers. As a result, cross-cultural trade and shopping practices have become a continuous part of everyday life in the German–Polish borderland. It seems that in this particular case of cross-border mobility, border restrictions generated curiosity and opportunities which people seized to experience and get to know the unfamiliar or previously known other side of the state border. It must be noted, however, that this contribution focuses on people aware of and not indifferent to crossing state borders. Others may regard the border as a dividing line between countries, rather than an opportunity for cross-border practices (Thuen 1999). Differences associated with state borders may then evoke feelings of discomfort or indifference and prevent people from engaging in cross-border mobility.

There is a certain degree of both familiarity and unfamiliarity when people undertake cross-border practices (see also Spierings and Van der Velde 2008, 2013; Szytniewski and Spierings 2014). Border stories and experiences of family members and friends who had already crossed the state border informed both other Polish traders and German shoppers about cross-border petty trade and shopping opportunities in the neighbouring country. People's journeys to West Berlin and cross-border shopping visits to Słubice not only generated curiosity but also familiarity with border-crossing practices, altering the mental distance between the hometown and the places where cross-border petty trade and shopping took place. As a result of social networks, people 'know' to some extent what to expect and use this knowledge – consciously or unconsciously – when making up their mind whether to engage in cross-border practices or not. Feelings of (un)familiarity can thus contribute to overcoming the mental border threshold. Moreover, as trade and shopping practices were not unfamiliar practices for either Polish traders or German shoppers, border crossings did not necessarily cause feelings of unease and avoidance, but instead were regarded by many as new adventures and opportunities. Polish border crossers for instance had already become familiar with cross-border trade and shopping practices during socialist times and continued with these familiar practices but at unfamiliar locations and with new customers. Even though the novelty and curiosity of the unfamiliar faded over time (Baláž and Williams 2005), the otherness of places continued to mobilize people to cross state borders. While differences in price and availability of goods used to be important reasons for German cross-border shoppers, the experience of taking a break from the everyday seems to matter more and more. Consequently, people seem to be attracted by the possibility to enter a space of difference, where they may experience differences in surroundings, atmosphere and culture.

Moreover, mobility in the German–Polish borderlands contributed to a regular experience of familiar strangeness. Polish trade tourists brought something new and unfamiliar to the streets of West Berlin and soon became familiar strangers not only to their customers but also to border guards in both East and West Berlin. The same may be said for the continuous flows of German customers at the border bazaars in Poland and their relationship with the Polish market vendors. People have become accustomed to cross-border trade and shopping – they recognize people and places across the state border, but they do not necessarily familiarize themselves entirely with the otherness of these places or people. As a result of national differences resulting from different political, socio-economic and cultural narratives on both sides of a state border, there is always something unknown present in the border-crossing experience. Although a degree of unfamiliarity exists, at the same time people also become attached to the familiar features of cross-border encounters across the state border and notice when their familiar strangers are not there or when a place has changed. A combination of familiarity and unfamiliarity remains and contributes to the attractiveness of cross-border trade and shopping practices.

Conclusion: A Dynamic Approach to the Mental Border Threshold

In the case of the German–Polish borderlands, curiosity, adventure, otherness and anticipated opportunities play an important role in cross-border trade and shopping practices. It appears that feelings of unfamiliarity do not necessarily hinder people from crossing state borders, but are associated with opportunities that may be found in places other than in the hometown. When deciding on places for trading and shopping, people consider – consciously or unconsciously – their mental borders. Familiarity – through social networks and experiences – and unfamiliarity – combining expected and unexpected encounters and opportunities – may influence people's choice of destination and trajectories and thus the decision to be mobile or stay immobile. Moreover, the degree of (un)familiarity may for a large part be influenced by travel restrictions and border policies. The closing and (re)opening of the state border between Poland and (East) Germany reduced familiarity with people and places across the border on the one hand, but also brought new opportunities for petty trade and shopping practices on the other, contributing to circular mobilities and cross-border encounters in different times, directions and places. Even though the initial novelty and curiosity fades, a certain degree of unfamiliarity with regard to otherness across state borders has remained, which even now enhances the attractiveness of cross-border trade and shopping in the German–Polish borderland.

The mental border threshold is thus more dynamic than it initially appears. Although state borders can considerably influence cross-border mobility, mental borders are constantly open to change and do not have to begin or end at a state border. Different others and different places may be encountered on an everyday basis, through personal experiences and formal and informal information flows, influencing people's (un)familiarity with regard to cross-cultural differences and similarities. Being indifferent therefore does not always mean being unaware of cross-border practices. Knowledge, practices and cross-cultural interactions can contribute to different ways of assessing different people and places – in borderlands and elsewhere. Consequently, both rational and emotional motives influence people's mobility. As mental borders are not restricted to state borders, people find opportunities in places where cultural, social and economic differences meet. To what extent should we then speak about overcoming a mental border threshold? It seems that physical and mental borders are simultaneously at work, continuously altering the mental distance between people and places, between 'us' and 'them' and 'here' and 'there'. They reflect perceived differences and similarities and the ways these are activated and acted upon. The mental border threshold is therefore continuously constructed, reconstructed and deconstructed in the everyday life.

References

Agnew, J. 2008. "Borders on the Mind: Re-Framing Border Thinking." *Ethics and Global Politics* 1 (4): 175–91.

Anderson, J., and L. O'Dowd. 1999. "Borders, Border Regions and Territoriality: Contradictory Meanings, Changing Significance." *Regional Studies* 33 (7): 593–604.

Anderson, J., L. O'Dowd, and T.M. Wilson. 2003. "Culture, Co-Operation and Borders." *European Studies: A Journal of European Culture, History and Politics* 19 (1): 13–29.

Baláž, V., and A.M. Williams. 2005. "International Tourism as Bricolage: An Analysis of Central Europe on the Brink of European Union Membership." *International Journal of Tourism Research* 7 (2): 79–93.

Baloglu, S. 2001. "Image Variations of Turkey by Familiarity Index: Informational and Experiential Dimensions." *Tourism Management* 22 (2): 127–33.

Bauman, Z. 1995. "The Stranger Revisited, and Revisiting." In *Life in Fragments: Essays in Postmodern Morality*, 126–38. Oxford: Blackwell.

Bell, M., and G. Ward. 2000. "Comparing Temporary Mobility with Permanent Migration." *Tourism Geographies: An International Journal of Tourism Space, Place and Environment* 2 (1): 97–107.

Castells, M. 2005. "The Network Society: From Knowledge to Policy." In *The Network Society: From Knowledge to Policy*, edited by M. Castells and G. Cardoso, 3–22. Washington, DC: Johns Hopkins Center for Transatlantic Relations.

Chessa, C. 2004. "State Subsidies, International Diffusion, and Transnational Civil Society: The Case of Frankfurt-Oder and Slubice." *East European Politics and Societies* 18 (1): 70–109.

Conradson, D., and A. Latham. 2005. "Friendship, Networks and Transnationality in a World City: Antipodean Transmigrants in London." *Journal of Ethnic and Migration Studies* 31 (2): 287–305.

Cresswell, T. 2010. "Towards a Politics of Mobility." *Environment and Planning D: Society and Space* 28 (1): 17–31.

Czakó, Á., and E. Sik. 1999. "Characteristics and Origins of the Comecon Open-Air Market in Hungary." *International Journal of Urban and Regional Research* 23 (4): 715–37.

Duncan, J. 1993. "Sites of Representation: Place, Time and the Discourse of the Other." In *Place/Culture/Representation*, edited by J. Duncan and D. Ley, 39–56. London: Routledge.

Economy Watch. 2013. "Poland Export Volumes of Goods Only (Percent Change) Statistics." Accessed 22 February 2013. http://www.economywatch.com/economic-statistics/Poland/Export_Volume_Goods_Percent_Change/.

Egbert, H. 2006. "Cross-Border Small-scale Trading in South-Eastern Europe: Do Embeddedness and Social Capital Explain Enough?" *International Journal of Urban and Regional Research* 30 (2): 346–61.

Ernste, H. 2010. "Bottom-up European Integration: How to Cross the Treshold of Indifference?" *Tijdschrift voor economische en sociale geografie* 101 (2): 228–35.

Faist, T. 1998. "Transnational Social Spaces out of International Migration: Evolution, Significance and Future Prospects." *European Journal of Sociology* 39 (2): 213–47.

Gabriel, Y., and T. Lang. 1995. *The Unmanageable Consumer: Contemporary Consumption and Its Fragmentation.* London: Sage.

Jagetić Andersen, D. 2013. "Exploring the Concept of (Un)familiarity: (Un)familiarity in Border Practices and Identity-Formation at the Slovenian-Croatian Border on Istria." *European Planning Studies* 21 (1): 42–57.

Jaješniak-Quast, D., and K. Stokłosa. 2000. *Geteilte Städte an Oder und Neiße: Frankfurt (Oder)-Słubice, Guben-Gubin und Görlitz-Zgorzelec, 1945–1995.* Berlin: Berlin Verlag.

Kristeva, J. 1991. *Strangers to Ourselves.* New York: Columbia University Press.

Marcińczak, S., and M. van der Velde. 2008. "Drifting in a Global Space of Textile Flows: Apparel Bazaars in Poland's Łódź Region." *European Planning Studies* 16 (7): 911–23.

Massey, D. 2005. *For Space.* London: Sage.

Massey, D.S., J. Arango, G. Hugo, A. Kouaouci, A. Pellegrino, and J.E. Taylor. 1998. *Worlds in Motion: Understanding International Migration at the End of the Millennium.* Oxford: Oxford University Press.

Moscovici, S. 1988. "Notes towards a Description of Social Representations." *European Journal of Social Psychology* 18 (3): 211–50.

Newman, D. 2006. "Borders and Bordering: Towards an Interdisciplinary Dialogue." *European Journal of Social Theory* 9 (2): 171–86.

Paulos, E., and E. Goodman. 2004. "The Familiar Stranger: Anxiety, Comfort and Play in Public Spaces." In *Proceedings of the SIGCHI Conference on Human Factors in Computing Systems*, 223–30. Austria: Association for Computing Machinery.

Pearce, P.L. 2005. *Tourist Behaviour: Themes and Conceptual Schemes.* Clevedon: Channel View Publications.

Prentice, R. 2004. "Tourist Familiarity and Imaginary." *Annals of Tourism Research* 31 (4): 923–45.

Riggins, S.H. 1997. "The Rhetoric of Othering." In *The Language and Politics of Exclusion: Others in Discourse*, edited by S.H. Riggins, 1–30. London: Sage.

Schutz, A. 1962. *Collected Papers I: The Problems of Social Reality.* Edited by M. Natanson. The Hague: Nijhoff.

Sik, E., and C. Wallace. 1999. "The Development of Open-air Markets in East-Central Europe." *International Journal of Urban and Regional Research* 23 (4): 697–714.

Spierings, B., and M. van der Velde. 2008. "Shopping, Borders and Unfamiliarity: Consumer Mobility in Europe." *Tijdschrift voor economische en sociale geografie* 99 (4): 497–505.

_____ . 2013. "Cross-Border Differences and Unfamiliarity: Shopping Mobility in the Dutch-German Rhine-Waal Euroregion." *European Planning Studies* 21 (1): 5–23.

Stokłosa, K. 2003. *Grenzstädte in Ostmitteleuropa: Guben und Gubin 1945 bis 1995.* Berlin: Berliner Wissenschafts-Verlag.

_____ . 2012. "Neighborhood Relations on the Polish Borders: The Example of the Polish-German, Polish-Ukrainian and Polish-Russian Border Regions." *Journal of Borderlands Studies* 27 (3): 245–55.

Szytniewski, B., and B. Spierings. 2014. "Encounters with Otherness: Implications of (Un)familiarity for Daily Life in Borderlands." *Journal of Borderlands Studies* 29 (3): 339–51.

Thuen, T. 1999. "The Significance of Borders in the East European Transition." *International Journal of Urban and Regional Research* 23 (4): 738–50.

Van der Velde, M., and T. van Naerssen. 2011. "People, Borders, Trajectories: An Approach to Cross-Border Mobility and Immobility in and to the European Union." *Area* 43 (2): 218–24.

Van Houtum, H. 2002. "Borders of Comfort: Spatial Economic Bordering Processes in and by the European Union." *Regional and Federal Studies* 12 (4): 37–58.

Van Houtum, H., and A. Strüver. 2002. "Borders, Strangers, Doors and Bridges." *Space and Polity* 6 (2): 141–6.

Van Houtum, H., and M. van der Velde. 2004. "The Power of Cross-Border Labour Market Immobility." *Tijdschrift voor economische en sociale geografie* 95 (1): 100–107.

Wessely, A. 2002. "Travelling People, Travelling Objects." *Cultural Studies* 16 (1): 3–15.

Chapter 3

Aspirations to Go: Understanding the Bounded Rationality of Prospective Migrants from Ghana

Lothar Smith

On a Sunday morning, in mid-December 2002, the author attended a Pentecostal Church service in the heart of the city of Accra, Ghana. While making his way out of the church he was stopped by a group of younger church members who were standing chatting. They asked him why he had come to their church, as he was the only white person there. His response was something non-committal, along the line of it being a good and valuable experience for a foreigner. At this they enthusiastically replied that this was also why they had stopped him, as they wished to join him when he went back to his country; they also wanted to 'experience' that place, i.e. go there for some time. Having been subject to requests to sponsor the travels of young Ghanaians to Europe many times before (i.e. provide funding), ranging from veiled, polite comments, to specific requests for a letter of recommendation, to outright demands for assistance, the first reflex was to say something non-committal and move on. However, this time the occasion provided a welcome opportunity to ask this group of young adults about the nature of their desire to migrate. To that end the author asked: 'You said that you wanted to come to my home, but do you actually know what my country is? Do you know where you would be migrating to?' The question provoked clear puzzlement within the group, but after some thought one responded: 'Oh, but anywhere outside Ghana is fine.' To this the others nodded in full agreement.

The complete indifference of these youngsters towards the specifics of a migration destination was surprising, and far removed from the general notion that people are well-prepared when embarking on long-distance travels. In that vein the response of the youngsters could easily be perceived as insubstantial and unrealistic, reflecting their dreams and wishes rather than a solid plan for the future. Yet the conversation stuck, and over time it provoked various other questions, notably considering more carefully whether their intentions were actually unrealistic, and indeed, if not narrowing their choice to one specific country was perhaps quite rational, reflecting a maximum scope to achieve their aspiration to go abroad?

The title of this chapter purposely alludes to the concept of 'bounded rationality' in order to embed the aspirations of young Ghanaians to migrate in certain theoretical assumptions on migration decision-making, notably the degree to which people are, or need to be, informed. What, however, is bounded rationality? Massey et al. (2008, 139) argue that the concept of bounded rationality is key in helping to avoid 'hyper-rationality' forms of thinking about the choices individuals make in engaging in migration. They explain the concept as follows:

> In thinking about migration ... the set of potential destinations is usually restricted and often people do not seriously consider moving at all (Speare 1971, 130). Only under very specific conditions people switch frames, overcome the enormous hurdle of inertia, alter habitual behaviors, and put, as a result, migration on the decision-making agenda and engage in difficult cost-benefit considerations.

This citation explains the role of boundedness of rationality in decision-making processes related to migration. Yet it appears to stand somewhat at odds with the sketch with which this chapter opened in two ways: Firstly, Massey et al. argue that there is a general indifference to migration, that is, until it rises to the fore due to changes of certain circumstances. Yet in countries in which migration features prominently in daily lives, such as the Philippines, Mexico but also Ghana, the so-called 'culture of migration' has resulted in a situation in which little such indifference, that is unawareness of the migration option, still exists. Instead there is a general awareness of migration as a valid and realistic avenue when setting out household and/or individual livelihood strategies. Secondly, Massey et al. make the point that once migration has been conceived of as an imminent step, the set of potential destinations also narrows down, an argument that seems logical. Yet the young Ghanaians outside the church maintained quite the opposite argument, in their aspirations to migrate the exact destination was not considered to be particularly relevant. So what is happening here? Do these youngsters fail to recognize how they are bounded by certain institutional constraints, or are they, quite conversely, very aware of these limitations?

The empirical point of departure of this chapter is the range of aspirations of would-be migrants residing in Accra, Ghana. The rationale of their aspirations and the related choices are thereby linked to the conceptualization of migration as a process by which different kinds of thresholds are overcome (Van der Velde and Van Naerssen 2011; Van der Velde and Van Houtum 2004), as set out in the introductory chapter of this book. Importantly such a threshold approach provides the opportunity to focus on the conception of migration, that is, on the aspiration of migrants, and not only on migration as a process that is already happening or completed. To that end this chapter especially focuses on the rationalities of would-be migrants in Accra and relates this to the indifference threshold in particular, but also to the other two thresholds, where appropriate (Van der Velde and Van Naerssen 2011, 191).

Chapter Outline

This chapter seeks to flesh out conceptual differences in terms of the diversity of perceived underlying rationalities for migration. To this end, young adults based in Accra with a clear interest in the potential of migration were selected as central cases. Thereby the focus is on the role migration culture plays in potential migrants' choice of destination. The bias towards destinations in the northern hemisphere, that is Europe and North America, is clear, despite Ghana having a long history of major intraregional migration movement, notably to and from Nigeria. This focus on the Global North relates especially to perceived differences in economic prospects, sometimes taking on mythical proportions, despite various counter cases of failed returnee migrants. However, this preferred choice of destination also has important financial consequences and is likely to produce many organizational complexities.

This chapter primarily draws on data collected during the Ghana TransNet research programme, which ran between 2002 and 2007. In this programme the role of transnational ties between Ghanaian migrants in Amsterdam, the Netherlands and members of their social networks based in their hometowns in the Ashanti Region (Kabki 2007) and in Accra, Ghana's capital city (Smith 2007), was explored through a concurrent, mixed methods approach of the transnational networks that connected the three locations (Mazzucato 2009). Subsequent visits to Accra between 2007 and 2012 helped maintain contacts with certain key respondents in Accra and stay abreast of further developments in their lives.

This chapter is organized as follows: The next section sets out the theoretical underpinnings of this chapter, focusing on the role of concepts like thresholds to migration and therein the role of aspirations and their rationales, the definition of trajectories, the influence of a culture of migration, and finally the role of a migration industry. Thereafter the role of migration as a potential step is presented through portraits of four young, aspirant migrants residing in Accra. In the final section the value of the concept of bounded rationality as related to the threshold approach is returned to in order to explain aspirations towards international migration.

Theoretical Background

Following the attention already given to the threshold approach as a whole in the opening chapter of this volume, this chapter will focus particularly on the mental border threshold, given its centrality in this chapter. However, as (anticipated) decision-making of the would-be migrants studied also raises some points linked to the locational and the trajectory threshold, these are also briefly discussed. With the assumed logic that migration only follows on an actual awareness of migration as an option, the mental threshold is discussed first, followed by the locational and trajectory thresholds.

With the *mental threshold* the question arises; how is migration conceived of as an option in the first place, and as an option for what, that is, to what end? Is it seen as a necessary move in the face of worsening local circumstances? Is it seen as a strategic option from a range of livelihood activities that can be pursued? Is it envisaged as an entrance into a new lifestyle that could never be achieved by remaining immobile? In that sense, the question arises; is it seen by the prospective migrant as a personal way to see more of the world than what is offered by the media, internet or through personal accounts of experiences by those who have been abroad? Any aspiring migrant's response could, of course, be a composite of these underlying motivations. But the gist of the threshold approach is that it avoids *a priori* conceptions of migration rationales being economic alone and rather sees it as part of the portfolio making up the totality of a livelihood trajectory.[1] Given the fact that in the course of this livelihood trajectory perceptions may change, or new opportunities unfold, it is then also better not to work with fixed notions and instead to speak of a shift in mentality, from indifference (as a non-attitude) to active consideration of migration as a feasible option. The threshold approach is conducive for more complex approaches to migration. This is, for instance, reflected in the role of the so-called 'migration industry' for aspiring migrants. Lamented in public discourse as an undesirable practise that no longer fits with 'modern Ghana' (see for instance various forum entries on the website *Modern Ghana*[2]), the industry remains very much in place, as prospective migrants are easily lured towards the expertise of 'connection men'. These are in essence information brokers and network players who claim to have fast access, with guaranteed results, to secure visas for their clients (Burrell 2009, 151). In some cases these brokers may also extend their role to other activities which can vary from facilitating cheap airline routes to arranging overland routes in situations where a visa could not be secured. Whilst the term 'connection man' may evoke a sense of dealing with 'shady' characters, many families who are eager to send one

---

1   Please note that the concept of a livelihood trajectory is a historical-sociological concept and should not be confused with the trajectory threshold concept. The latter, as set out in Chapter 1, and also explained later in this chapter, is purposely spatial.

2   See for example the article 'Can Ghana Ever Overcome the Activities of "Connection Men" in Passport Acquisition?'(*Modern Ghana*, 20 May 2013).

of their members abroad seek out someone they consider sufficiently trustworthy. This may be achieved by attending the same church, or by being residents of the same neighbourhood, and/or knowing that this person has successfully helped others in their network before.

The value attributed to the trustworthiness of such a broker provides two important insights: First it explains how the phenomena of the migration industry, which often comprises of networks of individual brokers, can become encapsulated in the social networks of would-be migrants. Second, it explains how the very existence of the migration industry, and these brokers more particularly, derives from the persistence of certain unfavourable macroeconomic, institutional and societal conditions hindering spontaneous migration. Thus the mental state of indifference can be considered as a two-pronged mental step: (1) to conceive of migration as an option in the first place and (2) to consequently fathom its meaning and thus relevance upon which the next steps, part of the locational and trajectory thresholds, can be explored.

With regard to the *locational threshold* one central question of relevance to this chapter is the relevant factors which influenced prospective migrants' choice to migrate, in particular the key thresholds migrants had to overcome in the conception and first onset of their migration trajectory. At what stage does migration begin as an actual project? This question is important, as migration may be preceded by a long run-up period. During this period potential migrants deal with various, often changing, personal conditions and ideas – their own and those of others – that affect how they regard whether, why and where they should migrate. The locational threshold thus conceives of migration as essentially multidimensional. It takes as its point of departure that the perception of the end location and insight into the trajectory that needs to be followed need to be considered together and not in isolation to understand how a decision to set off is made.

The third dimension with which to explore migration projects, and what influences them, is the *trajectory threshold*. In exploring the geographical trajectories that were envisaged by aspiring migrants, the key issue of rationality in migration decision-making is raised again. This links up to the concept of bounded rationality discussed in the introduction of this chapter. To reiterate the point: this boundedness relates to an often incomplete overview of the end destination and what can be achieved there. This may actually be a strategic choice to avoid becoming too discouraged from setting off altogether or deciding which route to take. Regarding the latter, for many migrants their migration is a project whose contours only become clear as they progress along the route. The ability of potential migrants to discuss the underlying rationality of their choices thus needs to be understood as confined (Simon 1982). It also needs to be understood that the process of migration may produce continual changes and adjustments due to all kinds of factors encountered which influence the trajectories taken and destinations chosen (Schapendonk 2011). Indeed in relation to this Faist (2000) points to the fact that the greatest danger lies in the researchers' quest to derive rational criteria from their respondents to explain their migration movements and decisions. The key issue is that these studies are often conducted *ex post*, that is, after the migrant has reached a certain destination. Yet if a study were conducted *ex ante* the migration of these very same respondents, it would become clear that they are much less able to explain their strategies and rationalize any of the choices made along the way. Such rationalization of migration crucially also assumes defined moments of departure and arrival and suggests the migration process is linear and progressive (Czaika and Vothknecht 2012). Yet neither linearity nor a fixed progression in thought and choice can be assumed in migration trajectories, as Schapendonk's (2011) recent study on what he calls 'turbulent trajectories' of migration so clearly shows for sub-Saharan migrants who aspired to reach Europe. In his work he makes clear how there is a continual interplay between actual migration behaviour, making a next step, and the mental state and logic of the migrant.

Explaining the Aspirations of Would-Be Migrants in Ghana

Following the theoretical framing of the threshold approach as related to aspirations to migrate, this section provides an empirical analysis of four case studies of aspiring migrants. The underlying questions are: When, and how, did international emigration come to be conceived of as an option in their lives, and what was it an option for? Was emigration therefore seen as an alternative or more secure form of livelihood? And if so, for whom? Was it also, or perhaps rather, envisaged as an entry to a new lifestyle that was less likely to be achieved by remaining immobile? Indeed, was the choice to migrate perhaps (also) seen as a way of seeing more of the world?

A response to these questions from a given set of migrants will, of course, show a composite of motivations. However, the crux of the matter here is to avoid taking a too narrow, functionalist focus when considering the decisions made with the activities pursued, notably zooming in on economic-oriented rationales as has sometimes been the case in livelihood studies (Rakodi 2002; Kaag 2004). Instead it could be preferable to take as a starting point the shifts in mentality, from complete indifference (as a non-attitude) to active consideration of migration as a goal.

This section presents four cases of respondents who, from different points of departure, explain how aspirations of migration develop. These cases give insight into the constellation of processes leading to the aspiration to migrate and the people involved in these migration projects, thereby influencing the trajectory they follow, the intended final destination, how their trip is financed, and so on. The order of respondents also reflects a shift from perpetual and relatively abstract pondering over migration, to actual experiences of international migration and the meaning of this, to a respondent who is meticulously and slowly crafting and planning his route.

Grace: The Shifting Importance of Migration as a Goal

Grace is a young woman who grew up in the town of Nsawam, which lies 30 km outside Accra, on the road to Kumasi. Following the completion of her secondary education in Nsawam she found employment in Accra. After first working in a household appliances shop in the city centre she was able to secure work as distributor of consumable goods on behalf of a wholesaler to traders around the city of Accra. When she was retrenched from this job some years later, she managed to find employment with an assurance company selling life insurances to government workers, both in and outside Accra. Beyond this she also runs a small 'drinking spot' (a bar) in Nsawam, together with her mother. She was able to invest in the expansion of this business through support of Asantewaa, a good friend based in Amsterdam. During a phone call Grace had asked whether Asantewaa could 'think of her' to allow her to expand her 'spot' by investing in a strong liquor license for the bar which she expected would strongly boost its overall sales. Asantewaa agreed that she would consider Grace's situation and a short time later Grace received a few hundred euro from her. Although during the phone call nothing had been explicitly discussed with regard to whether, and if so in what way, this money would be repaid, Grace felt that Asantewaa was just helping her establish the business and that it was thus not her intention to take any share of the profit afterwards or ask for the money back. Instead she felt that her migrant friend had helped her because Grace was her only very close friend in Ghana, a friendship which stemmed from their youth as they had grown up together.

Beyond Asantewaa Grace also has family members and two other good friends living abroad. With all these friends and family abroad she indicates that she also wants to leave: 'Yes! I am praying for it, I am ready!' This, she explains, is not only for her own economic prosperity, but also to be with her best friend in Amsterdam again. It would seem that she even has her suitcases already, as there are two brand-new looking suitcases stacked on the top of a wardrobe in her small room. However, it turns out that these are not meant for travelling, but rather for her future husband to present to her matrilineal family to complete the dowry claim allowing her to marry him.

While Grace claimed to be ready to leave at any time, the reality was that she has remained in Ghana over the past ten years which would have been the best period in life to go abroad. Various factors explain why she never left Ghana. First, she met her present husband whom she married in 2011, a clerk working with a local bank in Nsawam. This, as the second reason, also provided her with a steadier source of income than her own trading activities, reducing her incentive to look for better ways to generate income, such as through migration. Third, in the same year that she married she gave birth to their first child and felt she needed to remain close to them. This is a personal choice, as often parents do leave their young children in the care of a family member while they are abroad, often for years. All in all Grace now feels few urges to go abroad, at least for the moment.

Nigel: The Critic of Success through Migration

Nigel is a young man who completed his bachelor degree in geology in Kumasi in 2003, which was around the same time that he became a respondent in the research project. His parents live in Tema, the harbour town neighbouring Accra, whilst his girlfriend had been living in Amsterdam for some years already. Nigel had himself been abroad on a number of occasions. During his studies he went to New York three times for about four months on each occasion during his summer vacations. His father sponsored these trips. However he was able to repay him each time he returned to Ghana as on each trip he worked for about 16 hours a day at the Newark Airport of New York as a baggage handler. Nigel: 'I am proud to mention this, as it was only a temporary job.' The second aspect was important to him, as it contrasted with the situation of other Ghanaian migrants he observed on the streets of New York, Nigel: 'Some were even walking around in their Ghanaian slippers, as if they were still in Ghana!' To his mind these migrants had not managed to make it at all as they still struggled to make a living and thus were basically still wearing the same clothes they had taken with them from Ghana.

Following his own experiences he strongly recommends others in Ghana to get the experience of being abroad for some time instead of persisting with dreams of going abroad more permanently. In addition, he argues, such a shift in migration duration would also be good for the Ghanaian economy as people with such short-stay experiences will come back to Ghana, concentrating on opportunities to work or start up their own business here. They will now have their own view of how it is 'over there' rather than observe how those who have gone abroad come back to Ghana in new clothes, with new watches, mobile phones, proudly driving cars to their newly constructed houses. This all suggests that being abroad always brings economic success. In Nigel's opinion these kinds of observations also sustain the urge of young people to consider going abroad as their best option and not try their chances in Ghana instead. Nigel says: 'Many people [in Ghana] think that over there gold is on the streets, but the reality is that of the 10,000 Ghanaians abroad it is only a small number who really make it … Sure, some are able to build houses here, but can they really come back?'

In that sense he is happy that Ghanaian students get the opportunity to do summer jobs in the United States and Europe as then they can see that it is not easy being abroad, which, in his view, may reduce the brain drain. He adds:

Presently it takes about US $6,000 to get a person to go abroad! This money could have been well used to establish a healthy small business here as one third of this amount already buys you a communication centre. But such an investment is not the same thing, as people don't get access to the profit you make with this business while with a migrant you can expect to receive remittances from them.

If he were to have the money at some point and someone came to him with the request to go abroad to look for work he would want to hear a very clear argument why this person needs to go abroad whilst possibly neglecting good opportunities for gaining an income in Ghana which would require much less investment.

Nigel does have plans to go abroad again himself, but this would be specifically to the United Kingdom and only to further his studies to a master's degree level. With a smile Nigel adds that it is important to 'aim at the sun, and if you can't reach there, then go for the moon'. He explains that what he means is that if you want to go abroad, like he does, it has to be for a particular and *a priori* defined purpose. Thus in his case he wants to go to the United Kingdom to further his education. If, however, it turns out that this is not possible then he will look for other opportunities, whether there or in Ghana. For this masters degree he will need about £10,000 to pay for the fees. Nigel: 'This is big money!' To this end he is trying to secure a grant. Should this fail, then he also knows some people who will be interested in sponsoring him and providing financial support. Notably, he expects that he can rely on his father as he has some savings. He adds how he hopes that his father will provide this money as a gift, not as a loan. However, he will readily accept the second option too, should this be what his father prefers.

In January 2012 Nigel explains in a phone conversation how he has gained a well-paid job with a mining company in Ghana, and is now leading a comfortable life. This strongly contrasts with the experiences he had when he was abroad and with the experiences related to him by his girlfriend in the Netherlands. Thus he also makes it clear that at this point he will not consider going abroad again for any long period of time. Indeed, he clearly considers questions on the topic of migration irrelevant at this point. This, he adds, of course does not exclude him from going abroad for visits and holidays. The last comment epitomizes a clear shift in perception from seeing migration as a potential instrument of economic advancement to one in which visits abroad are done for purposes of cultural enrichment or otherwise are clearly linked to specific educational and/or work-related needs.

Peter: The Object of Migration

Peter is 21 years old and the eldest of four children of a female migrant. She had been living and working in Amsterdam for six years at Peter's first encounter with the research project in 2003. His mother had gone abroad with the help of her brother, James. He had decided to sponsor her trip as he felt that this would provide her with much better financial prospects. When she left he had also provided accomodation for her two eldest children. Peter sees it as a real privilege to stay at his uncle's place and also gain work experience in his business. Peter: 'We are fortunate that our uncle came to pick us in Kumasi.' At that time he had not been abroad, although he would much like to go there to see his mother and also to look for work. He sees this as a 'wild plan', however, and first wants to establish 'a foundation' of work experience in Accra before he goes abroad. To this end he claims that he really benefits from being an apprentice in the sliding doors company owned by his uncle. In three years he expects to have gained enough experience to try to start a business of his own. However, for the capital he would then still need to go abroad to save sufficient money to set up the business.

Peter turned out to be difficult to speak to in the subsequent months as he would often cancel meetings at short notice, usually claiming that he was too busy. Conversations on the phone revealed nothing of the turn of events around Christmas 2003. Appearing at his uncle's house with Christmas gifts for various family members, including Peter, the message was conveyed that the gifts for Peter were better posted to him as he was abroad. It then emerged that as of a few weeks ago Peter was no longer living in Ghana. Like his mother he had also been sent abroad by his uncle James. Talking through this sudden set of events with James, who was also a respondent in the research, he explained how, in contrast to the prospects for his own son (the same age as Peter), for Peter opportunities in Accra would always be limited as he was, first and foremost, a self-made man. And because his education was limited to the completion of secondary school, opportunities for him to earn a decent living in Ghana that would enable him to build a house, support a family, and so on, would always be limited. As Peter was still a young man, and as he had already shown an interest in going abroad, James felt that this was the best moment to give him that opportunity and he had sponsored his trip to the Netherlands: 'He can then decide whether he wants to work there for some time and see things, and when he wants to come back. That is his individual decision over which I have no influence.' Thereby he makes it clear that he does not expect to see any financial repayment from Peter for helping him in this way.

Being unable to speak to Peter again after his sudden migration, it is interesting to consider comments from his mother in Amsterdam to Mazzucato, the fellow researcher in the Ghana TransNet research programme, and by his younger sister in Accra. According to Peter's mother, the move from Accra to Amsterdam was one Peter himself had really wanted to make for some time, being keen to leave his old life behind. Yet his sister's comment was somewhat different. When she was asked how Peter was doing in Amsterdam she responded: 'He said that he was fine, except that he was still not happy with the way he had left Ghana suddenly.' She explained that he had particularly struggled with not being able to inform anyone of this plan. Indeed, in the month prior to his departure Peter had thrown a party for his friends to celebrate his birthday, which his sister had organized for him. Knowing what was ahead, she had felt that for him it had secretly been a farewell party. Yet even within the family no one had known that his departure would be so sudden. It materialized that James, their uncle, had indeed acted quite promptly, being able to suddenly secure a visa and ticket for Peter to board a plane, all within the space of 24 hours. Therefore, in order to avoid any legal and financial mishap, his uncle had instructed Peter not to inform anyone, even including his own girlfriend, that he was about to leave the country.

Ringo: The Long-Term Planner of Migration

After calling the mobile phone number provided by a migrant cousin in the Netherlands a meeting is arranged with Ringo on the corner of the Takoradi tro-tro (bus) station, in western Accra, in July 2003. Ringo turns out to be a friendly young man of 23 years who moved to Accra from Kumasi three years ago, once he had completed his secondary schooling. Initially he had moved in with his elder brother Raphael, who worked as a soldier in the Ghanaian army, and who had 'sent for me to come' as he would have better financial prospects in the informal economy of Accra than that of Kumasi or their rural hometown. Once he had settled in Raphael proceeded to support Ringo in starting a business of his own: selling goods on the street to passers-by at the Takoradi bus station. This business had grown over the years to what it was now.

Now, after so many years of 'hustling' on the streets of Accra, Ringo wants to go abroad to join his family members there. More precisely he indicates that he is preparing to go to the Netherlands, as this is where most of his family is. From the stories he hears from the various family members based in the Netherlands he speaks to on the phone, or when they come to visit Ghana, he has come to understand that the Netherlands is much better than Ghana for earning a good income. At the same time he also understands from them that in order for this success to become real he will need to be a hard worker. However, he has calculated that he could earn much more in the same amount of hours in the Netherlands than in Ghana. Indeed, he argues that life in the Netherlands may be very hard for migrants, yet at the same time the money that can be earned will be much more. He has heard that in the worst situation he can still earn 100 euro per month. If 50 euro is needed for all his expenses then this still leaves him with a profit of 50 euro, which is more than he can get here in Accra. He explains that he has come to know of these levels of income and expense from family members who have gone abroad and from some young men who work at the market near him who left Ghana eight years ago and have come back again. He adds that after working abroad for five years they have returned and each has managed to build a house in Accra in this period of time.

He is hoping that his elder brother will be both able and willing to sponsor him to go abroad. If Raphael does help him get overseas, will there be any expectation that Ringo repays him (reciprocity)? Ringo: 'Yes, I will remit every month so that Raphael can profit too from my going overseas.' Sending these remittances to him will also mean that Ringo is no longer relying on Raphael to help him, as is the case at present.

In a later conversation Ringo explains that those who are already abroad might also be willing to sponsor his trip. That this is not some random statement becomes clear when Ringo responds quite frankly to the question of whether an aunt based in the Netherlands would be willing to support him in his livelihood by sponsoring his trip: 'Yes, and I have asked her this very thing.' She had become his foster-mother following the death of his own mother (many years ago) and thereafter another aunt (who had become his foster mother following the death of his mother). He explains that now his aunt based in the Netherlands is trying to get the right documents to enable him to come over. However, the financial losses she had incurred with the funeral of her sister, his prior foster-mother, had set her back a little in carrying out these plans. In understanding why she would take Ringo abroad, and not some other family member, she responded that she should try to help at least two family members from different branches of the family to go abroad. Once abroad they, in turn, would be able to provide for their family branches, but also try and get someone else to come to the Netherlands. This is a clear case of chain migration.

In January 2012 there is an opportunity to meet with Ringo again after many years. It turns out that he started to run his own taxi three months previously. At the meeting he is not only happy to be in touch again, but also clearly proud to show off his new business. Ringo explains that the car had come from Raphael, his brother, who was in the United States for army training. They had pooled their financial resources in order to buy and import this car to Ghana. The long-term intention is to accelerate the pace at which Ringo is able to secure enough money to go abroad too, specifically to the United States. Originally the plan had been to sponsor him to come to the Netherlands, 'but that plan has been cancelled' because of lasting financial issues of his foster-mother, the aunt who had intended to sponsor him.

In explaining the current rationale for going abroad Ringo mentions that his two children are now aged 13 and 14 and attending junior secondary school. In the endeavour to go abroad he will take into account their age and basically wait for the moment they have completed their schooling. At that point he considers them grown up, and then he will also be 'free' to travel. At the same time, notably if he gets the opportunity to leave earlier for the United States, he will find a suitable woman to look after them. This will not be their biological mother, as neither he nor the children have any contact with her.

For his trip abroad Ringo indicates that the costs for all the paperwork, excluding the ticket itself, may amount to US $10,000, especially if the paperwork is to be completely legal. In his view the reason why the costs are so high is due to the long chain of middlemen one's papers have to pass through. As each seeks to derive some profit from this process the overall costs go up considerably. Yet, whilst false papers can be bought for half the money, he has not considered this option, as he feels that the chances of being caught in the process are simply too great. Furthermore, he adds, upon arrival one would face marginal employment prospects, which would also negatively affect the chance of saving money. Hence he will aim for a fully legal approach, even if this costs twice the amount.

Yet, who is trustworthy enough to rely upon to get this documentation, particularly given the fact that he is in a vulnerable position, certainly when the process passes down the chain of people involved? Initially Ringo explains that he trusts God to put the right people in his path. However, he adds there is a need 'to move with people in order to get to know them', that is, trust needs to develop out of a series of conversations he has with those who might help him with the legal paperwork. He also conducts thorough but inconspicuous checks of the credentials of these people: checking where they live, who their families are, what their networks of friends look like, and so on. Only then is it possible to be assured of the trustworthiness of the immediate connection man as well as those he has linkages with.

Ringo intends to stay in the United States for six to nine years. This, he reckons, will give him sufficient time to save enough money with which he can then set up his own business back in Ghana. He has in mind starting up a belt factory, thereby intending to use local hides and other material, and employing local people to sell these belts around Ghana. He has the feeling that there are few such businesses, particularly if the belts are original and considered of high quality because they are made with local resources. As Ringo puts it: 'When you invest you must do this in something you know well. And for me I know all about the belt business.'

His fellow trading companions and friends on the Kaneshi corner, whom he has known for a long time, all have no notion of his migration plans and the subsequent project. Already some of them look at him with certain jealousy for becoming a taxi driver in a relatively short time. As Ringo puts it: 'There are some who want to "pull the man down", they enjoy you not having success if they also cannot have it.' With this he means that many of his contacts, notably his fellow traders from the times he was plying his business on the street, but also his closer friends such as those coming from his hometown, are jealous of his successes so far. He fears that they might try to jeopardize his chances of migrating, should he disclose his plans to them. It is for this reason that the discussion takes place in a 'drinking spot' a little away from his friends.

Synthesis

This chapter has sought to give insight into the mental process of identifying migration as a possibility in the first place, and thereafter its concretization in terms of defining a destination and the trajectory to get there. The threshold approach, with its focus on processes affecting whether, when and where to migrate has helped to understand the decision-making process in all its dimensions. However, it also gives space to explorations of the relationship between borders (in the classical geopolitical sense, but also as a set of guiding social institutions), personal mental processes, and control mechanisms arising out of social networks. In the prior empirical section the relationships between these different elements were laid out for four young potential migrants, as related to varying notions on the role and influence of migration in their lives. Indeed, this also highlighted differential prospects for building up livelihoods and careers in Ghana, versus prospects available in destinations in the Global North.

What induces Ghanaians to develop aspirations to migrate abroad? For the Ghanaian respondents of this research this aspiration related to the knowledge they had, often through members of their social networks, on the suitability of various destinations. Therefore, alongside economic dimensions, social connections to specific locations were mentioned, albeit relating more to their connections with migrants they knew in those places or countries, rather than to an intimate knowledge of a local culture and the host society. In that sense both push and pull factors distinguished in the threshold approach under the locational threshold take on a more sophisticated meaning. Thus for the pull factor it is not the attraction of labour employment opportunities and other economic gains alone that motivate a prospective migrant, but the connection to migrants already there. For the push factor the need for a social dimension next to a (macro)economic dimension can also be argued: aspiring migrants may choose to purposely focus on the positive 'sounds' and signals coming from migrants abroad, rather than recognize that not all who set off actually make it to some foreign destination – as testified by a steady flow of returnee migrants to Ghana. In relation to the concept of bounded rationality the point to be made here is that given the enormity of the migration project when conceived as a totality, aspiring migrants may strategically choose to only focus on one aspect at the time, but also to ignore certain evidence that may adversely affect their willingness to act on their desires to migrate.

A further connection can be drawn between the locational and the trajectory threshold where this concerns the goal of aspiring migrants to reach *Schengen*. At face value such aspirations seem to indicate an unawareness of the sheer distance to this envisaged destination, but also what this perceived destination actually is. Yet, however tempting it might be, it would be wrong to dismiss these abstract ideas of destinations as fantasies for two reasons: First, many migrants purposely maintain a certain vagueness in the final destination they seek to reach in order to make this dream achievable. Making a list of all the issues they might come across along the way would make the idea daunting and result in them not setting off at all, even though this was their strong desire. Again this shows how the boundedness of rationalities not only pertains to what is actually known, but also to what is considered relevant moment to moment. Second, the concept of Schengen that aspiring migrants conceive of in their plans, may be abstract to its own citizens, yet it is a relevant geopolitical construct to these migrants, as this is the place they need to reach, more than the particular countries, places and cultures which make up Schengen.

Turning back to the migrants' aspirations, it can be discerned from each of the accounts of the four respondents how the migration horizon continually shifts. This is notably with regard to the level of perseverance to (still) migrate, which is linked to an awareness of opportunities elsewhere and what is required to get there, but also to social and economic developments taking place in the meantime. Thus Nigel, the university alumnus, imparted a rather personal perspective on what migration should mean for people following his observations of struggling Ghanaians abroad. He thus argued for a relative, and comparative understanding of the merits of migration. To that end the opportunity and actual costs of migration also came up in the other case studies, including the manner in which the destination and the trajectory were defined. Thereby the variation in definitions of the trajectory, from an ad hoc decision to long-term preparations taken for a specific trajectory, was noticeable from the accounts provided by respondents. In the case of Peter, his family had prompted his migration to the Netherlands. They saw this as a necessary step for him to take to be able to make certain progress in his life. Whilst this was a rationale he also accepted, the moment of migration was not quite of his choosing. His case provides a healthy qualification in any discussion on the agency of (prospective) migrants, but also attests to the relevance of looking at migration as a decision seldom made by migrants alone, as New Economics of Labour Migration theorists suggested some decades ago.

Overall it can be concluded that the four cases have articulated the relevance of the *threshold of indifference*, in some cases also interlinked to the *locational threshold*, to explain aspirations of potential migrants based in Accra.

Postscript

In January 2012 the author visited Ghana for a few weeks and was able to spend time with some of the respondents. Whilst in a taxi en route to a respondent the author had the opportunity to talk to the taxi driver for quite some time. The taxi driver explained how he had come to Accra from the eastern region roughly ten years earlier. He had been working there as a farmer, but improvements to the farmlands of his family would require a major capital investment. Hence he had decided that he should go to the city to try to save some money before going back to his hometown and take over management of the family lands, making these more commercially viable. Whilst he related his plans he suddenly looked at his passenger, whose origins he could only guess, and added he might also wish to go abroad, although only if he could work for an agricultural company. To his mind this would not only help him save money faster than he was able to in Accra, but it could also give him new insights to take back with him to Ghana and apply to his own agricultural activities.

This account draws clear links with that of the church visit which opened this chapter. Both accounts show the sense of opportunism with which international migration is considered as one of various avenues to achieve certain financial goals. The two accounts also help to explain how many younger people think about migration, namely as a perpetual grey zone of opportunity, the precise contours of which only become clearer after each hurdle in the process is overcome. This was exemplified in the variations found amongst the four respondents in their approach to migration and especially showcases the very dynamism of immobility and mobility as overlapping phenomena in the trans-world of migration-aspiring, young city dwellers in Ghana.

References

Burrell, J. 2009. "Could Connectivity Replace Mobility? An Analysis of Internet Café Use Patterns in Accra, Ghana." In *Mobile Phones: The New Talking Drums of Africa?*, edited by M. de Bruijn, F. Nyamnoh and I. Brinkman, 151–70. Bamenda: Langaa.

Czaika, M., and M. Vothknecht. 2012. "Migration as Cause and Consequence of Aspirations." IMI Working Papers Series 57. Oxford: International Migration Institute.

Faist, T. 2000. *The Volume and Dynamics of International Migration and Transnational Social Spaces.* Oxford: Oxford University Press.

Kaag, M. 2004. "Ways Forward in Livelihood Research." In *Globalization and Development: Themes and Concepts in Current Research*, edited by D. Kalb, W. Pansters and H. Siebers, 49–74. Dordrecht: Kluwer Academic Publishers.

Kabki, M. 2007. "Transnationalism, Local Development and Social Security: The Functioning of Support Networks in Rural Ghana." PhD diss. Leiden: African Studies Centre.

Massey, D.S., F. Kalter, and K.A. Pren. 2008. "Structural Economic Change and International Migration from Mexico and Poland." *Kolner Z Soz Sozpsychol.* 60 (48): 134–61.

Mazzucato, V. 2009. "Bridging Boundaries with a Transnational Research Approach: A Simultaneous Matched Sample Methodology." In *Multi-Sited Ethnography: Theory, Praxis and Locality in Contemporary Social Research*, edited by M.-A. Falzon, 215–32. Farnham: Ashgate.

Rakodi, C. 2002. "A Livelihoods Approach: Conceptual Issues and Definitions." In *Urban Livelihoods: A People-Centred Approach to Reducing Poverty*, edited by C. Rakodi and T. Lloyd-Jones, 3–22. London: Earthscan.

Schapendonk, J. 2012 "Turbulent Trajectories: African Migrants on Their Way to the European Union." *Societies* 2 (2): 27–41.

Simon, H.A. 1982. *Models of Bounded Rationality: Behavorial Economics and Business Organization*. Cambridge, MA: MIT Press.

Smith, L. 2007. "Tied to Migrants: Transnational Influences on the Economy of Accra, Ghana." PhD diss. Leiden: African Studies Centre.

Speare, A., Jr. 1971. "A Cost-Benefit Model of Rural to Urban Migration in Taiwan." *Population Studies* 25 (1): 117–30.

Van der Velde, M., and H. van Houtum. 2004. "The Threshold of Indifference: Rethinking Immobility in Explaining Cross-Border Labour Mobility." *Review of Regional Research* 24 (1): 39–49.

Van der Velde, M., and T. van Naerssen. 2011. "People, Borders, Trajectories: An Approach to Cross-Border Mobility and Immobility in and to the European Union." *Area* 43 (2): 218–24.

Chapter 4
Rational Routes?
Understanding Somali Migration
to South Africa

Zaheera Jinnah

Contemporary Somali migration has been framed by the political and economic conditions induced by the collapse of the central government in Somalia in 1990 and the ensuing civil war. Yet, the understanding of Somali migrants in this chapter does not depart from the onset of war but rather from the concept of *buufis*. This is a Somali word, translated as 'to inflate or blow' (Zorc and Osman 1993) and used by Somalis in Kenya and elsewhere to refer to the longing to go abroad (Horst 2006). It is also used by Somalis to describe the state of depression that many associate with longing to migrate. The rationale of Somali migration to South Africa is constructed both from the logical advantages of the latter's legal policy framework in the region, and the less logical mythical expectations of work, trade and onward migration that the country seems to offer, created by and transmitted through social networks. The lack of an encampment policy in South Africa, and the notion of free movement and trade for asylum seekers and refugees are strong motivators for migration to the tip of the continent. In addition, the increasingly anti-immigrant and anti-Islamic sentiments in parts of Europe and North America has been fuelling more nuanced, strategic decision-making amongst Somalis who are in search of social and economic opportunities and a sense of belonging. At the same time, the practical support offered by and narratives of a 'land of opportunity' told and carried by transnational social networks in Kenya and Somalia serve as additional impetus in the decision-making process to realize a better life abroad.

Drawing on original empirical research this case study seeks to understand how Somali migrants and refugees create, and invest in, passages of hope and profit to make the journey to South Africa. This chapter applies the threshold approach, which is taken as a theoretical framework that aims to understand the role that personal and geographical factors and processes play in influencing mobility and immobility. In particular, it seeks to understand how the locational threshold, and to a lesser extent the trajectory threshold, are useful paradigms in better understanding the migration project of this group and how these approaches can be extended, adapted or better understood in light of the empirical work presented here. The central argument of this chapter is that in instances of forced migration, people exercise strategic and subjective rationality in the migration decision-making process. More explicitly this case study demonstrates that formal and informal socio-economic and political characteristics of South Africa, subjectively perceived and informally carried through social networks, play a significant role in drawing Somalis to the country.

The outline of this chapter is as follows: First the empirical context of the study is given followed by the theoretical framework of the locational and trajectory thresholds including the methodology of the empirical research. Next a series of narratives are presented in which this approach is fleshed out through case studies of Somali migrants in Johannesburg. Finally the key points of analysis and limitations arising out of embedding this case study in the threshold approach are discussed.

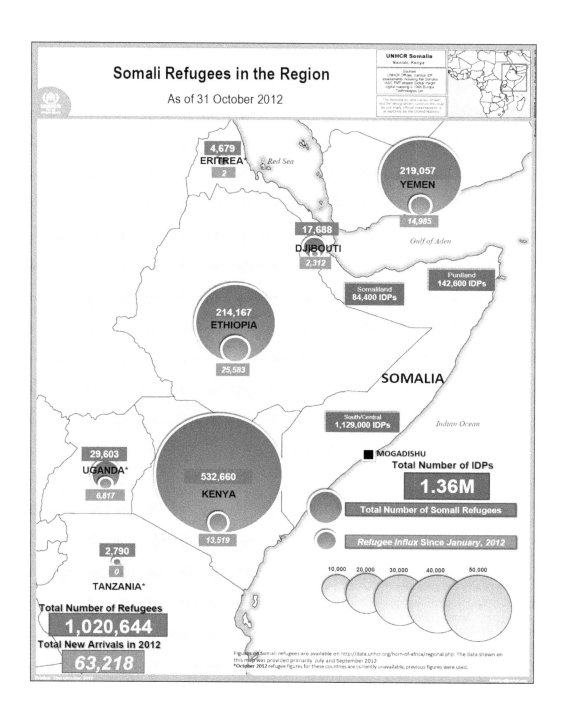

Figure 4.1   Somali forced migrants and IDPs

Source: available at http://www.unhcr.org/5077ccd318.html.

Empirical Context

Mobility and migration are a characteristic of Somali life dating from nomadic practices to modern-day realities. As pastoralists and herdsmen, Somalis historically moved in search of greener pastures to cultivate or for their stock to feed on. Others who were traders, were drawn to fluid caravan routes and far-away destinations to source and sell their ware. As part of the Arabian trade networks, Somalis adopted a hospitable view of outsiders who they regarded as their link to commercial and spiritual resources. Following the civil unrest and conflict that has plagued Somalia for most of the latter part of the twentieth century, mobility took on an additional function: a means to escape war and political and economic uncertainty. Current estimates show that Somalia has a population of 9,557,000 people (UN 2013),[1] of which roughly a quarter is believed to be living in the diaspora (Lewis 2008), including 1,130,939 refugees originating from Somalia (UNHCR 2014).

Although in the last two decades much attention has been paid by scholars and policymakers to those seeking political refuge and economic opportunities outside the country (Buijs 1993; Horst 2004, 2006; Hopkins 2006; McMichael 2002), more than half of the mobility has been within Somalia (see Figure 4.1),[2] an aspect which has been often overlooked (Noor 2007). By contrast, official figures of Somalis in South Africa do not exceed 30,000 (UNHCR 2012). This strengthens the argument made elsewhere in this book that most people, even in daring circumstances, do not move across borders (Van der Velde and Van Naerssen, 2011; see also Smith et al. 2010).

Nevertheless, for those that have moved, and have felt forced to move abroad, the decision as to why, where and how they move remains pertinent, and needs to be elaborated at multiple levels and through multiple paradigms. It is easy to consider all movement from Somalia as a homogenous flow of displaced people, given the macro-level impact of economic and political factors shaping the country during this time. However, such a perspective undermines the individual decision-making, the particular needs of the group and the family, and the agency of the migrant within the complex, contoured, and often composite patterns of mobility (Jinnah 2010).

Drawing from a 20-month ethnographic doctoral research project undertaken between 2009–11 in Mayfair, a suburb nestled close to the central business of Johannesburg that is home to an estimated 12,000 to 18,000 Somalis in the city, this section demonstrates some of these complexities to show that the threshold approach is useful in better understanding the ways in which migrants conceive, perceive, and undertake their transnational journeys. In this chapter three narratives are presented to illustrate how this approach helps in exploring the subjective meanings that migrants make before, during and after migration. They also demonstrate, in different ways, the deep personal meanings attached to decision-making throughout migration and settlement.

Somalis began arriving in South Africa in the early 1990s, following the collapse of Syed Barre's regime in Somalia and the prospect of new opportunities in South Africa. Initially they settled in Mayfair, which had a significant South African Muslim population and was established as a place for trade (Jinnah 2010). Gradually Somalis began venturing further afield, and over the next two decades created economic opportunities for themselves through small-scale retail businesses in cities and in townships (former black areas located on the fringes of major cities and towns) in almost every major and secondary city in South Africa.

---

1   The last national census in Somalia was undertaken in 1975.

2   According to a 2012 UNHCR snapshot of Somalia, there are 1,356,845 internal displaced people or IDP's versus 1,077,048 refugees.

Most Somalis in South Africa hold refugee permits (UNHCR 2005; FMSP 2006). This allows them to live, work, trade, study and move freely within the country. However, despite this provision to ensure their protection, they face considerable risks from xenophobic-related violence by the public and state bodies (Misago et al. 2009). Partly to counter this threat, and partly to preserve a sense of community, Somalis have established several community organizations in South Africa (Jinnah and Holaday 2010; Polzer and Segatti 2011; Johnson 2010).

Although South Africa has not had a long and entrenched history of Somali migration, it has received steady flows of refugees and migrants from that country since the early 1990s. The perceived new opportunities for work, trade, settlement and resettlement, and the political stability that it posed in the post-apartheid era drew many migrants from across the continent, including Somalis. When South Africa confirmed a non-encampment policy for asylums seekers and refugees in 1998, the move was widely applauded by forced migrants. However, in time, many Somalis have found that self-settlement in competitive and hostile urban centres has resulted in a great deal of economic and physical vulnerability and risk (WRC 2011; Landau and Gindrey 2008; Misago et al. 2010).

In the last two decades, there have been four main waves of Somali migration to South Africa (Jinnah 2010). The first wave refers to what is often called the pioneer stage amongst migration scholars (Massey 1990) where a small group of Somalis, mainly single men and small-size families, came to South Africa in the early 1990s, fleeing the political conflict in their home country. At the time South Africa was still under apartheid rule, and rights for (non-white) foreigners were not existent. In South Africa, Somalis settled, as mentioned, in Mayfair, Johannesburg, which was home to a large Muslim community of Indian origin, who provided shelter and assistance to Somali refugees at a personal, private and public level through religious and charitable institutions. The latter were in some cases already working directly in Somalia or were raising money to send to Somalis. Therefore there was some knowledge amongst locals of the political situation in Somalia and empathy for Somalis entering the city (Jinnah 2013). The early Somali migrants slowly established themselves by working for, and/or trading with local Muslim-owned businesses. In time Mayfair also proved to be a strategic base for the economic and social settlement of Somalis. This was for a number of reasons. Firstly, Mayfair was situated close to the bus and railway services linking Johannesburg to the rest of the country and the region. Secondly, it was close to key economic hubs in the city such as the central business district, and other wholesale and trading areas where Somalis could find work or buy goods for trading (such as the Chinese malls).

The second stage of Somali migration to South Africa, from the mid-1990s to 2000, occurred as a result of the effects of social networks. Somalis at home had begun hearing of the possibilities of work, business and self-settlement in South Africa and decided to join the small community there. The groups largely consisted of men who were related to or knew someone who had travelled to South Africa. Around this time, a few factors spurred an increase in those seeking to come to South Africa. Firstly, a few Somalis in South Africa had resettled in the United States, or Scandinavia, sparking a desire to come to South Africa as a means to travel onward to Western countries. Secondly, the first group of Somalis (who had saved money from petty trading and working) slowly started opening small businesses in Mayfair. They, in turn employed other Somalis creating the impression that business would be possible in South Africa. Later businesses were also opened in smaller towns in Gauteng and other provinces and in the townships in Western and Eastern Cape and Gauteng, which became the target of xenophobic-related violence and intimidation.

The third phase of migration to South Africa occurred as a result of the Ethiopian incursion of Somalia in 2000. This movement was more diverse and intense than the previous two waves.

Younger men and women of all tribes joined the mainly older and predominantly Ogaden Somalis who were in South Africa at the time. This changed the profile of Somalis in South Africa and also created tensions amongst Somalis due to tribal, gendered and generational differences. Whereas the first groups of Somalis were seen as hardworking and honest, who respected cultural and religious norms, newer migrants became known amongst the established Somalis as young and reckless who would rather hang out on the streets and chew *kat*[3] than work or trade for a minimal income. Between 2000 and 2005, as a result of the ongoing war of more than a decade in Somalia, and the breakdown of families and livelihoods through migration, economic restrictions and conflict, social norms at home began to shift. This resulted in many more women starting to leave Somalia in their own right, with their children (as divorced or widowed mothers) or as single women looking for work to support their extended family at home (elderly parents and younger siblings and children). This added to the diversity of Somalis in South Africa and would later shape social and gender norms in the host country, as women began working, trading and asserting economic independence.

The fourth phase of migration, like the second, was an intense movement of a number or people. It was caused by the drought and famine in Somalia in late 2011. The international media coverage of the famine in 2011 renewed NGOs support and public interest in the Somali issue which raised significant funds for Somalis at home and also created a more sympathetic public discourse for Somalis in South Africa.

Theoretical Context

The threshold approach (Van der Velde and Van Naerssen 2011) regards migration as a process, conceived, undertaken or disregarded by individuals who often rely on incomplete information. In part this explains the continued mobility across borders at a global scale despite stringent and severe attempts by states and regional blocks to restrict, limit or stop this movement. In South Africa for example, waves of deportation of Zimbabweans has not stopped the flows of migrants from that country (Amit 2012). At the same time, the reliance on subjective knowledge also explains immobility; even when people face severe political and economic conditions in their home countries they sometimes are unlikely to make the decision to move if they fear a loss in networks, or notions of belonging.

The approach defines thresholds as 'barriers or hurdles' that inhibit mobility, recognizing that physical borders may not be the most significant of these for migrants. One aspect which the approach does not develop fully is the interaction between structural factors and personal ones. How do potential migrants understand and engage with hurdles to international mobility, whether there are personal limitations or state-imposed restrictions? How are these perceived differently, if at all? In sum, what is a threshold and how is it conceived and understood by different types of (forced) migrants? In the analysis to follow, this is unpacked further using the empirical findings as a means to deepen this theoretically.

Secondly, the subjective realities of South Africa carried by social networks are explored within the locational threshold. Here particular attention is paid to the opportunities and risks presented by a potential host country, and the likely impact this has on making the decision on where to move.

---

3 *Kat* or *meera* is an herbal substance with narcotic properties commonly used in eastern Africa and parts of the Gulf. Its legal status differs from country to country. It is illegal in South Africa but legal in Somalia, Yemen and the United Kingdom.

What is of importance here is that this decision is largely a subjective one, which is made relying on incomplete or distorted information rather than information obtained through formal channels of settlement or migration. This supports the question of whether the actual location, or indeed the trajectory, is as important as the perception of these in informing potential migrants' decision-making processes.

As stated earlier, the core material for this chapter is drawn from original ethnographic fieldwork conducted in Johannesburg from 2009 to 2011. During this time, in-depth interviews were conducted with key informants from Somali business, community and political structures, South African civil society organizations, repeated interviews with 60 Somalis, and participant observation in the houses, streets and shops of Mayfair, an area of the city in which most Somalis live and work. Respondents were selected through a combination of purposive and snowball sampling techniques. Interviews and field notes were transcribed and analysed using thematic and narrative analysis. The section below presents some of the narratives emerging from this fieldwork. The use of narratives is opted for to both understand the subjective realities of migrants and to provide case studies that illustrate some of the nuances and provide a basis for the analysis that follows.

Three Narratives

Fartuun: 'I just want a good life'

Fartuun began thinking of moving out of Somalia in 2007 after the assassination of two of her colleagues. At the time, she was a journalist-in-training in Baldweyne, central Somalia, reporting for an American radio company. It was this that made Fartuun and her colleagues a constant target by Islamists, many of whom considered those working for Western companies, and particularly for American ones, as traitors or, even worse, as spies. Fartuun had never considered leaving Somalia before her colleagues were killed. She was born just before Syed Barre, the Somali dictator who ruled the country for more than two decades (1969–91), was overthrown in 1991, and her young life had been marked by the political upheaval in her country. This was the only life she knew and one that she accepted, for she believed that even amidst the chaos and violence, she would be able to make a life for herself. More importantly Somalia was home: her friends and family were there, she knew the language and customs and she felt that she belonged there. Although some of her half-siblings were living in Kenya and Norway, migration was not an option for her as she had a decent job and was able to live what she calls a 'a normal and happy life'. In a way she showed a clear indifference towards migration.

However, that all changed one fateful day in November 2007, when two of her colleagues were killed by members of Al-Shabab. Fartuun was forced to go into hiding and eventually had to leave the city and resettle in Mogadishu, where she stayed with family for a few months. During this time she realized that her dream of being a journalist was no longer possible in Somalia. At this point she started thinking of migration as a way out of her personal circumstances – safe but without work. Eventually the decision to move was made in order to find opportunities for study, work and 'live a good life', as she puts it. Yet the decision to leave Somalia was not an easy one to make, she needed to consider whether the protection that migrating would offer would offset the betrayal she felt in leaving behind her mother and younger siblings. In the end, migration was an option for her because of two additional related factors: firstly the company she worked for arranged a route out of Somalia for her, and secondly it offered her a job (albeit an informal, temporary one) reporting on the diaspora. Armed with the economic muscle that this gave her, Fartuun left Somalia in 2008 and settled temporarily in Kenya. Although she had paid work from an American news company, she did not have the documentation to stay in Kenya. Rejecting the option to live in a refugee camp, she self-settled as an undocumented migrant.

Although she was safe here, she was dissatisfied living on the 'fringes of society' without being able to enjoy full access to educational and work opportunities. She did not feel that she could find the life she wanted in Kenya. In bustling Eastleigh, a suburb of Nairobi in which many Somalis live, she heard from other Somalis of a country where Somalis are easily granted refugee status, able to work or study, and perhaps even be offered resettlement in the United States or Europe. For Fartuun this new country – South Africa – represented her dream of a better life and she once again made the decision to move. This time though preparations for the long and costly journey to South Africa (she would have to travel by road as she had neither the money nor the documentation to leave Kenya legally) would take her several months, and at times she even considered cancelling when she encountered set-backs in her travel plans. Eventually though her desire to escape the stagnating life in Nairobi propelled her to finalize her plans and leave.

The journey was a long and difficult one: passing through four countries in two months, travelling clandestinely in the hands of several agents who used back roads and cut through ravines to avoid being detected by authorities, sleeping in buses, going without food or water for hours and even days. She hadn't heard of these difficulties from the Somalis who had made the journey to South Africa; they had emphasized the life after the journey, and she had thus been unprepared for it. Several times she felt like turning back and would have if she had the money and know-how to do so.

Once in South Africa though, she realized that life was better than what she had experienced in Somalia or Kenya. She obtained a legal asylum seeker permit on arrival, through the regular formal channels, which enabled her to move freely in the country with documentation and gave her the right to work, study or trade; shortly thereafter she had a full refugee permit.[4] She resumed her reporting duties for the American company and easily integrated into the Somali community in Mayfair, which, although dominated by the Ogden tribe to which she did not belong, was fairly receptive to all Somalis. After three years in the country, Fartuun says, 'Life in South Africa is beautiful if you have money,' but for a poorly paid journalist who has a large family to support at home, she is unable to enjoy most of the resources around her. In particular she is unable to access the costly tertiary education that she covets. Hence, once again she is thinking of moving, this time as a resettled refugee by the UNHCR (United Nations High Commissioner for Refugees) to the United States, where she hopes she will find her better life. This dream and journey too are laced with sentiments she has heard from those who have already undertaken it. In the United States, Fartuun has heard that education and housing for refugees is free, that she will be able to have language classes and eventually send for her family to be reunited with her. She is uncertain of the exact conditions of each of these, or of the time frames associated with them. The details are irrelevant, rather the excitement in the voice of relatives who have made the journey and the constant updates of work, study and cars on Facebook from friends who are in the United States are testament that a better life awaits her.

---

4   The Department of Home Affairs, the national state entity mandated to regulate non-nationals, grants most Somalis asylum seeker and refugee permits.

Aweys: 'I always wanted to leave'

In contrast to Fartuun, Aweys cannot imagine a time when he did not want to leave Somalia. Meeting for the first time in downtown Mayfair in 2010, he was unhappy, sitting forlornly on a kerb outside a Somali coffee shop, chewing *kat*. He said matter of factly, 'I am in *buufis* and I was always in *buufis*.' It is easy to understand why Aweys was in *buufis* in Somalia: he had no job, no education and no money, and unlike Fartuun did not have any strong emotional bonds to his family. 'There was nothing to live for,' he said, 'People were killing each other around me and I was going mad.' Also, he had few resources that would facilitate migration.

Then an opportunity presented itself to him: his brother, who was living in a small city in South Africa, where he ran a shop, was killed in a xenophobic-related incident, leaving behind a wife and a young child. In line with tribal customs, Aweys was sent by his family to marry his widowed sister-in-law. 'I was so happy when I heard I was going to South Africa, I only thought that I am (finally) leaving Somalia,' he says. And, 'I didn't consider what life I was coming to.' Without much money and no documentation his journey would be a risky and difficult one, yet the knowledge of this did not deter him, unlike Fartuun. About the long journey by sea from Kismayo to northern Mozambique, in a makeshift vessel, and then on foot through the wilderness to enter South Africa, Aweys says: 'I felt I was in the air, nothing troubled me.' Once in South Africa though, he realized the extent of the fate he had been forced into: being married to a woman he did not love, looking after a business he did not care for and which did not excite him, and stuck in a small city miles away from other Somalis. He found life stifling and wanted out. Challenging the wrath of his family, he divorced his wife, left the business and moved to Johannesburg. Now he sits all day chewing *kat*, jobless, almost penniless and dreaming about a better life abroad. For Aweys migrating was imagined to be a route to a better life, and the reality of the journey or the location did not play a significant role in the decision to migrate. Neither was migration seen an as a linear, once-off project, rather it was embedded in a longer life project.

Hajiro: 'I want to be free'

'I didn't have a happy childhood,' says Hajiro, a stout woman in her 40s who has a stern look on her face. This opening statement sets the tone for the interview, in which Hajiro describes her desire to live a free life. She narrates the many problems she faced growing up in a poor household: 'My father was old and couldn't work, so my mother was selling clothes to feed us. I went to school until form 1,[5] and then I had to leave as my parents couldn't afford the school fees.' With limited education, and meagre economic resources at home, Hajiro was coerced into a marriage by her aunt. She was 16. The marriage was an unhappy one and Hajiro's husband turned abusive. Yet she had two children with him, as social norms dictated. Facing a bleak existence, she turned to her father and asked for his assistance. 'He saw how unhappy I was and he made me get divorced,' she explains. But the divorce would cost her their children as her ex-husband retained custody of them.

Although happy to be free of the marriage, Hajiro suffered from stress and depression. It was this stress which was rooted in her everyday physical environment that made her consider migration as an option. She recounts:

_____

5   Form 1 refers to the first year of secondary education. At this stage students have completed seven years of primary schooling.

I thought that I needed to get away from my life, my family, and my worries. I thought that I need to find peace as a person, peace from the war in Somalia, peace as a woman from the stresses of my life. I am a Muslim woman but I have rights, so many Somali men think that Islam lets them treat women badly, and this is not the truth. So I went to Tanzania and stayed there for a while.

After a few months, she was joined by her younger sister. They were free of the restrictive social norms they had experienced as women in Somalia but still faced obstacles as they were undocumented. 'We had no papers,' Hajiro says, 'You can't be free without an ID. We were scared of [being arrested and] ending up in the refugee camp.'

She heard through social networks that South Africa had no refugee camps and that Somalis were able to obtain documentation relatively easily. This information prompted Hajiro and her sister to join a group of Somali migrants and travel to South Africa. She remembers the journey:

The journey was terrible: at the Mozambican border the Somali agent paid money to a South African woman who took us across the border but kept us locked in house. They took all our money, and our jewellery. When they let us go we made our way to Johannesburg and then to Mayfair [but] with nothing [money or clothes]. I thought I have to work a little harder for the freedom I wanted. I got my refugee card and my refugee ID, so now at least I am someone even if I am only a refugee. I can sell things and be my own person. South Africa is safe from war but it has crime. It is funny that I can't find the peace and freedom I want. I can be free here but I can't travel out of the country.

Like many refugees in South Africa, Hajiro needs to apply for a travel document at the UNHCR, a process that is lengthy, and bureaucratic. And meanwhile, in South Africa, she finds that the close density of Somalis in Mayfair results in many social norms being produced and reproduced, often to the detriment of women. 'I have rights as a refugee but Somalis here still think they can tell you what to do because you are a woman,' she says. Initially Hajro dreamt of leaving Somali to escape traditions and a lifestyle that she found suffocating. She associated migration with making a new start in life in a new place, yet as a migrant she has experienced new forms of exclusion and discrimination in Tanzania and later South Africa. Her migration was in essence her life improvement project.

Her narrative raises the point of how embedded repel and keep factors are in personal lives in both the host and home societies. Although migration processes occur in a broader political climate, the subjective reality is as important in shaping decisions around moving or staying as Hajiro's narrative testifies.

Discussion

These narratives illustrate that rationality as it is often conceptualized in migration theory, needs to be differently understood. It is a personal decision-making process that is rooted in subjective realities that derive from the ability to mobilize information and resources at hand rather than as an objective exercise taken by a rational individual who relies on formal, complete knowledge. This confirms the argument made earlier in this book. There are three distinct arguments that this contribution raises in relation to this.

First, in all the narratives, despite the presence of strong overarching political and economic conditions that could justify movement, it were personal, subjective realities that were the real or final driving force in considering migration. Migration was conceived as a journey toward self-improvement, an option to overcome personal difficulties or to meet personal goals rather than as a response to general environmental conditions. At the same time, a combination of factors in South Africa shaped movement or furthered mobility amongst the respondents. It was the ability to have documentation, to live and trade freely and perhaps to move onward that made migrants consider South Africa as destination country.

Whilst this decision is easily understood given the encampment policy in much of eastern and southern Africa, it is less logical to apply when Somalis are choosing for a host country with non-encampment policies such as Egypt, which is closer to Somalia and with which there are stronger linguistic and religious ties than with South Africa. In parts of Europe and North America, refugees are entitled to social welfare benefits and take advantage of study and employment opportunities. So can it only be understood from a personal perspective that South Africa was perceived as a better country to migrate to? The second argument that these narratives raises is that of the role of social networks: through (incomplete) transnational flows of information, many Somalis heard of the relative tightening of immigration and refugee laws and the anti-Islamic sentiments in Europe and North America on the one hand, and on the other, of the opportunities to work and trade and circumvent some of the immigration restrictions that they faced in eventually getting to the West by moving, if only temporarily, to South Africa.

The third point is that these narratives suggest that the locational and trajectory thresholds are more relevant to some than others. Although the cost and length of the journey from home to host country are important considerations, its actual determining value differs. For Aweys for example the harshness of the route was perceived as an easy threshold to overcome given his desire to leave Somalia, whilst for Fartuun and Hajiro, the difficulties encountered en route questioned their decision to migrate. For each of these migrants the reality encountered in the first host country propelled them to move again. Thus the amount of resources invested in the migration project determine to an extent the strength of each threshold.

Conclusions

The central objective of this chapter was to position an empirical case study of Somali forced migrants in South Africa within the conceptual framework of the threshold approach in understanding decision-making towards migration (Van der Velde and Van Naerssen 2011; Smith et al. 2010; Van der Velde and Van Houtum 2004). In so doing it aimed to extend the theoretical and methodological rigour of the approach by raising questions about its applicability in understanding mobility. In light of this, three points are raised in this conclusion.

First, the threshold approach provides a useful lens to explore (ir)rationality in migration. Even within particular and dominant macro political and economic factors, it is the realities of one's own world and the strength of one's perceptions and dreams that condition decision-making in the migration process. For the migrants in this chapter, personal, subjective realities provided the real driving force in considering migration. That is, knowledge of migration routes was seldom complete, even for those holding professions which might be expected to provide access to such information, as in the case of the young journalist Fartuun.

Second, and in a similar vein, the role of social factors within migrants' lives needs to be understood at multiple levels and ways: all the migrants spoke of the networks and ties they had in destination countries and their reliance on these as pull factors. Aweys's case though highlighted how a pull factor can also shift to a repel factor. Thus the changing nature and divergent outcomes of social networks are also important considerations in migration projects.

Finally, the (personal) weight of repel and keep factors are likely to shift over time. Although macroeconomic and political factors in Somalia have fluctuated in the last two decades, and distinct increases in flows of migration have been associated with severe conditions (Jinnah 2013), there has also been distinct impetuses for mobility arising from personal experiences. This is evident in Fartuun's narrative. Again, the interplay between the imagined and the real, the personal and the public, is worthy of more attention as this approach develops.

References

Amit, R. 2012. *Breaking the Law, Breaking the Bank: The Cost of Home Affairs' Illegal Detention Practices*. ACMS Research Report. Johannesburg: University of the Witwatersrand.

Buijs, G. 1993. *Migrant Women: Crossing Boundaries and Changing Identities.* Oxford: Berg.

Czaika, M., and M. Vothknecht. 2012. "Migration as Cause and Consequence of Aspirations." IMI Working Papers Series 57. Oxford: Oxford University Press.

Faist, T. 2000. *The Volume and Dynamics of International Migration and Transnational Social Spaces*. Oxford: Oxford University Press.

FMSP (Forced Studies Migration Programme). 2006. *Migration and the New African City: Citizenship, Transit, and Transnationalism*. Johannesburg: University of the Witwatersrand.

Hopkins, G. 2006. "Somali Community Organizations in London and Toronto: Collaboration and Effectiveness." *Journal of Refugee Studies* 19 (3): 361–80.

Horst, C. 2004. "Money and Mobility: Transnational Livelihood Strategies of the Somali Diaspora." Global Migration Perspectives 9. Geneva: Global Commission on International Migration.

_____. 2006. "Buufis amongst Somalis in Dadaab: The Transnational and Historical Logics behind Resettlement Dreams." *Journal of Refugee Studies* 19 (2): 143–57.

_____. 2008. *Transnational Nomads: How Somalis Cope with Refugee Life in the Dadaab Camps of Kenya*. Oxford: Berghahn Books.

Jinnah, Z. 2010. "Making Home in a Hostile Land: Understanding Somali Identity, Integration, Livelihood and Risks in Johannesburg." *Journal of Sociology and Anthropology* 1 (1–2): 91–9.

_____. 2013. "New Households, New Rules? Examining the Impact of Migration on Somali Family Life in Johannesburg." Special issue, *QScience 2013, Family, Migration & Dignity.*

Jinnah, Z., and M. Holaday. 2010. "Migrant Mobilisation: Structure and Strategies for Claiming Rights in South Africa and Kenya." In *Mobilising Social Justice in South Africa: Perspectives from Researchers and Practitioners*, edited by J. Handmaker and R. Berkhout, 137–76. Pretoria: Pretoria University Law Press.

Johnson, J.G. 2010. "Transnationalism and Migrant Organizations: An Analysis of Collective Action in the Johannesburg/Pretoria Area." PhD diss. Johannesburg: University of the Witwatersrand.

Landau, L.B., and V. Gindrey. 2008. "Migration and Population Trends in Gauteng Province 1996–2055." Migration Studies Working Paper 42. Johannesburg: University of the Witwatersrand.

Lewis, I.M. 2008. *Understanding Somalia and Somaliland: Culture, History, Society.* New York: Columbia University Press.

Massey, D.S. 1990. "Social Structure, Household Strategies, and the Cumulative Causation of Migration." *Population Index* 56 (1): 3–26.

McMichael, C. 2002. "'Everywhere is Allah's Place': Islam and the Everyday Life of Somali Women in Melbourne, Australia." *Journal of Refugee Studies* 15 (2): 171–88.

Misago, J.P., L.B. Landau, and T. Monson. 2009. "Towards Tolerance, Law, and Dignity: Addressing Violence against Foreign Nationals in South Africa." Johannesburg: International Organization for Migration, Regional Office for Southern Africa.

Misago, J.P., V. Gindrey, M. Duponchel, L. Landau, and T. Polzer. 2010. *Vulnerability, Mobility and Place: Alexandra and Central Johannesburg Pilot Survey*. Johannesburg: Forced Migration Studies Programme.

Noor, H. 2007. "Emergency within an Emergency: Somali IDPs." Forced Migration Review 28. Oxford: Refugee Studies Centre.

Ngwato, T.P., and Z. Jinnah. 2013. "Migrants and Mobilisation around Socio-Economic Rights." In *Socio-Economic Rights in South Africa: Symbols Or Substance?*, edited by M. Langford, B. Cousins, J. Dugard, and T. Madlingozi, 389–420. New York: Cambridge University Press.

Polzer, T., and A. Segatti. 2011. "From Defending Migrant Rights to New Political Subjectivities: Gauteng Migrants' Organisations after May 2008." In *Exorcising the Demons Within: Xenophebia, Violence and Statecraft in Contemporary South Africa*, edited by L.B. Landau, 200–225. Johannesburg: Wits University Press.

Simon, H.A. 1982. *Models of Bounded Rationality: Behavioral Economics and Business Organization*. Cambridge, MA: MIT Press.

Smith, L., M. van der Velde, and T. van Naerssen. 2010. "Across the Threshold: States, Communities and Individuals in a Methodological Approach to Migration and Transnationalism." Paper prepared for the International Conference on "Migration Methodologies: Researching Asia", 8–9 March. Singapore: National University of Singapore.

UN (United Nations). 2013. "2013 World Statistics Pocketbook Country Profile: Somalia." Accessed 10 March 2014. http://unstats.un.org/unsd/pocketbook/PDF/2013/Somalia.pdf.

UNHCR (United Nations High Commissioner for Refugees). 2005. "2005 UNHCR Statistical Yearbook. 2005." Accessed 10 March 2014. http://www.unhcr.org/4641bebd0.html.

_____ . 2014. "2014 UNHCR Country Operations Profile: Somalia." Accessed 10 March 2014. http://www.unhcr.org/pages/49e483ad6.html.

Van der Velde, M., and H. van Houtum. 2004. "The Threshold of Indifference: Rethinking Immobility in Explaining Cross-Border Labour Mobility." *Review of Regional Research* 24 (1): 39–49.

Van der Velde, M., and T. van Naerssen. 2011. "People, Borders, Trajectories: An Approach to Cross-Border Mobility and Immobility in and to the European Union." *Area* 43 (2): 218–24.

WRC (Women's Refugee Commission). 2011. *No Place to Go But Up: Urban Refugees in Johannesburg, South Africa*. New York: WRC.

Zorc, R., and M. Osman. 1993. *Somali-English Dictionary with English Index*, 3rd ed. Maryland: Dunwoody Press.

Chapter 5

Thresholds in Academic Mobility: The China Story

Maggi W.H. Leung

> Globalization, a key reality in the 21st century, has already profoundly influenced higher education … For some the impact of globalization on higher education offers exciting new opportunities for study and research *no longer limited by national boundaries* … With 2.5 million students, countless scholars, degrees and universities *moving about the globe freely* there is a pressing need for international cooperation and agreements. (Altbach et al. 2009, iv–v; emphasis added)

Education studies expert Altbach and his co-authors began their critical report on the key trends in global education prepared for the UNESCO with the above framing, putting an accent on the high level of freedom, mobility and the demise of boundaries in the increasingly globalized academic field. Indeed, as academics, researchers and students have become incorporated into the development visions of universities and other research institutions, and more broadly urban centres, subnational regions and states near and far from where these talents (in-the-making) are currently located, these (potentially) mobile individuals are increasingly represented as footloose, brain *sans frontières* who can spread their wings and cruise across all boundaries, scouting for the best deals in the wide, wide world. This contribution unsettles this representation and illustrates how academic mobility is in fact a regulated terrain. It engages with the concept of thresholds (Van der Velde and Van Naerssen 2011) to produce a grounded analysis of the decision-making processes of Chinese scholars and students. Narratives of the research participants illustrate that rather than 'moving about the globe freely', these students and scholars in fact perform mobility, having to adhere to the influence of a range of macro (structural), meso (household, university and other institutions, social network) and micro (individual) level factors anchored in diverse places (home, work, China, overseas and the virtual space). In addition, the impact of coincidence will be discussed, drawing attention to the often unplannable character of mobility.

This chapter draws on the results of two recent research projects. The first one (fieldwork conducted in 2009 and 2010) investigates the motivation for and impact of international academic mobility among Chinese scholars of postdoctoral level or above who have conducted research visits (of minimum three months) in Germany. The study implemented a two-page email/postal survey to gather general and quantifiable data from 123 Chinese scholars. Findings of the survey provide an overview of important trends and patterns. They are also used to generate the question guide used for subsequent in-depth interviews. A total of 64 scholars participated in the interviews either in person or by telephone. All these interviews were conducted in Chinese. Expert interviews with six representatives of German organizations involved in academic mobility programmes for Chinese scholars and three with Chinese officials were also conducted in German and Chinese respectively. All interviews were audio-recorded and fully transcribed for analysis. Participant observation and quasi focus group discussions with small groups of 'mobility alumni' were also carried out. The second project is a small-scale research conducted in 2013. Using a biographical approach, Heuts (2013) conducted eight biographical interviews with Chinese students pursuing higher education in Utrecht, the Netherlands, to study their motivations for and experiences of studies overseas. Interviews were conducted in English, audio-recorded and fully transcribed for analysis.

In the following, an overview of academic mobility in China as a regulated terrain will first be provided. Drawing on fieldwork research findings, the next section will proceed to apply the threshold approach to tease out the various forms of stimulants and impediments in the Chinese academic mobility field. Examples will be provided to illustrate how the three types of thresholds, namely (1) the indifference threshold; (2) the locational threshold; and (3) the trajectory threshold are manifested in the Chinese academic mobility field and in turn shape the mobility intention and experiences of the mobile scholars and students. The chapter will then conclude with some implications of analysing academic mobility with the threshold approach for unsettling some conventional ways of thinking about this form of mobility.

Academic Mobility in China: A Regulated Terrain[1]

The 'academic mobility for development' strategy has become a common option among policymakers who are looking for quick fix solutions for unmet labour demand in targeted economic sectors as well as challenges posed by demographic changes (Leung 2013b). Depending on the context, academic mobility has been assigned a wide variety of missions and utility level beyond the immediate goals of developing human capital and advancing the production of knowledge in the academia. One of the prime examples of state regulation of academic mobility for development is China. At the 2012 International Education Summit, Jinghui (2012, 43), secretary general of the China Scholarship Council, emphasized the importance of the internationalization of higher education – of which academic mobility is a key element – for 'the nation's wisdom and personnel reserve' and how it has become 'one of the crucial factors to China's comprehensive national strength and international competitiveness'.

The current era of academic mobility promotion was initiated by Deng Xiaoping, the chief architect of the Open Door policy that commenced in 1978. Deng believed that scholars and students sent abroad would bring back advanced ideas and expertise needed for China's modernization and development. His conviction was concretized in an expansion of study abroad programmes. Subsequent political leaders have further developed foreign study and scholar exchange programmes in China.

In 2010, the Chinese state launched the National Outline for Medium- and Long-term Educational Reform and Development. A part of the policy document focuses on promoting educational exchanges and cooperation. It assigns a high importance to internationalization with concrete measures to introduce high-quality education resources from overseas, encourage Chinese-foreign cooperative education and joint research, and promote academic mobility. All the initiatives aim to train 'talented students imbued with global vision, well-versed in international rules, and capable of participating in international affairs and competition' (Jinghui 2012, 43).

China is now the world's largest source of overseas students, accounting for 14 per cent of the global total (*Caixin Online*, 25 September 2012). According to statistics from the Chinese Ministry of Education, almost 400,000 Chinese students went abroad in 2012, representing an increase of

---

1   In recent years, the Chinese state has been active in attracting international students and staff to China. This chapter focuses, however, on outgoing academic mobility among Chinese academics and students.

17 per cent from 2011. The most popular destinations for Chinese students are Australia, Britain, Canada and the United States (*China Daily*, 12 March 2013).

Since 1978, Chinese academics have gained geographical mobility and intensified their participation in the global knowledge economy. Partially due to its transient nature, systematic, internationally comparable data on the mobility of academic staff are not available. Certain is that academic mobility is still a privilege among relatively few and its symbolic capital is high (Leung 2013a). In Europe, only Germany, Finland, Portugal, Sweden and the UK (out of 27 EU member states) provided statistics relating to the numbers of Chinese academic staff in higher education. Table 5.1 shows the figures collected from the various sources. At least 6,697 academic staff of Chinese nationality worked in the EU between 2008 and 2009. Most of them were working at German and British higher education institutions (GHK Consulting and Renmin University 2011). The figures indicate the number of academic staff with Chinese nationality, irrespective of how long they have been in the specific country and where they had been before coming to that particular EU country, reflecting the general challenge in interpreting migration/mobility data. These data can therefore not be treated as an accurate measurement.

Table 5.1   Academic staff mobility from China to the EU

Country		2005	2006	2007	2008	
Finland		34	32	36	48	
Germany	Scientific staff of Chinese nationality	1,027	1,174	1,298	1,636	
	Chinese scientists on exchange programme	1,535	1,678	1,779	2,199	
Country		2004/05	2005/06	2006/07	2007/08	2008/09
Portugal[a]		10	12	11	12	9
Sweden		18	18	18	26	35
UK						2,770[b]

Notes: [a] Gender ratio approximately 50/50; [b] Of whom 1,780 male and 990 female.
Source: GHK Consulting and Renmin University 2011, 48.

According to the Chinese Ministry of Education, academics going to Europe are generally aged between 30 and 40 years, male and from disciplines related to engineering. They are mostly affiliated with institutions located in Beijing or Shanghai. Academic mobility of Chinese scholars to Europe is steadily increasing and these scholars mostly travel to Europe for short-term periods (three months to one year) (GHK Consulting and Renmin University 2011). Mobility patterns vary greatly depending on academic disciplines, the existence of exchange programmes, academic networks and sponsorship among other factors. While the US is commonly referred to in published studies, anecdotal references and our own fieldwork as the 'number-one destination' among many Chinese academics, there are no comprehensive data available to verify this preference. National-level data can neither capture the intranational geographical pattern of mobility of academic staff, which is particularly important in understanding the impact of academic mobility on particular laboratories, research groups, departments, universities or consortia, nor the institutional-level disparity in academic performance and expertise production.

Taking a Closer Look: An Analysis with the Threshold Approach

This section engages with the threshold approach to identify the various thresholds that work (or not) in intersection in defining students' and scholars' intention and act to become mobile.

Indifference Threshold

According to Van der Velde and Van Naerssen (2011), the indifference threshold is the hurdle preventing individuals from recognizing a mobility possibility or unwillingness to perceive it as such. The concept offers a sedentarist perspective – assuming that people 'in nature' stay put and they have to overcome a series of thresholds to become mobile. Contrasting to this assumption, this contribution underlines the dwindling indifference to mobility in the Chinese (like many other) academic field. In particular, this research brings forth the rise of a culture of mobility in the Chinese academia and society in general and illustrates that a range of institutions play a significant role in nurturing this culture, lowering the indifference threshold in academic mobility – though by no means evenly for every individual in the academic field.

'Mobility has come to epitomize a new and modern China' is the key message of Nyíri's (2010) insightful account of the evolving meaning of mobility in contemporary Chinese society. Ever since the late 1970s, the Chinese state has successively loosened its restrictions on internal and international migration, other forms of human mobility such as tourism, as well as the mobility of commodities, money and information, among others. This transformation has been fuelled by the relaxation and promotion of a variety of human mobilities and the increasingly positive portrayal of migrants and other mobile persons including Chinese and foreign mobile students and academics.

Indeed, at high policy levels, students and overseas scholars are often conferred to as the key to China's future. In May 2010, then President Hu Jintao announced plans to upgrade the country's 'software' basics such as health care, education and academic freedom to entice skilled Chinese overseas to return, would be installed as an important part of the up-coming 10-year plan. In addition, he reiterated that talent 'is the most important resource' for China's continued development (*China Daily*, 7 June 2010). At the 2010 Business Establishment Week for Overseas Chinese Students, Wang Xiaochu, vice minister of Human Resources and Social Security, announced that 'students abroad are in a unique position to help China's modernization and the government attaches great importance to luring them back' (*People's Daily Online*, 30 June 2010). Statements announced by key political figures at important occasions, like the ones cited above, are examples of soft, discursive governing technologies implemented to reinforce the 'academic mobility = heroic acts ⇨ national development' narrative, instil the motivation to move and moral responsibility to return or at least engage in the diaspora knowledge network.

In addition to the Chinese state, other actors are also influential in lowering the indifference threshold in the Chinese academic mobility 'market'. International education now represents a major export sector for many popular study abroad destinations across the world, adding billions to domestic economies like those in the US, the UK, Canada and Australia.[2] Chinese students have become the targets in this booming global industry. Numerous actors and institutions both in the public and private sectors, based in China and overseas, are involved in this vibrant and competitive market. Overseas education fairs are organized regularly with high visibility, often approved and supported by high-level policy bodies such as the Chinese Ministry of Education. One recent example is the 18th China International Education Exhibition Tour held in March 2013 in seven Chinese cities: Beijing, Shenyang, Xian, Shanghai, Hefei, Fuzhou and Guangzhou. The same event held in 2012 hosted 485 various types of educational institutions from 35 different countries and regions, and attracted over 40,000 visitors.[3]

Studying abroad has become an increasingly common goal for children from families with a background of higher education. In recent years, this culture of student mobility has expanded to a wider population, signified by the increasing number of students from working-class families are also studying abroad in recent years (*Caixin Online*, 25 September 2012). The normalization of this mobility culture is also reflected by its high visibility in daily media. The *Guangzhou Daily*,[4] for instance has a three-full-page 'Emigration and Overseas Study' section which appears at least four days a week, devoted to articles and advertisements related to outward student mobility. Needless to say, information on studying abroad is easily available on the web, provided by public[5] and private[6] sectors as well as via personal channels such as diverse social media (for instance RenRen and Qzone) where young people (and their parents) exchange experiences and advice. Chinese student mobility is a lucrative and booming market that has been groomed by an array of actors and institutions, located near and far from the students potentially going overseas. The desire to study abroad is also closely linked to the lust for other forms of international mobility (for instance tourism and outmigration) brewing quickly and ever more widely in China (Leung 2009). Viewing the contemporary mobility turn in China, one can confidently conclude that a bundle of social, cultural, economic and political forces have worked together to lower the indifference threshold for young people and their parents to pursue their educational mobility projects.

The general indifference thresholds among academics have also sunk over the recent decades. However, while the desire for academic mobility has soared, it is not equally accessible to all. As mentioned earlier, Chinese academics going to Europe are predominantly male from disciplines related to engineering. There are many factors that lead to the gender and disciplinary biases in Chinese academic mobility (Leung 2013b). One of the determinants is the different perception of the possibility to be mobile, which is related to the notion of indifference thresholds, as reflected in the research findings. The price of mobility is different among male and female academics. Li Meimei[7] (female, 40s), professor in urban planning considered herself to be 'lucky', comparing

---

2   For instance, international students and their dependents contributed US $21.8 billion to the US economy in tuition and living expenses in 2011 (McMurtrie 2012).

3   As indicated on the website of the China International Education Exhibition Tour, 17 April 2013 (http://www.cieet.com/en/dslj.asp, site discontinued).

4   http://gzdaily.dayoo.com/ (accessed 8 December 2014).

5   See for example the public web portal on study abroad operated by the Ministry of Education: http://ufind.cscse.edu.cn/Portals/AbroadEducation/PageThreeColumnMainLeft.aspx?tabid=5 (accessed 17 April 2013).

6   See for example a privately operated web portal on study abroad: http://edu.qq.com/abroad/ (accessed 22 October 2014).

7   Pseudonyms are used throughout the chapter.

her experiences with her female peers. Li had witnessed the divorce among a number of women colleagues 'because their husbands could not take it [having their wives being abroad]'. She states a gendered 'price for mobility' that is deeply rooted in the Chinese culture:

> If men go to Germany to study or work, they can bring their wives with them. That is acceptable in the Chinese society. Men take charge of the outside, women take care of things inside the family [*nan zhu wai, nü zhu nei*] – that is Chinese tradition. It is not really probable that a man accompanies his wife to study or work.

For the sake of the family, therefore, it is not uncommon for qualified female academics to forego their mobility desires. While not having been completely immobilized, Hu Lin (female, 40s), associate professor in electronic information, recalled her decision against further mobility possibilities:

> The biggest loss was the separation from my child, leaving him at home. Nobody took care of him. And then my son's eyesight got worse. He became short-sighted. And his school performance also deteriorated. Actually there was a possibility for me to extend my research stay from one year to at least two years then … But I did not dare to take the offer. Because I did not dare to leave my child behind longer … My husband was very busy. And if my child went on like that [performing less well at school and becoming short-sighted], my family would be finished.

Besides gender bias, there is also stark disciplinary inequality in academic mobility. Whether academic mobility belongs to the 'normal' element of one's career differs greatly between scholars engaged in strategic research areas and those marginalized. Yuan Mijiao (female, 40s), specialized in rural and gender development studies spent some years in Germany with German funding during her postgraduate years and is now a senior lecturer at a university in one of the poorest provinces in China. When asked if she would be interested in going abroad again as a visiting scholar, she said that, even though the Chinese state was investing huge amounts in promoting academic mobility and international exchange, 'Research like ours [not belonging to any strategic areas] is seldom funded. We struggle enough to keep our ordinary teaching and research activities going.' One can therefore conclude that there are multiple 'cultures of academic mobility' in China (like elsewhere) where structural biases define differentiated indifference thresholds. Academics of varied gender, disciplinary and geographical backgrounds, institutional affiliations and so on recognize mobility as a desirable and/or realistic option in an uneven way.

Locational Threshold

Having traversed the indifference threshold, if it at all exists, aspired international students or mobile scholars would encounter locational thresholds. Concurring with general observations, (potentially) mobile students and academics make their academic mobility plans that reflect the combined effects of micro (for instance individual behavioural factors), meso (for instance family, community and social network factors) and macro (for instance immigration and international education policies) structures. On a global level, the US (19 per cent of the global total), UK (11 per cent), Australia (8 per cent), France (7 per cent) and Germany (6 per cent) are the five top destination countries among international students (UNESCO 2012).

Such global figures level out a high level of complexity. For one, generalized locational data cover tremendous intranational differences in international student mobility. In the UK, for instance, although international students are located throughout the country's higher education sector, they are particularly concentrated in specific institutions, especially in London (Madge et al. 2009). Sometimes, the decision to go to London (or anywhere else) is as academic-founded as economic. Real estate agencies are witnessing an increasing number of Chinese parents buying homes for their children when they study overseas. In the UK, for instance, buying property provides more than accommodation for the children. Owning a real estate means saving on university accommodation fees, and potentially making a profit from capital appreciation or earning income from renting out the property after their children have moved on.

Since the vast majority of the mainland Chinese students are single child at home, the accompanying *peidu mama* phenomenon ('study mamas', women who accompany their children to receive education overseas) has also taken shape, especially in Singapore where a significant number of Chinese mothers move with their children to receive primary and secondary level education. In her work on transnational families straddling Canada and Hong Kong, Waters (2005) has insightfully illustrated the particular and crucial role that education and children play within a wider family project of capital accumulation. When moving young people to receive overseas education has missions beyond the study itself and direct impact on others (mostly mothers) in the families, the locational threshold of certain study-related mobility trajectory can only be understood taking these other stakeholders and factors into account. Stretching the original concept of the New Economics of Labour Migration (NELM) somewhat, one can also regard such student mobility as a strategic act within larger social units, in this case the household or family, seeking to optimize their livelihoods and reduce their vulnerability to (financial and/or political) risks, and not merely an individual decision.

Acadamic staff mobility displays yet another set of patterns. Where scholars pursue their academic visits they are influenced heavily by the structural and institutional opportunities such as the existence of exchange programmes, academic networks and sponsorship. These in turn vary significantly across disciplines and geographical space. Recurrently, participants of the current research explained their location decisions with such institution-related reasons. 'My department in China has a long-standing relationship with this department' and 'My senior colleague [or supervisor] suggested me to come here' are some of the commonly stated justifications. Repeating the stretching exercise performed above with the concept of the NELM, one can also observe the need to rescale academic staff mobility as a strategic act performed within a large social unit, namely the mobile academic's research group, department and university. International partner research structures function like hubs in a corridor along which researchers but also ideas, knowledge, research materials, money, work culture and practices travel (Leung 2011). These institutions stimulate or hinder the mobility of particular (groups of) scholars to optimize not only the mobile academics' but perhaps more importantly, the institutional collective performance in their respective field. As central as the perception of the moving individuals, their affiliated institutions are crucial factors when we consider the locational threshold of specific academic mobility flows. Universities in China, like most others worldwide, are now engaged in institutional network-building. A close look at the list of any institution would reveal certain spatial imbalance and exclusivity. Almost always having an impact on funding, the system of designated partnerships regulates the directionality and intensity of academic mobility among students and staff.

While partnership among higher education institutions is often branded with the term 'strategic', the selection of the partners is sometimes also a result of small, but arguably important, coincidence. At a recent workshop on the internationalization of higher education in Brussels, a colleague from a Swedish university who is responsible for his university's internationalization policies explained to the author how so often seeds of institutional partnership are sowed at the dining table or during coffee break when two colleagues, university chancellors happen to meet each other at international meetings or other similar occasions – in fact, just like the one we were at. Indeed, coincidence works often to move the locational threshold in mobility plans and their execution. When asked why she chose to conduct her research visit at her host Aachen University (RWTH), which is by the way very renowned in her field internationally, Lin Xuefei (female, 30s) explained that 'it was not intended but a good coincidence.' She recalled her decision process with a sense of humour:

> I was looking for a host university for my fellowship application. I went to the website with all the German universities and Aachen was on top – the list was sorted alphabetically [interviewee laughed]. So I clicked on the link and looked for the department. Only after that did I learn that it is a good department in the field. I contacted one of the professors and he was so welcoming and that's how it happened.

The impact of coincidence in shifting a certain location threshold is apparent in the decision-making process among many of the interviewees. In order to understand the sometimes surprising way a particular mobility trajectory takes shape, therefore, we need to pay attention to both institutional/macro-level factors as well as the path dependency of 'small(er) events', which are generally sidelined in academic inquiries. Findings of a recent study on the motivations and experiences of Chinese students in Utrecht (Heuts 2013) also underline the coincidental character of mobility. To cite one example, John, an economics student (23 years old) in Utrecht recalled how the Netherlands entered his options for overseas study:

> A lot of my classmates also decided to go abroad to study further. We also discussed with each other and then someone mentioned the Netherlands is a good destination. And I was very surprised actually, because before that I didn't even think about the Netherlands as an option. I didn't know the education here. This person gave me some brief points about the Netherlands, like they have a good education system. After that I did some research on the internet and thought: 'Well, why not?' (Heuts 2013, 22)

Subsequently, another unplanned event took place. John met a Dutch person, who happened to work as an information broker to the Netherlands, at a family party:

> In Shanghai I met a Dutch person, from Leiden. She also gave me a lot of help. I met her on a family party. She stimulated me to go specifically to the Netherlands. Before that I had a concern about the level of English of the Dutch people. After I met her, I totally convinced myself and my parents the Dutch are good in English. (23).

The impact of one or a series of coincidental or unplanned (small) events does not end in shaping one's mobility decision or trajectory. A cumulative causation effect often takes shape through social networks (Findlay et al. 1994; Fussell 2010; Heering et al. 2004). John continued reconstructing the chain of mobility events in which he ended up playing an instrumental role:

> My mother is a piano teacher. One of her students also came to the Netherlands to study at Tilburg University after me. I think he followed me. I think he is like me, he never thought about studying in the Netherlands before, but when he heard about my story, like me he did a lot of research

afterwards and he decided to study in the Netherlands as well. So it works the other way around as well, I motivate people to study in the Netherlands too. It's contagious. It's like a trigger. (Heuts 2013, 23)

Similarly, hearsay and social (including professional) networks play a crucial role in shaping the locational threshold among aspired mobile academics. Many (previously) mobile scholars refer their students and colleagues to contacts they have established at places where they have worked or at least have good contacts, hence reinforcing the strength and impact potential of the particular chains or webs of relationships that are anchored in certain sets of localities (Leung 2011).

Last but not least, the role of intermediaries like agencies in shifting locational thresholds should not be overlooked (see also Agunias 2009; Findlay and Li 1998; Radcliffe 1990; Mahmud 2013). As mentioned before, international student mobility is a booming industry in which an array of actors and institutions are involved. The active engagement of student recruitment agencies in China was discussed earlier in the chapter. Agencies do not only facilitate clients' mobility plans. They also steer them. Karin, an international communication and media student in Utrecht (20 years old) originally wanted to study in Finland. But because of some complications with her agency and the fact that her cousins had studied in the Netherlands and transmitted some knowledge about the country and the education system to her, she eventually agreed with the agency's recommendation to study in the Netherlands (Heuts 2013, 22).

As shown above, the locational threshold of a mobile person is shaped by multiple actors, institutions and macrostructures, as well as coincidence. The interplay between the respective influences is highly contextual in any specific mobility trajectory.

Trajectory Threshold

Decision-making does not stop after one makes a decision to move and where to move. Rather, it is an ongoing process, conducted en route, after the initial locational decision has been made and mobility plan (partially) worked upon or even after the initial embarkment. Due to a whole array of intersecting sociocultural, economic, political and security factors that crop up along the way, migrants or other kinds of mobile persons often arrive at 'another' destination or series of destinations. Van der Velde and Van Naerssen (2011) refer to this as 'trajectory threshold'. In their conceptualization, trajectories are deviations of initial routes. This contribution operates, however, from a more dynamic perspective through which mobility is seen as a process rather than a line between the origin and destination points. Thinking about mobility and trajectory this way acknowledges those individuals who aspire geographical mobility but do not necessarily have one fixed destination to start with. Rather one might make one step or decision after another while moving or have multiple plans for moving and staying. If trajectories are not perceived per se as deviations of original, well-planned routes, they can be conceptualized, perhaps closer to reality, as a series of locational decisions. The route is always in the making and there are no definite destinations to speak of. Having said that, we ought not think of the trajectory threshold as 'just' an accumulation of locational thresholds, which arguably is a rather static and binary (here or there) concept. The strength of the trajectory threshold concept lies in its recognition and emphasis on the dynamic, often unpredictable and extremely contextualized nature of the mobility process.

The following section illustrates how an array of factors and processes contribute to redirect the mobility routes of many Chinese scholars, hence bringing forth the manifestation of the trajectory threshold in the Chinese academic mobility field. In order to lure Chinese former students, many of whom academics and researchers, overseas back 'home', many schemes have been set in place at different governance levels in China. The Thousand Talents programme was established late 2008, with incentives sometimes reaching one million US dollars to entice the crème de la crème of China's best academics overseas. Such perks have worked to effectively change the mobility plans of many 'brains in diaspora'. However, as opposed to entrepreneurs who are much more willing to return to China 'permanently' (at least in a more permanent manner), 'returning' academics and scientists opt mostly for short-term visits and are reluctant to make a permanent return (*New York Times*, 21 January 2013). As transnationalized and/or translocalized professional and personal lifestyles become more common, there is a need to adjust the notion of trajectory beyond stressing the unpredictability and fluidity of migration/mobility routes to addressing the multiplicity and simultaneity of mobility trajectories. Among transnational academics (like people in other sectors ranging from international consultants to rotating care workers), their mobility trajectories are more complex than the imagery of a singular line with zigzags suggests. Rather, they take the shape of bundles of lines with multiple zigzags of their particular dynamic directionalities, spatiality and temporalities of mobilities, immobilities and moorings.

The availability of scholarship, fellowship and exchange opportunities play a determining role in charting the mobility trajectories of scholars and students. As Wu Shen (male, 30s), a research fellow in engineering science, traced his route:

> One thing after another, I ended up here in Germany. I left China for a fellowship in Paris, and there I met a colleague who mentioned to me about the Alexander von Humboldt fellowship. I applied and got it. So now I am here. I don't know where I will go next. But I would like to stay overseas for a while longer.

Among young (visiting) academics, it is quite common that their professional and personal lives make a quick and sharp turn when the supervisors take up a new job and move to another institution, country or even continent. This 'secondary mover' phenomenon is particular prevalent among female academics. Ackers's (2000) work shows that female mobile researchers in double-career partnerships in the EU move predominantly as 'tied' movers, typically following a male partner. The dependency of some academics on others in aspiring for and practising mobility often spans time and space. The following multigenerational mobility chain illustrates the need to think beyond one person, one trajectory, one period of time in our conceptualization of mobility and its associated impact.

Professor Schmidt (now in his mid-60s) is a German social scientist specialized in North American topics who worked in an American university from 1983 to 1984. The one-year experience was very positive. That motivated him to act as a hospitable host when Shen (now in his late 60s), a visiting Chinese professor, settled in the office next to him in the late 1980s. He recalled how his connections with the subsequent generations of Chinese scholars took root:

> It was actually all a coincidence, not because I was interested in China. [Shen] sat next to me … and we talked to each other and I took care of him a little, we ate together a few times and then later on, they asked me when they wanted to come, and I supervised them. Often in groups of six or seven … and then he invited me to China. It happened this way.

Little did Schmidt expect that his academic mobility experience would lead to a chain of mobility events that would eventually also contribute to the flows of knowledge and research practices that would 'travel in space' to an academic field spanning Germany, China and beyond (for instance the US). Since Shen's first return to China, he has supervised a large number of young scholars,

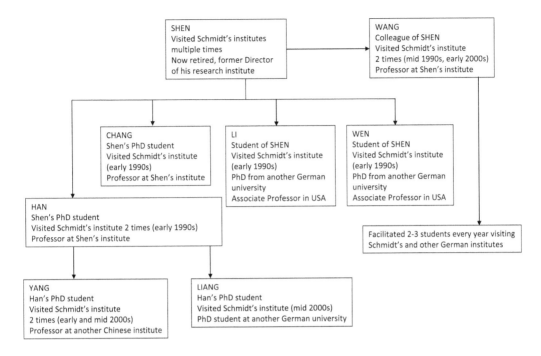

Figure 5.1   Schmidt/Shen mobility trajectory chain

Source: the author.

some of whom have also followed their mentor's footsteps and visited Schmidt's institute. Figure 5.1 maps the sequences of selected events and actors involved in this academic mobility trajectory chain that spans considerable spatial and temporal distances, illustrating the apparent working of path dependency in the seemingly free-floating transnational academic mobility field.

As the higher education and research field becomes more globally connected, it is increasingly common for academics to be associated with colleagues, research groups, universities, institutions and private corporations in different places, within a country but also internationally, at the same time. Mobility trajectories of these transnational academics are manifold, dynamic and anchored simultaneously in multiple places, being moulded by factors spanning micro, meso to macro levels. The modification of any of the factors and their constellations would introduce new dynamics to the trajectory threshold of a specific person associated with the mobility web.

Conclusions

In order to check the mileage of the threshold approach, this chapter has applied the approach to a study of academic mobility, focusing on the Chinese case. It is important to note that the approach was first developed to conceptualize more permanent labour migration through which migrants are assumed to be searching for a second 'home' while the case of Chinese academic staff mobility concerns mobility projects that are meant to be temporary – at least on the onset. Hence, some fitting was necessary in the application.

One important disjuncture is noted on the usefulness of the concept of the indifference threshold. While the original approach denotes a sedentarist perspective in which people are seen as rooted and mobility demands a series of threshold-jumping, the Chinese academic field the author has studied is characterized by a rising culture of mobility where the indifference threshold has quasi vanished, at least among some groups. Despite the arguably rather fundamental difference in the mobility contexts under study, the approach provides a frame and a set of interlinked vocabularies for the analysis of the decision-making processes among Chinese students and academics studied here, teasing out the differentiated impact of factors that promote or hinder their academic mobility. Rather than considering academic mobility solely as a dependent outcome of structural factors, the approach underlines the intersections between agency of individuals and regulating structures and institutions. This chapter has illustrated how mobility decisions are always a product of multiple factors anchored at diverse interrelated levels, as well as the shifting and superseding of a variety of thresholds. While the key elements of the threshold approach are recurrent questions in contemporary migration and mobility research, the approach offers a structure, an umbrella of a sort, to frame mobility inquiries and warrants a (more) comprehensive research set-up, which explores issues regarding how people become mobile (or not), where they go and how their routes eventually take shape.

Through applying the threshold approach, this chapter has arrived at a few insights that have implication in the ways academic mobility can be conceptualized and studied. First, the importance of thinking about academic mobility beyond the individual level is highlighted. Like many other forms of mobilities, the (im)mobility of potential students and academics is often an outcome of determinants anchored at a collective level. Echoing the essence of the NELM thinking, this chapter has illustrated how some students pursue their education abroad as part of a familial/household economic accumulation strategy, while female academics often have to forego their mobility desires because of their familial responsibilities. By situating our (academic) mobility inquiries at the household level, we reclaim space for analyses with a gender lens that has been underused in the extant scholarship. The particular and important role of young people and children, as well as the position of overseas education in household mobility projects also receive thereby their deserved attention.

In the realm of academic mobility, it is imperative to expand the conventional categories employed in the classical NELM framework, namely households or families, and consider the academic collectives in promoting and impeding the mobility of certain individuals and groups. Indeed, in research on academic mobility, there has been a consensus that the role of mesoscale institutions is little understood and hence deserves further research (Raghuram 2013).

Second, when one sieves through the many narratives collected in both of the research projects, a high level of path dependency can be observed. A particular mobility project/trajectory should not be considered in isolation. More sensitivity in recognizing and thinking about linkages between previous and subsequent actions and decisions when studying sequences of events is called for. Zooming into the way a mobility trajectory unfolds, as a process rather than a line linking two points, reveals how trajectories are dynamic and often seem to be unplannable. This does not mean, however, that they are fully coincidental and therefore unexplainable. To understand the outcomes of individual trajectories, we need to take seriously the big and small events that affect decisions and therefore the mobility routes, moorings and life courses of the enacting and other affected individuals. Breakthrough moments in mobility trajectories can be affected by unexpected events and the entanglement of trajectories in a mobility web – or multiple mobility webs – in which an individual is embedded.[8]

---

8   The author thanks Joris Schapendonk for helping put these thoughts in words here.

Last but not least is a plea for more concerted efforts in expanding the spatial and temporal reach of the trajectory thinking, by considering the linkages across trajectories. While the present conceptualization has pushed us to recognize the dynamic and ad hoc characteristic of the route-finding and -making processes, it has not explicitly called for attention to the interplay of multiple, simultaneous and entangled trajectories enacted by individuals, families and other relevant collectives (for instance academic institutions in the case of academic mobility). In other words, in order to develop the concept further, there is a need to add dimensions to a one dimensional, lineal imagery of mobility trajectory – may the lines be full of turns and knots. Considering mobility trajectories as webs and/or bundles that extent into multiple time-spaces will contribute to a more sophisticated understanding of decision-making process and impact of a specific mobility trajectory (or a set of related mobility trajectories). By setting our lens to see how mobility trajectories of different nature, orchestrated and undergone by different individuals (as complex subjects with multiple identities) and collectives (can) relate to, stimulate or hinder each other, we can traverse our own intellectual thresholds in appreciating the relationalities across conventional categorizations. In turn, we problematize the common dividing/dichotomising thinking regarding tourism versus academic mobility, investment migration versus work migration, brain drain versus brain gain, places of origin versus destination, voluntary versus involuntary mobility, mobility versus immobility and past, present and future mobility. Hence, rather than artificially isolating one thread of mobility experiences/outcomes out of organic bundles and webs, a more integrative approach should be attempted. This will help underline connections among different mobility trajectories that, at first sight, might be anchored at different time-spaces, and undergone by/affecting disconnected and distant individuals and communities. Mobility has been an element of humanity since time immemorial, critical study of which is, however, still a nascent field and there is much more to be done. The threshold approach offers a way forward.

Acknowledgements

The project on Chinese scholars in Germany was funded by the Research Grants Council (Hong Kong) (HKU 746909H, 2009–10) and the Alexander von Humboldt Foundation. The small project on Chinese students in Utrecht was conducted by my former student Hugo Heuts. I thank him for letting me use his transcripts. I am grateful to all research participants who took part in these two projects. Joris Schapendonk provided very helpful comments on the earlier draft.

References

Ackers, L. 2000. "The Participation of Women Researchers in the TMR Marie Curie Fellowships." Accessed 25 April 2013. ftp://ftp.cordis.europa.eu/pub/improving/docs/women_final_rpt_3march2000.pdf.

Agunias, D. 2009. "Guiding the Invisible Hand: Making Migration Intermediaries Work for Development." Human Development Research Paper 2009/22. New York: United Nations Development Programme.

Altbach, P.G., L. Reisberg, and L.E. Rumbley. 2009. "Trends in Global Higher Education: Tracking an Academic Revolution." Report Prepared for the UNESCO 2009 World Conference on Higher Education. Paris: UNESCO.

Findlay, A.M., and F.L.N. Li. 1998. "A Migration Channels Approach to the Study of Professionals Moving to and from Hong Kong." *International Migration Review* 32 (3): 682–703.

Findlay, A.M, F.L.N. Li, A.J. Jowett, M. Brown, and R. Skeldon. 1994. "Doctors Diagnose Their Destination: An Analysis of the Length of Employment Abroad for Hong Kong Doctors." *Environment and Planning A* 26 (10): 1605–24.

Fussell, E. 2010. "The Cumulative Causation of International Migration in Latin America." *Annals of the American Academy of Political and Social Science* 630 (1): 162–77.

GHK Consulting and Renmin University. 2011. *EU-China Student and Academic Staff Mobility: Present Situation and Future Developments; Joint Study between the European Commission and the Ministry of Education in China.* Brussels: GHK Consulting.

Heering, L., R. van der Erf, and L. van Wissen. 2004. "The Role of Family Networks and Migration Culture in the Continuation of Moroccan Emigration: A Gender Perspective." *Journal of Ethnic and Migration Studies* 30 (2): 323–37.

Heuts, H.C. 2013. "International Student Mobility as a Socially Embedded Process: A Case Study on Chinese Degree Students in Utrecht." BA Thesis (unpublished). Utrecht: Utrecht University.

Jinghui, L. 2012. "National Priorities in Promotion and Internationalization of Higher Education: Recent Developments and Future Trends in China." Speech delivered at the 2012 International Education Summit. Washington, DC: Institute of International Education. Accessed 4 September 2014. http://www.iie.org/Videos/Corporate/G8-2012/Overview-China.

Leung, M.W.H. 2009. "Power of Borders and Spatiality of Transnationalism: A Study of Chinese-Operated Tourism Businesses in Europe." *Tijdschrift voor economische en sociale geografie* 100 (5): 646–61.

_____. 2011. "Of Corridors and Chains: Translocal Developmental Impact of Academic Mobility between China and Germany." *International Development Planning Review* 33 (4): 475–89.

_____. 2013a. "'Read Ten Thousand Books, Walk Ten Thousand Miles': Geographical Mobility and Capital Accumulation among Chinese Scholars." *Transactions of the Institute of British Geographers* 38 (2): 311–24.

_____. 2013b. "Unraveling the Skilled Mobility for Sustainable Development Mantra: An Analysis of China-EU Academic Mobility." *Sustainability* 5 (6): 2644–63.

Madge, C., P. Raghuram, and P. Noxolo. 2009. "Engaged Pedagogy and Responsibility: A Postcolonial Analysis of International Students." *Geoforum* 40 (1): 34–45.

Mahmud, H. 2013. "Enemy or Ally: Migrants, Intermediaries and the State in Bangladeshi Migration to Japan and the United States." *Migration and Development* 2 (1): 1–15.

McMurtrie, B. 2012. "China Continues to Drive Foreign-Student Growth in the United States." *The Chronicle of Higher Education*, 12 November. Accessed 17 April 2013. http://chronicle.com/article/China-Continues-to-Drive/135700/.

Nyíri, P. 2010. *Mobility and Cultural Authority in Contemporary China.* Seattle: University of Washington Press.

Radcliffe, S.A. 1990. "Between Hearth and Labor Market: The Recruitment of Peasant Women in the Andes." *International Migration Review* 24 (2): 229–49.

Raghuram, P. 2013. "Theorising the Spaces of Student Migration." *Population, Space and Place* 19 (2): 138–54.

UNESCO (United Nations Educational, Scientific and Cultural Organization). 2012. "Global Flow of Tertiary-Level Students." Accessed 18 April 2013. http://www.uis.unesco.org/Education/Pages/international-student-flow-viz.aspx.

Van der Velde, M., and T. van Naerssen. 2011. "People, Borders, Trajectories: An Approach to Cross-Border Mobility and Immobility in and to the European Union." *Area* 43 (2): 218–24.

Waters, J.L. 2005. "Transnational Family Strategies and Education in the Contemporary Chinese Diaspora." *Global Networks* 5 (4): 359–77.

Xiang, B. 2012. "Predatory Princes and Princely Peddlers: The State and International Labour Migration Intermediaries in China." *Pacific Affairs* 85 (1): 47–68.

Chapter 6

Gendered Thresholds for Migration in Asia

Ton van Naerssen and Maruja M.B. Asis

All over the world, national labour markets are involved in processes of transformation that affect sectoral employment and the participation of women. In 2009 there were as many women as men above the age of 15 years (2.5 billion each), but among these only 1.2 billion women were employed as opposed to 1.8 billion men. Nevertheless, in the three decades between 1988 to 2008, the gender differentials in employment narrowed. Also, over time there has been a decrease in the distance between countries with low levels and countries with high levels of female economic participation (ILO 2010).

Accompanying these trends, the number of female labour in international migration has also increased. Overall, women represent half of the migrant population in the world but contrary to the past, many of them migrate independently nowadays (IOM 2010). *Feminization of labour migration* refers to the increasing number of women who migrate individually, distinguishing this trend from the older dominant flow of women migrating for family reunification. The feminization of labour migration underscores that since the 1970s, an increasing number of women cross borders with the aim to support themselves and their families. In other words, women increasingly choose to be income earners for themselves or their family, and so break the traditional pattern of the male migrant as the provider of household income. The main destination countries for Asian female labour migrants are the more developed countries in East, West and Southeast Asia. In several of these countries, the number of female labour migrants has in fact reached 70 per cent of the total number of labour migrants. For example, between 2006 and 2007, 69 per cent of labour migrants from Indonesia were female (IOM 2010, 8).

The push and pull factors of migration of women have their own gender-specific characteristics. The processes, motivations and social norms governing men's and women's cross-border mobility are different and this requires a gender perspective on the relationship between migration and the meaning of borders: borders are gendered. Similarly, the various reasons explored in this contribution are geographically also gendered in different ways. The participation of women in migration flows varies from country to country. For instance, in the case of the Philippines, it is estimated that 56 per cent of the migrants are female but only 19 per cent in the case of Senegal and 13 per cent in the case of Albania (Robert, forthcoming). The costs, risks and benefits of migration also differ by gender. The violence that women migrants are exposed to, both on the way to and across the borders, is different from the violence men may experience.

In this chapter, a gender lens is used to consider the threshold approach to decision-making by labour migrants. Van der Velde and Van Naerssen (2011) argue that socio-spatial thresholds impact differently on people who migrate and cross international borders. If one feels at ease in the place one lives – since one has a pleasant job that provides a sufficient and stable income, disposes of sufficient social capital and is free from social and political pressure – there usually does not exist a compulsion to move to another place and/or cross borders. Thus, circumstances like these are conducive to keep people in their home area. This also occurs when on the other side of the border the economy is weak and/or insecurities exist. These conditions will repel people from crossing the border. On the other hand, crossing borders becomes attractive when social and economic conditions at home deteriorate, thus pushing people to move, whilst at the same time, better conditions abroad pull potential migrants across the border. The willingness to migrate implies that the first threshold, the disposition towards migration, is passed. Next comes the issue of where to go precisely. Other opportunities and barriers will determine the choice of destination. These include the (un)availability of sufficient financial resources to pay private intermediaries to facilitate the border crossings and/or the (non-)existence of social networks to support the arrival in a foreign environment. After these barriers that constitute a 'locational threshold' are passed and a destination is chosen, the geographical route to the area can prevent migration: the distance, travel costs and the dangers along the route constitute a third threshold, that is, the 'trajectory threshold' that forces the potential migrant to look for other destinations or ultimately prevents him/her from leaving his/her home country.

Recognizing the importance of macrostructures, this chapter explores three sets of factors as they shape and are shaped by gender considerations: (1) labour markets; (2) sociocultural norms and practices: and (3) migration policies. These sets of factors impact on the thresholds differently for women compared to men, whether they want to leave or stay (the threshold of indifference), the direction of the migration if they decide to leave (the locational threshold) and the different routes to the chosen destination (the trajectory threshold). Special attention will be paid to migration laws and policies, since a specific approach should not only impact on our view about the subject matter but also on policies, be it of the government or civil society actors, including migrant associations. These issues are examined in the context of Asia, a macroregion in the throes of intense intraregional migration and where gender matters significantly in migration debates. Although the examples in this contribution are mainly from South and Southeast Asia, the same arguments may apply to other regions in the world as well.

The Global Labour Market

All around the world, the division of labour in households is, to different degrees, gendered. Men are considered as the major income earners and women as primarily responsible for reproductive tasks, even when the latter become the major contributors to family production and/or income. When women are active in the formal labour market, they usually perform different tasks and work in different sectors compared to men: the labour market is thus, gender segmented. According to the report *Women in Labour Markets* of the International Labour Organization (ILO 2010, 16):

> In terms of occupations ... nearly two-thirds of women in manufacturing are categorized as labourers, operators and production workers while only a few can be found in the administrative and managerial positions predominantly held by men. Women workers are usually employed in a limited number of industrial sectors: more than two-thirds of the global labour force in garment production is female-accounting for almost one-fifth of the total female labour force in manufacturing. With respect to employment status, the majority of family workers are female and, it may be added, often unpaid. These patterns suggest that women and men in the labour market are employed in different sectors and that, even when they are in the same sector, they

carry out different tasks at different levels of responsibilities. This phenomenon, sex segregation in the labour force, makes female and male workers work in 'compartmentalized' activities that usually lead to different rewards and different career opportunities even though workers may have comparable labour market attributes.

The working conditions of female employment are on average inferior to those of men and also more insecure, since they often concern informal and/or flexible work: part-time, temporary, casual, work in the home and subcontracting. According to the ILO report, there are some modest signs of progress, since the share of women working in the categories of vulnerable employment declined from 55.9 to 51.2 per cent between 1999 and 2009 (the male share fell from 51.6 to 48.2 per cent). Nearly one fourth of women remain in the category of unpaid home/family workers, implying that they receive no direct pay for their work. The ILO report concludes that gender inequality at work remains with regards to the quality of employment. Moreover, there is a clear segregation of women in sectors that are characterized by low payment, long working hours and often informal work arrangement.

The ILO report is based on national statistics and research results presented by macroregions (for instance North Africa, the Middle East and sub-Saharan Africa). Gender selectivity and segmentation of the labour markets are global phenomena and they increasingly impact on the migration of women who are in search of paid work across the border. The demand for female workers in the labour market of migrant-receiving countries is an important pull factor in migration. Push factors in sending countries also play a role, such as poverty, a high level of unemployment of both men and women, gender inequalities (women have more difficulties finding work), political and economic instability, violence in general and gender-based violence in particular. Labour migration increasingly constitutes an important part of female cross-border migration but the sectors migrant women are engaged in are usually as gender-specific as in their home countries.

From a region sending international migrants to the more developed regions in the Global North, Asia itself has transformed into a hub of international migration from the 1970s onwards, following the infusion of petrodollars in West Asia and the rise of newly industrialized countries in East and Southeast Asia in the 1980s. In the 1970s, India, Pakistan, Bangladesh, South Korea, the Philippines, Thailand, Indonesia and Sri Lanka (the last two joined relatively late) responded to the labour needs of the Gulf countries. Except for South Korea, which became a labour-importing country by the 1990s, and Thailand, which became a sending as well as a receiving country, these countries later also sent their nationals to work in Japan, Hong Kong, Taiwan, Singapore and Malaysia. A few more countries also embarked on labour migration programmes from the 1990s: Vietnam, Nepal, Cambodia, Laos and Myanmar.

The high demand for domestic workers contributed to the feminization of migration to the Gulf countries. With nationals not attracted to employment in the domestic sector, these countries had to rely on a foreign workforce to meet the labour needs. Not all sending countries readily deploy women migrants to the Gulf. In general, South Asian countries with the exception of Sri Lanka, send mostly male workers to this region. Sri Lanka, Indonesia and the Philippines stand out for the large proportion of women among their overseas worker population who are mostly engaged in domestic work. When East and Southeast Asian labour markets also turned to foreign workers to meet their labour needs, the demand for domestic workers once again fuelled female labour migration. Until 2004, the demand for 'entertainers' in Japan contributed to female migration and many of them come from the Philippines (through legal channels) and Thailand (through irregular channels).

It is important to note that in many of the destination countries, women are preferred to men because of their (assumed) precision at work, tolerance or patience and non-political engagement. Hence, they typically work in the export industry (garment, shoes and electronics), in the tertiary sector (as cooks, waitresses, entertainers and sex workers) and, in accordance with the traditional reproductive tasks of women, in the care sector (as domestic workers, nannies, caregivers for the elderly and nurses), that is, as part of the 'global care chain' (Ehrenreich and Hochschild 2002; Pérez Orozco 2007).

Female migrants are thus typecast in domestic (and entertainment) work and even when professional jobs are available, these are mostly in nursing and health. This leads to the observation that female migration basically responds to the demand for care work or reproductive work while male migration responds to the demand for work in the productive sector(s). A typical example is Indonesia. The majority of male migrants work in the construction, transport and oil industry in the Gulf States and in construction and agriculture in Malaysia. The majority of female workers migrate to the same countries but are overwhelmingly employed as domestic workers (Ananta 2009).

The demand for female migrant workers in the care sector is explained by the entry of local women into paid employment in the rich and emerging economies. Since there are many households whereby partners work outside the home, the care for the young, the physically handicapped and the elderly has become a problem. In large emerging economies such as Brazil, China and India, there is a pool of cheap local labour to fill the need for care workers, which in turn results in huge internal migration. In the rich countries of the US, Europe and East Asia, however, labour needs to be imported. Between the 1970s and 1990s, in the 'Little Tiger' economies of East Asia, a rapid growth in female employment occurred. The female labour participation rate of 64 per cent in 2008 was substantially higher than in the US and the EU, where it was only around 50 per cent (ILO 2010). The high level of female employment in East Asia is explained by their initial reliance on export industries where women were the preferred workers. Moreover, the investments in human capital (education) were high so local women entered the labour market as highly paid professionals as well (see, for example, Lee 2008). These changes generated a flow of female workers from the Philippines, Indonesia, Sri Lanka and Bangladesh to Hong Kong, Singapore, Taiwan, Malaysia and – to a lesser degree – South Korea.[1] Again, the constitution of the labour market is gendered since the demand for male workers is caused by other tendencies, more specifically the boom in the housing and construction markets, the need for modernization of urban infrastructure and the scarcity of local workers in agriculture.

Sociocultural Norms and Practices

Sociocultural norms and practices, which are often related to religion, create environments that encourage, facilitate or hinder the motivation to migrate. Major factors that impact on the mobility of female labour force are the limited access to education, the attitude of national institutions and policies towards women, and, more in general, the state of the national or regional economy. These factors are interrelated, for instance, women's level of education directly impacts on their level of economic and social development and vice versa. These same factors play a role in the international labour migration of women. In nearly all countries, the social barriers for women to

---

1   South Korea does not have a formal system for recruiting domestic workers – the domestic workers in the country are from China and mostly ethnic Koreans (Lee 2008).

migrate are greater than for men, but the degree may vary by country and region. In this section, we will focus on the sociocultural and religious environment of potential female migrants that legitimizes or obstructs female cross-border migration.

The role of social institutions in legitimizing migration is important. In particular, family structures determine if an individual has the power to decide on migration and the spaces of negotiation among the members of the family or household. Is it the individual who decides or the household, as explained by the New Economics of Labour Migration? Who are the members of a household and what structures the intrahousehold relations, the hierarchies and gender relations? Intrahousehold decision-making does not, however, occur in isolation. The role of kinship, clan or tribal relations and other social relations and networks surrounding the households are equally important. However, the agency of the individual should not be underestimated either. In the Philippines, for example, women have been found who migrate against the will of their families, among them women who leave to escape from the surveillance or control of the family and community.

Family structures are important in many aspects of life. At the level of the smallest unit, the household, Robert (forthcoming) gives the example of divergent household structures in the Dominican Republic. In the south of the republic, the dominant ethnic and cultural patterns are of African descent. They are characterized by men–women relationships that tend to be uncompromising, marital instability and strong ties between mother and children (stronger than between couples or father–child). Therefore, according to Robert, the model of the man as breadwinner of a nuclear family is not valid. In these circumstances, women can decide autonomously to migrate but they have to rely on kin members to take care of their children. On the other hand, in Las Placetas, a community located in the centre of the Dominican Republic, with a population that is predominantly of mixed race and ethnicities, the patriarchal family model is overriding and hence, women have limited decision-making power.

In the case of the extended family pattern, a difference can be made between the patriarchal extended family and the 'flexible' extended family. The power of decision-making is more shared in the latter case. Here the existence and influence of the extended family, and particularly the support of female kin, provide greater opportunities for women to migrate. For the married migrant women, their children are taken care of by their sisters, mothers, mothers-in-law or other family members, even when their husbands remain at home. In the patriarchal family, whether nuclear or extended, women's autonomy in decision-making is out of the question, because men are afraid to lose control on their women, thereby making it difficult for women to migrate. Moreover, the couple will be blamed for behaviour that is considered indecent, as migration exposes the women to the outside world (Abadan-Unat 1977; Desai and Banerji 2008).

In traditional Islamic countries, the mobility of women is often limited. However, care should be taken not to stereotype the so-called 'female Islamic behaviour' in countries where Islam is the dominant religion. For example, in Indonesia, the largest Islamic country in the world, women have considerably more negotiation power within the households – they are usually the managers of household finances – and more space to move compared to women in Saudi Arabia and the Gulf States. Also within countries, a variety of interpretations of dominant values and norms on behaviour exist, which varies according to class, ethnicity, caste (if so), religion and geographical region. Keeping this diversity in mind, Bangladesh will be presented as a case study on how crossing international borders tend to be conceived and interpreted as 'hostile and dangerous' to women.

Bangladesh has a population of around 160 million inhabitants with an average population growth of 2.5 million a year. International migration has kept the country's unemployment rate virtually unchanged since the 1980s, even though the growth rate of its labour force is nearly double that of its population growth. There are 7.5 million documented migrant workers from Bangladesh all over the globe. Many more, however, remain undocumented (Siddiqui 2008). An overwhelming majority of the migrants are men. Women more so than men go through informal, private sector channels rather than the government channels to obtain jobs, visas and support in facilitating their journeys. The proportion of women migrants relative to men remains minute: according to the International Organization for Migration (IOM 2010) only 4 per cent of the total documented migrant population are women. The official 'invisibility' of migrant women is also due to the fact that until 2006 the government prohibited women below 35 years of age to go overseas for work (Siddiqui 2008). The ban against female migration, however, did not stop women from migrating for work. On the contrary, many continued to migrate through private channels, which in turn aggravated migrant women's vulnerability and insecurity.

Whilst it is true that the Bangladeshi migration policies are not in favour of female international migration, the major reason for women's small representation in international migration is their embeddedness in a sociocultural environment that restricts the mobility of women, as a result of social taboos with *purdah* (seclusion) as the most extreme.[2] Obviously, the major push factor for labour migration is poverty and the major pull factors are to support families and eventually to save for one's own business. According to Dannecker (2005), who did qualitative research among Bangladeshi female international migrants, families only supported female migration in two cases: either because there is no male member in the household or the financial possibilities of the family only allow sending a daughter or wife (the lower wages paid to female labour migrants abroad lead to lower recruitment costs). A female migrant brings shame to the family, especially if the woman concerned is unmarried. Thus, only male migrants receive social approval. The migration of a female family member is not at all considered as a success story, neither for the migrants nor for their families. Dannecker (2005, 660) notices:

> Whereas households of male migrants proudly present the symbols of the successful journey or stay of their family members, for example, a new tin roof, a television or a huge picture of their relative abroad, these symbols are hardly to be seen in households of female migrants. These families are very reluctant to speak about their female family members abroad, or even try to hide the fact that female members have lived or currently live and work abroad.

The impact of sociocultural circumstances and institutions also applies to receiving countries. For example, the large number of Filipino domestic workers in Spain and Italy can be explained by the fact that all three countries are predominantly Roman Catholic. To a certain degree, this may be true. However, there are also Moroccan, Albanian and African women working as domestic workers in these countries, much more than in north-western Europe. As a matter of fact, having domestic workers is a tradition and a sociocultural trait in the southern countries of the EU, more so than in the north where in-house domestic helpers are rare.

The spatial trajectories to and from other countries are also clearly gendered. Labour migrants en route have to take the risks of being cheated or physically violated but male and female migrants encounter different problems. Generally, women are much more vulnerable than men but some aspects of vulnerability are socially constructed. For example, the approach taken by some Asian

---

2   Like in other Asian countries, female mobility was facilitated by the flourishing of the garments industry in the 1980s, an example of how economic needs and opportunities can modify cultural norms regarding mobility.

governments in dealing with female migration is typically patriarchal, such as the tendency to consider women travelling abroad on their own as being inappropriate. Consequently, women end up paying bribes, suffer physical violence, are being trafficked, have their money stolen upon return, et cetera.

To counter these abuses and to protect overseas workers, in 1999 the Indonesian government opened a special terminal for their migrant workers at the Soekarno-Hatta International Airport near Jakarta. However, this had led to new abuses. Silvey (2007) describes how female returning migrants had to pay fees for various compulsory services: transport to the special terminal, a porter to help them with their baggage and for documentation. Their family members also had to pay a fee when they picked up these women at the airport and they had to pay again upon leaving the arrival gate. For these reasons, the terminal was closed in 2012. Due to the Indonesian government's concern about the decency of its female migrants, it decided to provide to/from airport transport. If no family member is there to pick up the female returnees, they are required to ride home in government-registered buses. Unfortunately, such a move also resulted in unintended consequences; stories of drivers extorting money from the returnees abound. In short, and relevant in the context of the threshold approach, both governmental actions and the migration industry of agencies and intermediaries demonstrate the gendered nature of trajectories in international labour migration.

The foregoing seems to suggest that these are some very significant factors that repel or keep women at home. However, this picture may be too pessimistic as there is another side to the story as well. Bach (2011) refers to a study among Nepalese women labour migrants who considered that their situation had improved after migration, regardless of the many obstacles faced. They went to countries as diverse as India, Hong Kong SAR, the UK and the US and accumulated assets that gave them more control over land, housing and other commodities which increased their autonomy. Female labour migration and the women's newly acquired role as remitters can be empowering, thus contributing to a greater sense of self or self-esteem and independence (Van Naerssen, forthcoming). That is why among this group of women, about 60 per cent wanted to work abroad again (Bach 2011). There are similar findings in the Philippines: overall, female migrants are positive about their migration experience.

Government Laws and Policies

This section focuses on the links between gender, migration and laws and policies. While the gendering of the exit and admission laws and policies is fairly established in the migration literature (Boyd 2005; Oishi 2005), it is important to examine the translation of these laws and policies into practices. This reveals unintended consequences, tensions with other laws or policies, or conflicts with other stakeholders, that impact differently on the chances of migration for women and men.

While all lower-income countries in Asia actively pursue overseas jobs for their nationals, only Indonesia, the Philippines and Sri Lanka allow the deployment of women migrants without much restrictions. Bangladesh and Nepal had an on-off policy with a recent relaxation of restrictions in Bangladesh (Kibria 2011). India maintains a policy of allowing women of 30 years and older to migrate for work and for less-skilled migrants. This differential approach to labour migration made Oishi (2005) to carry out a multilevel analysis (state, individuals and society, or macro, micro and meso levels, respectively) to examine why some Asian sending countries do not encourage sending female migrants (Bangladesh, India and Pakistan) while others do (Indonesia, the Philippines and Sri Lanka). She found that policies for male migration are economically driven, while those for women are value-driven, citing in particular the need to protect women. The three countries that restrict female migration cite concerns for the safety of women migrants. Oishi concluded that emigration policies at the state level, next to women's autonomy and decision-making at the individual level and social legitimacy of women's employment and mobility, contribute to conditions promoting female migration. Her fieldwork dates from the early 2000s. Since then in several countries some relaxation of restrictive rules took place, influenced by the advocacy of migrant and women organizations.

In stark contrast to the borderless world of capital and goods, in the (still) bordered world of migration, national laws and policies governing the exit or entry of people function as barriers and remain significant in filtering and separating those who are allowed to migrate. However, despite the daunting restrictions, there are loopholes, gaps or opportunities that can subvert the intent of laws and policies.

From the Origin: Exit Policies

The discrepancy between the explicit right to leave from and return to one's country and the silence on the right to enter another country in the Universal Declaration of Human Rights (Article 13) suggests that it is easier to leave a country than to gain admission in another. Because of their location at the point of origin, restrictive exit policies can stop, prevent or delay migration at the initial step. Travel restrictions or outright bans have been employed by repressive or authoritarian regimes as part of their system of governance. It may be recalled that prior to the introduction of market reforms in 1978, international migration from and to China was negligible. After 1978, international migration steadily increased as restrictions to leave and enter China were dismantled. North Korea and, until very recently, Myanmar keep a tight lid on international migration, such that leaving these countries tend to be unauthorized, and those who are caught can be punished severely.

In relatively democratic societies, exit policies tend to be open, although restrictions may be introduced as the need arises. As noted earlier, the enthusiasm with which some origin countries in Asia pursued labour migration was reserved only for men while women's access to jobs abroad was restricted in the name of protecting them from dangers while they are overseas. At least, this was initially the position of the governments as the restrictions were later lifted, and interestingly, the countries which allow female migration are currently reconsidering to reduce the deployment of women workers. Among the three South Asian countries, Bangladesh, India and Pakistan, the first one has a history of imposing a ban on women leaving as domestic workers (the migration of professional and skilled women is not considered problematic). Although India and Pakistan did not impose a ban, the conditions allowing female migration had the effect of banning women from migrating: India required a minimum age of 30, completion of matriculation and clearance from the Protector General of Emigrants and in the case of Pakistan the minimum age was 35 (Thimothy and Sasikumar 2012). Data on legally deployed labour migrants from these countries are supporting this since the migrants are overwhelmingly male (IOM 2010, 25).

While effective in curbing legal migration, concerns about women migrating through irregular or trafficking channels are rife. In other words, the very efforts aimed at protecting women have the opposite effect by pushing women to extralegal channels and rendering them more vulnerable to abuse and exploitation. The rationale for the restrictions is also criticized for its paternalistic view of women and a violation of their right to migrate and obtain gainful employment. If governments reconsider their position, these considerations cannot be discounted.

In addition, it is not far-fetched to propose that economic benefits of female migration also provide a compelling reason to relax or abandon restrictive exit policies. Over time, the ban or restrictions in the three South Asian countries have been relaxed or lifted. In response to pressures from civil society invoking women's rights to migrate, the Bangladeshi government started to relax the ban in 2002 and eventually lifted it in 2006 (Kibria 2011).[3] Although the basic conditions for female migration remain, India has stepped up efforts to promote the protection of women migrants in the domestic work sector. For those migrating to the Gulf countries, India has pushed for a minimum monthly wage of US $300–$350 and the issuance of the Smart Card (which contains basic information about the migrants) for first-time migrants. Recently, in 2009, Pakistan's national emigration policy to promote regular migration and the protection of migrants acknowledged the importance of female migration. Among others, it created a task force that examines the obstacles and prospects of female migration, look into skills development and promote safe migration. But the age requirement is unchanged and the report is silent on women's right to work (Thimothy and Sasikumar 2012, 83).

A recent player in international labour migration, Nepal, has changed its policy on female migration many times since 1997. When it launched the overseas employment programme in 1997, the government did not have a specific policy to restrict female migration. However, in 1998 the mysterious death[4] of Kani Sherpa, a Nepalese domestic worker employed in Qatar called attention to the risks and vulnerabilities of women migrants, which resulted in the imposition of a ban on female migration, particularly to the Gulf countries. The ban on female migration to any country was lifted as of May 2012 (Thimothy and Sasikumar 2012, 78). But less than a year later, increasing reports of Nepali women encountering problems in the Gulf countries prompted the Nepali cabinet on 9 August 2012 to impose a ban on women under the age of 30 migrating to work in the Gulf (Human Rights Watch 2012; Caritas 2012). The move is seen as discriminatory against young women and as ineffective.

Meanwhile, the Philippines, Indonesia and Sri Lanka, the three countries where female migration is significant, are constantly confronted with challenges in promoting the protection of women migrants in domestic work. Various measures such as, pre-departure information programmes, skills training, cultural orientation and on-site services have been attempted, but reported abuses persist (see for instance various reports by Human Rights Watch and the annual US *Trafficking in Persons Report*). The Philippines (Lebanon, Jordan and Syria)[5] and Indonesia (Saudi Arabia, Jordan, Kuwait and Syria) had resorted to bans because of abuse cases, but eventually, the bans were lifted when certain conditions were met. Sri Lanka also attempted to ban the deployment of women migrants, but for a different reason. In consideration of the welfare of the young children left behind, the Sri Lankan government proposed to ban the departure of mothers with young children (that is, children under the age of five), but the ban was withdrawn because of the protests lodged by local and international rights groups. As elsewhere, critics said the ban was misguided and not the answer to the problems of the families and children left behind by women migrants.

---

3   Kibria (2011) cites 2007 as the year when the ban was lifted.

4   Kani Sherpa died under similarly mysterious circumstances. While Kani's employer said she fell to her death from the top storey of the house, her family alleged that she was murdered after being raped (Sarkar 2011).

5   The ban on Saudi Arabia, imposed in July 2011, was lifted as of 1 October 2012, following Saudi Arabia's agreement to the US $400 minimum monthly salary for Filipino domestic workers.

Indeed, for the countries that imposed a ban, it only curbed legal migration but was ineffective in stopping migration through extralegal channels. For example, when the Philippines imposed a ban on domestic worker migration to Singapore in 1995 (following the diplomatic strain over the execution of the Filipino domestic worker Flor Contemplacion[6]), the migration of women continued, but through irregular means. The ban on domestic worker migration to Afghanistan, Iraq, Lebanon, Jordan and Syria has not prevented Filipino women from finding employment in these countries as domestic workers. And as feared, as unauthorized migrant workers or trafficked persons, many cases of abuse continue to be reported (Battistella and Asis 2011).

Indonesia, the Philippines and Sri Lanka are now aiming to reduce the sending of domestic workers due to persistent concerns with safety and well-being. In recent years, the share of women migrants in the annual deployments has indeed gone down, in Sri Lanka from 59 per cent in 2005 to 51 per cent in 2009 (IOM 2010, 24) and in the Philippines from 71 per cent to 52 per cent during the same period (Asis and Agunias 2012, 2). The Philippines introduced policy reforms on domestic worker migration in 2006 to promote better protection. The reform package includes the increase of the minimum age to 23, disallowing the collection of placement fees by recruitment agencies, requiring domestic workers to undergo skills training and language/cultural orientation, and raising the monthly minimum wage to US $400. The package has the implicit objective of reducing domestic worker migration. By increasing the salary threshold, it is hoped that the demand for Filipino domestic workers will be redirected to countries that pay higher wages and provide better legal protection, such as Taiwan and Hong Kong SAR. An assessment of the reform package reveals that domestic worker deployment declined temporarily (in 2007 and 2008) but started to pick up from 2009. Considering the many violations and ways of getting around the reforms (for details see Battistella and Asis 2011), it does not seem likely that the increase reflects increasing compliance with the new conditions.

The need to have a coherent and decisive policy on domestic workers is called for. Indonesia banned the sending of domestic workers to Saudi Arabia in 2011, following the execution by beheading of Ruyati binti Sapubi without informing Jakarta. The incident triggered Indonesia to work out a plan to stop the deployment of domestic workers from 2017; the preparation includes setting up skills development programmes to prepare Indonesians for jobs in other sectors (*Jakarta Post*, 29 May 2012). Similarly, the Philippines is considering to phase out the deployment of domestic workers by 2017. The plan, which is in its conceptualization phase, will not completely eliminate domestic worker migration as officials acknowledge that there are destinations and types of domestic work which are high paying and protective (*Inquirer*, 26 August 2012).

At the Destination: Entry Policies and Beyond

Present-day laws and policies governing the admissibility of non-nationals in destination countries in Asia are intimately linked with labour market needs. As was discussed earlier, the shifts in the labour market needs of the Gulf countries played an important role in transforming the gender composition of migrants from a predominantly male to a more gender-balanced population. The labour migration policies of the receiving countries in the Gulf and Asia are highly economistic, mostly oriented to the needs of the receiving countries. Unlike the temporary guest worker programme in Europe which was between governments, labour migration to the Gulf and East Asia is mediated by private recruitment agencies. The whole

---

6   She was found guilty of murder by the Singaporean court and hanged.

migration regime rests on the temporary labour migration specifically for migrant workers in less-skilled occupations. Those in highly skilled or professional occupations generally have the option to be joined by family members and obtain residence (and citizenship) in the destination country.

The differential treatment of the highly skilled and the less-skilled migrant workers indicates that other key social categories than gender – in this case, class – determine the impact of migration on migrants. Ethnic or national origin also figures in determining the source countries of migrant workers. Considering the many potential sources of workers, destination countries can decide from which countries to recruit needed workers. If one source country decides to impose a ban, destination countries can choose from various other countries to meet their labour needs. Or a destination country can ignore the exit policies of a sending country for as long as a worker meets the former's documentation requirements, a situation which leads to the migrant being rendered irregular from the perspective of the origin country but an accepted one from the perspective of the destination country. This is how many Filipino women secure employment as domestic workers in Singapore: they leave the Philippines as tourists, enter Singapore as tourists, and thereafter apply for a work permit (that is, with a prearranged employment) as domestic workers.[7]

Destination countries in Asia focus on the length of stay or residence of foreign workers. As mentioned earlier, despite varying histories and contexts, all destination countries in the region are adamant on limiting the stay of migrant workers in less-skilled occupations (and this applies to women and men alike) while open to granting family reunification, residence and citizenship to migrants in highly skilled or professional occupations. The limit on the stay of less-skilled migrant workers is typically two years (which may be renewed depending on the agreement of the worker and employer). Malaysia and Singapore do not allow migrant workers to marry a national; those who do will have their work permits revoked, which means repatriation for the worker. This policy has the effect of circumventing the possibility of foreign workers gaining residence through marriage with a citizen in the destination country and complicating the ethnic composition (and citizenship issues) of the receiving society.[8] Singapore requires foreign domestic workers to undergo a biannual medical examination; those who are found pregnant will be repatriated. The greater surveillance over migrant women's bodies was cited by Huang and Yeoh (2003) as one of the ways in which migration policies are gendered.[9] The policy's singling out foreign domestic workers and exempting foreign women in high-level occupations from restrictions and requirements also speaks of gender intersecting with class in their impact on migrants.

An unexpected development in Asia, particularly in East Asia (Japan, Taiwan and South Korea), is the rise in international marriages and the highly gendered feature of this phenomenon.

---

7   This example also highlights the lack of consensus on migration policies and procedures between countries.

8   Both Malaysia and Singapore are multicultural societies with populations of Malay, Chinese and South Asian origin. See Chee et al. (2012) for a discussion on the difficulties encountered by Indonesian migrants and Malaysian nationals who marry. In view of Malaysia's restrictions against such unions, couples have to navigate circuitous pathways to legalize their marriage or to find ways (including extralegal means) to enable the Indonesian partner to stay in Malaysia.

9   See, for example, Channel News Asia (2012). The report stated that the number of diseases and pregnancies among foreign domestic workers is very low, which was taken as evidence of the wisdom to continue the policy.

Before the 1990s,[10] international marriages in the region were mostly between Western men and Asian women, but in more recent decades, international marriages had become intraregional involving men from more developed countries and women from less developed countries in the region. Marriage migration, thus, increased the participation of women in migration in the region. In addition, the admission of women as spouses of their citizens and as mothers of future citizens has touched off discussions on the emerging cultural diversity of otherwise homogeneous societies. The phenomenon of marriage migration in South Korea has resulted in the introduction of policies and programmes towards Korea's transformation into a multicultural society.[11] Through marriage migration, women are able to access residence and citizenship more so than men. The participation of brokers has infused irregular practices and risks to marriage migration – and trafficking of women has entered the scenario. Thus, like labour migration and the concentration of women migrants in unprotected sectors, the possibility of trafficking in marriage migration brings out the possibility of thrusting women in risky and vulnerable situations.

Challenging Migration Laws and Policies

Although labour migration in Asia is strictly regulated, irregular migration, including traffic in persons, is significant. The barriers put up by migration laws and policies deter migration, but they are not sufficient to stop people from leaving or to keep people out. Irregular migration is symptomatic of policy limitations, labour needs and the assertion of the right to migrate. Defining irregular migration as a departure from the norms and regulations of the origin, transit or destination country and thereby defining irregular migrants as violators, privileges the state as the authority over migration matters. However, the state, through conflicting, incoherent or unrealistic laws and policies can contribute to irregular migration. The lack of or limited channels for legal migration amidst a high demand for workers can lead to high volumes of irregular migration, such as in the case of Malaysia where the number of irregular migrant workers surpasses the number of regular ones. Both legal and irregular migration are determined by factors operating in the origin and destination countries; both are mediated by legal and illegal brokers and intermediaries. Brokers and intermediaries have become key players in facilitating migration. Migrants diverge into legal or irregular pathways depending on their access to legal versus irregular brokers, including traffickers (Battistella and Asis 2003). The irregular or abusive practices of brokers can be addressed by increasing migrants' access to information, providing access to migration and more employment opportunities (Agunias et al. 2011).

The impetus for change in migration policies comes from various sectors whose divergent positions pose tensions and challenges to arriving at coherent, realistic and rights-based policies. In destination countries, employers' demand for workers may be challenged by civil society organizations (including trade unions) as an attempt to cut costs. Meanwhile, the public may be concerned with the social, cultural and political implications of migration. In origin

---

10   Earlier in the case of Japan.

11   Policies promoting multiculturalism, as promoted in South Korea, are not quite the same as the multicultural policies introduced in Australia, Canada and Europe. Programmes offered by multicultural centres focus on foreign women married to Korean men to help the former adjust to Korean society (for example by getting to know the Korean language, Korean cuisine, Korean culture and so on), an approach that is more reminiscent of assimilation than multiculturalism. Interestingly, there are no programmes for foreign men from other Asian countries who marry Korean women.

countries, labour migration policies can be criticized by civil society organizations as an easy but unsustainable strategy for development. The right to migrate can be used by governments to support or facilitate migration; the same argument can be used by civil society organizations to governments' decisions to restrict or ban migration for the purpose of protecting migrants. The perspective of migrants, the 'object' of government actions and advocacy efforts, should not be neglected.

Conclusions

Gender matters in international migration and is reflected in norms, policies and laws concerning mobility in general. Although there has been notable progress in broadening the gender perspective in migration, gender analysis of migration still tends to be, by and large, based on women's realities rather than on a more holistic view that include men's experiences as well. For some time now, despite the expanded definition of trafficking provided by the 2000 Trafficking Protocol, most research in this area has focused on trafficking in the sex industry. As a result, women and girls figured prominently in discussions about trafficking, leading to anecdotal observations as 'men migrate while women are trafficked' (D'Cunha 2002, 8). In recent years, research in other sectors has provided more evidence and a nuanced appreciation of the trafficking of women and men in the other sectors (see for example Pearson et al. 2006, vol. 1). Also, the tendency to highlight the experiences and conditions of workers in the sex industry inadvertently draws attention to the trafficking of primarily women and girls while the trafficking of men and boys in other economic sectors have been downplayed.

Whilst gender does matter in international migration, the previous discussion has also highlighted the intersection of gender with ethnicity and class, and how in some instances and contexts, gender may be more or less important as compared to other social constructs. However, in this contribution, gender is placed in the centre of interest and it is demonstrated that the various thresholds have differential meanings and impact on the decision to migrate by both women and men. The meaning of thresholds, in particular the mental threshold, differs by gender. In general, the thresholds are lower for men than for women. This is related to the three specific sets of factors distinguished: labour markets, sociocultural traditions and migration policies. Men are usually considered to be the main breadwinners, and traditionally there are less restrictions imposed on their mobility. From a broader perspective, this chapter shows that for the threshold approach to be more policy and politically relevant, it has to be refined and to incorporate the rich diversity of labour migrants.

References

Abadan-Unat, N. 1977. "Implications of Migration on Emancipation and Pseudo-Emancipation of Turkish Women." *International Migration Review* 11 (1): 31–57.

Agunias, D.R., C. Aghazarm, and G. Battistella. 2011. *Labour Migration from Colombo Source Process Countries: Good Practices, Challenges and Ways Forward.* Dhaka: International Organization for Migration.

Ananta, A. 2009. "Estimating the Value of the Business of Sending Low-Skilled Workers Abroad: An Indonesian Case." Paper presented at XXVI IUSSP International Population Conference, 27 September–2 October. Marrakech: International Union for the Scientific Study of Population.

Asis, M.M.B, and D. Agunias. 2012. "Strengthening Pre-Departure Orientation Programmes in Indonesia, Nepal, and the Philippines." Issue in Brief 5. Bangkok: International Organization for Migration.

Bach, J. 2011. "Remittances, Gender and Development." In *Gender and Global Restructuring: Sightings, Sites and Resistances*, 2nd ed., edited by M.H. Marchand and A.S. Runyan, 129–42. London: Routledge.

Battistella, G., and M.M.B. Asis, eds. 2003. *Unauthorized Migration in Southeast Asia*. Quezon City: Scalabrini Migration Center.

——. 2011. "Protecting Filipino Transnational Domestic Workers: Government Regulations and Their Outcomes." PIDS Discussion Paper Series 2011–12. Manila: Philippine Institute for Development Studies.

Boyd, M. 2005. "Women in International Migration: The Context of Exit and Entry for Empowerment and Exploitation." Paper presented in the Fiftieth Session, Commission on the Status of Women, High-Level Panel on the Gender Dimensions of International Migration, 27 February–10 March. New York: United Nations.

Caritas. 2012. "Banning Women from Work is Not the Answer, Says Caritas." *Caritas*, 31 August. Accessed 01 October 2014. http://www.caritas.org/2012/08/banning-women-from-work-is-not-the-answer-says-caritas/.

Channel News Asia. 2012. "Not Many Foreign Domestic Workers Have Diseases or Pregnancies." *Channel News Asia*, 15 May. Accessed 1 October 2012. http://www.channelnewsasia.com/stories/singaporelocalnews/view/1201247/1/.html.

Chee, H.L., B.S.A. Yeoh, and R. Shuib. 2012. "Circuitous Pathways: Marriage as a Route toward (Il)legality for Indonesian Migrant Workers in Malaysia." *Asian and Pacific Migration Journal* 21 (3): 317–44.

D'Cunha, J. 2002. "Trafficking in Persons: A Gender and Rights Perspective." Paper presented at the Expert Group Meeting on "Trafficking in Women and Girls", 18–22 November. New York: United Nations Division for the Advancement of Women.

Dannecker, P. 2005. "Transnational Migration and the Transformation of Gender Relations: The Case of Bangladeshi Labour Migrants." *Current Sociology* 53 (4): 655–74.

Desai, S., and M. Banerji. 2008. "Negotiated Identities: Male Migration and Left-Behind Wives in India." *Journal of Population Research* 25 (3): 337–55.

Ehrenreich, B., and A. Russell Hochschild. 2002. *Global Woman: Nannies, Maids and Sex Workers in the New Economy*. New York: Henry Holt and Company.

Huang, S., and B.S.A. Yeoh. 2003. "The Difference Gender Makes: State Policy and Contract Migrant Workers in Singapore." *Asian and Pacific Migration Journal* 12 (1–2): 75–98.

Human Rights Watch. 2012. "Nepal: Protect, Don't Ban Young Women Migrating to Gulf: New Ban on Young Women Migrants Discriminates, Not a Solution." *Human Rights Watch*, 14 August. Accessed 2 October 2012. http://www.hrw.org/news/2012/08/14/nepal-protect-don-t-ban-young-women-migrating-gulf.

ILO (International Labour Organization). 2010. *Women in Labour Markets: Measuring Progress and Identifying Challenges*. Geneva: International Labour Office.

IOM (International Organization for Migration). 2010. *Labour Migration from Indonesia: An Overview of Indonesian Migration to Selected Destinations in Asia and the Middle East*. Jakarta: IOM.

Kibria, N. 2011. "Working Hard for the Money: Bangladesh Faces Challenges for Large-Scale Labor Migration." *Migration Information Source*, 9 August. Accessed 1 October 2012. http://www.migrationinformation.org/Profiles/display.cfm?ID=848.

Lee, H.-K. 2008. "Migration and Development: Migrant Women in South Korea." In *Global Migration and Development*, edited by T. van Naerssen, E. Spaan, and A. Zoomers, 269–87. London: Routledge.

Oishi, N. 2005. *Women in Motion: Globalization, State Policies, and Labor Migration in Asia*. Stanford, CA: Stanford University Press.

Pearson, E., S. Punpuing, A. Jampaklay, S. Kittisuksathit, and A. Prohmmo. 2006. *The Mekong Challenge: Underpaid, Overworked and Overlooked: The Realities of Young Migrant Workers in Thailand*, vol. 1. Bangkok: International Labour Office.

Pérez Orozco, Amaia. 2007. "Global Care Chains." Gender, Remittances and Development Series Working Paper 2. Santo Domingo: United Nations International Research and Training Institute for the Advancement of Women.

Robert, E. Forthcoming. "A Gender Perspective on Migration, Remittances and Development: The UN-INSTRAW Experience." In *Gender, Remittances and Development in the Global South*, edited by T. van Naerssen, L. Smith, T. Davids, and M.H. Marchand. Aldershot: Ashgate.

Siddiqui, T. 2008. "Women Migration from Bangladesh: Present and Future." Paper presented at the Conference "Remittance: Does Remittance Really Empower Women", 1 November. Dhaka: Bangladesh Support Group.

Silvey, R. 2007. "Unequal Borders: Indonesian Transnational Migrants at Immigration Control." *Geopolitics* 12 (2): 265–79.

Thimothy, R., and S.K. Sasikumar. 2012. *Migration of Women Workers from South Asia to the Gulf.* New Delhi: Giri National Labour Institute.

Van der Velde, M., and T. van Naerssen. 2011. "People, Borders, Trajectories: An Approach to Cross-Border Mobility and Immobility in and to the European Union." *Area* 43 (2): 218–24.

Van Naerssen, T. Forthcoming. "Exploring Gender and Remittances." In *Gender, Remittances and Development in the Global South*, edited by T. van Naerssen, L. Smith, T. Davids, and M.H. Marchand. Aldershot: Ashgate.

PART II
BORDERS AND BORDERING

Chapter 7

(Im)mobility in Karelia:
A Space of Transforming Belonging

Alexander Izotov and Tiina Soininen

This chapter contributes to our understanding of EU's external borders and their regulatory mechanisms in international mobility. The Finnish–Russian border is used as a case to understand what prevents mobility at the individual level. The issue of 'latent' mobility (as in commonness, closeness or distance and familiarity) and actual mobility is problematized by viewing the border region as a space that is constructed out of social relations. The overarching threshold approach of Van der Velde and Van Naerssen (2011) is adopted as the analytical framework in order to understand these different aspects of mobility/immobility. At the individual level, they argue that there are three kinds of thresholds: the mental border threshold, the locational threshold and the trajectory threshold. Through the selected case of northern Europe and northern Russia, the authors will demonstrate how these thresholds influence mobility and immobility.

To operationalize the thresholds, several variables are used as proxies. For instance, the mental threshold is represented by social networks, personal biographical experience and motivation. The locational threshold is represented by the place of residence and, in particular, the distance to the border. The third threshold, the trajectory threshold, is attributed to the mode of transportation, as this indicates the passage between different places. Through the clarification of various thresholds, spatial behaviour is analysed at a particular external border of the EU and the aim is to describe how these different thresholds manifest themselves.

This chapter addresses mobility or immobility in a particular space of Finnish and Russian Karelia (see Figure 7.1). On the Finnish side, the focus is on North Karelia, and on the Russian side on the Republic of Karelia. North Karelia is often portrayed as an economically deprived, peripheral area in Finland. The Republic of Karelia, as one of the 21 republics of the Russian Federation, plays a key role in the north-western region of the country in terms of cross-border contact with the EU.

After a general section on the historical and current cross-border contacts in the Finnish–Russian border region, this chapter examines the results of a survey carried out in North Karelia. The survey was conducted in order to discover the travel interests and cross-border mobility motivations of the Finnish people in the borderland of both Karelias. The population is divided into three groups: mobile, potentially mobile and immobile, and it shows that these three groups differ in nature. The analysis draws on new ideas about how cross-border mobility actually takes place in everyday life of individuals. The main question is: What defines cross-border mobility from Finland to Russia in a context of two previously culturally close but later geopolitically divided regions?

The geopolitical location, the common border (the two countries share a border of more than 1,300 km long with a total of eight permanent border-crossing points for regular travellers), and the location remote from population centres make the position of Russian and Finnish Karelia of interest from both the northern European and Russian perspective. Since the early 1990s, when the border regime was liberalized, attempts to facilitate a gradual integration of these border regions

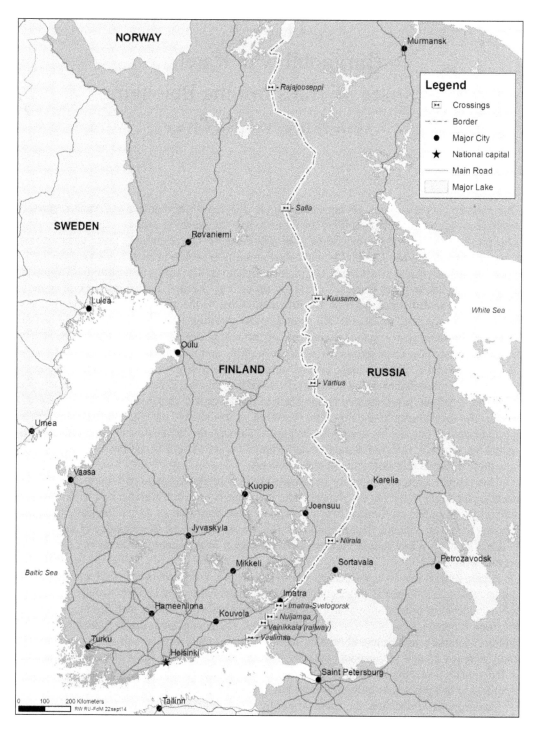

Figure 7.1   Finnish–Russian border regions and border-crossing points

have been undertaken. Cross-border economic and cultural cooperation and formal and informal contacts have grown at regional and municipal levels. Regional and local actors have already created transborder links based on market-oriented interaction and EU programme-based cooperation.

As Izotov and Laine (2013) argue, the practical problems stemming from linguistic complexity, differences in institutional frameworks and mismatches between the parties involved are commonly overshadowed, even greatly, by the rhetoric of success in the evaluation of Russian and Finnish experts (see for instance Romanova 2002; Liikanen and Virtanen 2006). Simultaneously, more cautious views of these processes exist, which take into account historically rooted geopolitical tensions, export policies and the achievements of cross-border cooperation in general (see Eskelinen 2011). It is also important to take into account the significant shift in the policies of the Russian federal government towards its border regions from the late 1990s to the early 2000s.

The contribution of this chapter to the discussion on cross-border mobility is based on two considerations. First, it concerns external borders of the EU. Secondly, this chapter deals with peripheral regions. Both facets demonstrate contrasts between, and various characteristics of, bordering and regionalization processes in the EU, and the varied circumstances which impact on decision-making concerning mobility and immobility. Programmes, such as the Euroregion Karelia,[1] which aim to create flexible joint economic structures and organize cross-border trade, tourism and cultural contact in the border regions of Finland and the Karelian Republic illustrate attempts of region-building on the edge of the EU. These measures have had different impacts on the actual cross-border mobility and integration of border regions in different locations.

This chapter presents a micro perspective on cross-border processes. It emphasizes the role of local actors in everyday life in the border areas. The focus on localities in the border region allows to investigate social networks and public measures that enable mobility in cross-border activity. This perspective, which emphasizes the role of human agency, is often ignored in theories that focus on historical macro-level processes (Hebinck et al. 2008). In this respect, the threshold approach is employed, which presupposes and refers precisely to an understanding of cross-border (im)mobility centred on spatial actors (Van der Velde and Van Naerssen 2011). The combination of people, borders and trajectories inherent to cross-border mobility allows us to understand how people decide to go and how their spatial behaviour in this process might be studied. This concept also envisages interrelations between places (localities, regions, countries) of origin and places of destination.

Cross-Border (Im)mobility in the General Context of Finland and Russia

The Finnish–Russian border is one of the oldest extant state borders in Europe. However, it is not a consistent entity; it has been constructed by historically varying circumstances. The oldest parts of the border were defined in peace treaties between the Kingdom of Sweden and the Grand Duchy of Muscovy in 1595 and 1617. Since then, several wars have taken place between the independent state of Finland and the former Soviet Union, including conflicts during the Second World War, which have moulded the structures of the border. Historically, the border area of Karelia has been

---

1   Euregio Karelia is a cooperation area and cooperation forum of three Finnish regions: Kainuu, North Karelia and Northern Ostrobothnia, and the Republic of Karelia of the Russian Federation. The aim of its foundation was to improve living conditions of inhabitants through cross-border cooperation. Euregio Karelia is a cooperation forum for the participants. It deepens cross-border cooperation, and introduces strategic and political guidance into the cooperation. Regional lobbying is an important task of the Euregio Karelia (see http://www.euregiokarelia.com/en/, accessed 25 September 2014).

the arena for a wide range of ethnic and cultural encounters as populations were moved across the newly created border (Kokkonen 2012; Paasi 1999).

During the Second World War, the previously Finnish region of Karelia was divided into two regions. A part of the former Finnish Karelia was annexed by the Soviet Union. The entire Finnish population moved to Finland. The ceded territory was turned into a closed border zone. Migrants from different parts of the former Soviet Union settled in the region (Izotov 2012). Consequently, the Finnish language and the shared Karelian cultural identity were also divided. Since a large part of the border area in the Soviet territory became a closed border zone, contact became limited. Only in the post-cold war era, at the beginning of the 1990s, did liberalization of the border regime take place, allowing inhabitants of border areas to visit their neighbours.

Since the opening of the border, there has been interest in creating transborder social and economic interaction in the area. The idea of cross-border 'integration' is currently a matter of public debate at the state level in Russia and at the supranational level of the EU. However, it is argued that residents on both sides of the Finnish–Russian border are involved in different processes of self-initiated and self-motivated integration and separation. Furthermore, macro-level factors, such as the history of separation and differences in identity and national institutions, as well as, more importantly, daily experiences on both sides of the border create the conditions under which citizens make their decisions about cross-border mobility. Academic opinions on the coherence of the region differ: Some Russian scholars have been rather optimistic in their assessments of the situation, proclaiming that, despite differences in living standards between the Finnish and Russian parts of Karelia, the region can be considered a transnational area characterized by common cultural and historical traditions (Romanova 2002; Roudometof 2005). Some Finnish scholars emphasize 'the otherness' between the two sides of the border. For example, according to Kiiskinen (2012), the border tends to retain its role as a cultural border at the state level and has not transformed into a platform for negotiating heritage and alternative belonging.

A growing number of visitors cross the border for work-related and business purposes, but the majority are tourists. Trends in border crossing show that the number of Russians in Finland has increased from a tiny percentage to about three quarters of the total number of visitors (OSF 2013). Mobility has increased most strikingly in Kymenlaakso and South Karelia, the southern parts of Finland where the impact of St. Petersburg as a large metropolis can be seen vividly (see Figure 7.2). At the crossing point in North Karelia (Niirala), the overall growth in mobility has been rather modest. It is also worth noting that mobility from Russia to Lapland mainly occurs via the southern crossing points.

Tourism from Russia in Southern Finland is often characterized as transit or gateway tourism, as people tend to travel onwards to Helsinki, Lapland or other European cities via the southern border-crossing points (Kosonen et al. 2005). Russian tourists form the largest group of visitors to Finland and also the fastest growing group in all border regions. Tax-free shopping is growing fast, as in the area of Joensuu in North Karelia. The year 2012 saw the highest growth rate since the border was opened, with an increase of 37 per cent compared to the year before. Altogether, tax-free shopping (of which 99 per cent is done by Russians) amounted to over 12 million euro. (Global Blue Records 2012). Tourism from Russia is seen as a major opportunity in Finland and the key to economic growth in the east of the country.

Economically, Finland offers many possibilities more than just tourism for the Russians. Expectations of a common labour market are particularly high, especially since in the near future, Finland is expected to face labour shortage, mainly in the public health sector. At the same time, Russia in the post-Soviet era has been experiencing a socio-economic crisis, which has led to dramatic increases in unemployment, poverty and social insecurity. These differences on both

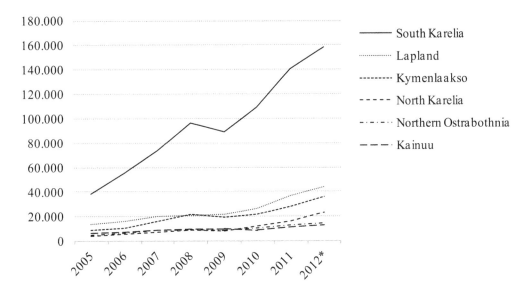

Figure 7.2   Number of incoming visitors from Russia to eastern Finland border regions, 1995–2012

Note: * Year 2012 numbers are a prognosis.

Source: OSF 2013.

sides of the border magnify the importance of balancing the development between the border regions of Finland and Russia, but at the same time this also raises fears of a 'brain drain' from Russia's border regions and of social dysfunctioning in the Finnish border regions, particularly in the form of an illegal labour force (Arosara 2004).

Case studies on the north-eastern communities of Joensuu (in North Karelia), Sortavala and Petrozavodsk (both in the Republic of Karelia), illustrate the complexities in cross-border (im) mobility. The shift from a closed border zone to a more open cross-border region has been one of the key factors in Sortavala's transformation as a Karelian town. While relatively distant from major European cities and transportation possibilities, the border town of Sortavala, located in the North Ladoga region (see Figure 7.1) at 64 km from the border, has nevertheless gained attention as a destination for Russian, Finnish and other visitors. The nearby border-crossing point of Niirala-Värtsilä has also become an important transit point between Russia and Finland. Since 1995, the volume of both goods and passengers passing through the crossing point has increased fivefold. In 2011, the number of border crossings rose to nearly 1.3 million (Rajavartiolaitos 2012). The town is very near to the border and therefore acts as a gateway to Finland. Shopping tourism, for instance is popular. Residents of the Finnish border area also enjoy visiting Sortavala. Meanwhile, as the transportation system between the two capital towns in the regions (Joensuu and Petrozavodsk) is poor, these towns have not benefited from the open border to any significant degree. The border continues to hinder economic and cultural contact (Laine 2013).

As in the past, the economic situation in both Karelias has suffered from the closed border, new features of common interest are gaining a foothold. For instance, projects led and funded by the EU operating on Karelian territory have resulted in some degree of EU employment among local residents. Also, since the late 1990s, a number of local private-sector enterprises in Sortavala have been involved in economic cross-border cooperation with Finnish partners. They produce low value-added goods for EU markets. Sometimes, these productions are subcontracted to nearby Finnish companies which have employees on both sides of the border (Eskelinen and Zimin 2004).

Karelia as a Space of Contention

The historical bordering and the image of the Finnish–Russian border are important in relation to cross-border mobility. In the case of the two Karelias, the historical image of the 'enemy' has made it difficult to transform relations into a 'neighbourly' one with common interests (see for instance Liikanen 2005; Hurd 2010; Raittila 2011). Furthermore, the first encounters of the Finnish people with their Russian neighbours, as well as their first experiences in trips to Russia, have created a picture of 'the other' that was economically dispossessed and poor. On the other hand, the Finnish still carry pre-WWII nostalgic images and childhood memories of the ceded territory. This drastic difference in images and the situation just after the opening of the border is firmly lodged in the minds of the Finns. Despite the subsequent economic improvement in this border region, these early images have persisted. As a result, the border still represents a gateway to keep the 'other' out. Regarding the public opinion, Lounasmeri (2011), referring to the Finnish media and public opinion, argues that the image of Russia in Finland continues to separate cultures on both sides of the border.

In addition, due to their traditional Western view of Russia, the Finns still consider the Russians as a threat, particularly in relation to immigration. The media tends to portray Russia as a super (imperial) power over and above its cultures. Thus, these media discourses, which tend to frame issues only in 'black-and-white' terms, not only reflect the Finnish public opinion but are also influenced by it. Contemporary civic and media opinion about Russia in Finland lacks placatory arguments present during the Soviet era (Kangaspuro 2011; Raittila 2011). On the other hand, the growing discursive tendency is to present Russia as an economic possibility (Ojajärvi and Valtonen 2011; Laine 2013). Thus, the historical development of bordering brings forth ideas about how localities are shaped and structured in time, and how historical transformations turn into bordering on an individual level (Paasi 2003, 2012).

Borders are now often understood as multi-faceted social institutions, rather than as markers of state sovereignty only (Scott 2006). State borders connect people rather than divide them. Cross-border cooperation often means new kinds of mobility, which does not presuppose moving from one country to another, changing citizenship and so on. This vision of borders is also based on the idea that 'the structure, functions and meanings of state borders seldom remain fixed or stable for long periods' (O'Dowd 2002).

The borders of everyday practices result in specific social outcomes for the people affected by them. The borders relate to people's border constructions, which create spaces of belonging, and these constructions may change. These shifting borders are grounded in ethnic and language-based discourses as well as in memories, which are in turn politically utilized in specific contexts (Hipfl et al. 2002).

In 2002, Van Houtum and Van Naerssen contributed to the geographical debate on bordering and 'the othering' processes by focusing on the issue of (im)mobility. The authors initiated debate on the issue of 'bordering' through immobilizing people. Although this chapter is not dealing with migration per se, it shows the importance of the idea of 'immobilizing' people in everyday cross-border mobility. Van Houtum and van Naerssen argued that territorial borders continuously fix and regulate the mobility or human flows and thereby construct or reproduce places in space. They pointed out that liberalization and cross-border integration, which characterize the process of globalization, coincide with the reproduction of mythologized or imagined borders of the past. Therefore, mobility and fix go hand in hand.

Further, 'bordering' can be conceptualized as space created not in some absolute, independent dimension but rather constructed out of social relations that incorporate time and change (Massey 1994, Urry 2002). Structural definitions, historically built power relations, social groups and attitudes of individuals are all interlinked in the structuring process of 'bordering'. It is not just how state power politics and governance determine the borders and how individuals would in turn adjust to these structures, it is also about how the individuals themselves create feelings of belonging, regardless of power politics. The historically built understanding of 'we and the others' allow contemporary acts of mirroring into the past and reflection upon the current needs in relation to the past. The past and the contemporary world with its new global sense of space and future are all present in these processes (Massey 1994, 2005; Moisio 2012).

The situation on the Finnish–Russian border has changed radically since the liberalization of the border regime in the late 1980s and early 1990s. During the Soviet era, the North Ladoga region was a closed border zone. After twenty years of cross-border interaction, there is now a new generation who can visit the neighbouring border regions regularly. For them, a border represents more of a contact zone than a dividing line.

The shared history of the two Karelias is another (perhaps debatable) argument in favour of the creation of transborder communities. Divided between two states, and yet sharing a common cultural background, Karelia today is looking for a new basis for mutual cross-border activities. On the other hand, the shifting border and shared feelings of belonging are also dependent on socially and historically built institutions and the peripheral nature of both the Finnish and Russian Karelian regions. The openness of the border is, in many ways, restricted by thresholds.

It is interesting to look at how the different perspectives on the images of a border, which are historically rooted, are mirrored in the actions and mobility patterns of the Finns into Russia, and into the Republic of Karelia in particular. In reality, the impact of 'bordering' on the development of the whole region might be transforming. There are limits on cross-border connections in everyday life and as fears about mobility grow, it may even become a barrier to balanced social development of the area. On the national level, the state is not really interested in creating longstanding possibilities of encountering the other. Based on these arguments, a theoretical approach can be drawn which conceptualizes the mobility patterns of individuals in relation to national and local public institutions, such as traffic on both sides of the border. The individual habits, wishes and actual actions are bound to what is possible or restrained by public measures.

Cross-Border Mobility from Finnish Karelia to Russian Karelia: A Context-Bound Phenomenon

The empirical data presented here was gathered in autumn 2012 in the Joensuu region. The local administration and VR Group, the Finnish national railway company, conducted a survey that targeted Finnish people living in the area. The purpose was to shed light on cross-border mobility patterns and interest among the population in travelling to Russia. Specific information was gathered about the level of interest in a regular passenger train connection between Joensuu and Petrozavodsk, the capital of the Republic of Karelia.

The survey was carried out through the website of the city of Joensuu. It was advertised in the local media and passed on in the email lists of the largest organizations in the Joensuu region. Altogether, there were 1,097 respondents, of whom 200 lived outside North Karelia. The respondents also included 200 representatives of the Russian-speaking minority. In general, the survey was skewed towards the immobile population (about 5 per cent of the total sample), and it cannot easily be extrapolated. Also, it is important to mention that there is no existing study on the general attitudes towards travelling to Russia among the North Karelian population with which these results can be compared to. However, the data are a first attempt to picture the region's mobility vis-à-vis Russia, and in this regard, the data are exceptional. This chapter uses the data for a descriptive analysis of the mobility patterns.

The survey consisted of five parts: (1) respondents' travelling behaviour to the Republic of Karelia in the last 12 months; (2) prospects of respondents travelling to the Republic of Karelia in the following year of 2013; (3) prospects of respondents travelling to the Republic of Karelia when a regular train connection would be established; (4) prospects of respondents travelling to the Republic of Karelia when visa-free travel would be launched immediately; and (5) general information about the respondents including car ownership, monthly household income, gender and so on.

Mobility Patterns

The data reveal three distinct groups of respondents: (1) those who had travelled to the Republic of Karelia and were keen to return the following year ($n = 547$); (2) those who have not travelled to the Republic of Karelia but were highly interested to travel there the following year ($n = 495$); and (3) those who had not travelled there and would not travel there in the following year ($n = 56$). These groups are labelled as 'mobile', 'potentially mobile', and 'immobile' respectively.

The socio-economic backgrounds of these groups differ slightly, which suggests that they are actually interdependent in many ways. In the next section, some important notions about the socio-economic groupings of Karelian cross-border mobility will be discussed since spatial division of social relations of mobility patterns brings forth notions of the spatiality of power structures (Massey 1994).

The gender divide between the groups as represented by the data is fairly consistent. It seems to indicate that, generally speaking, women may be slightly more mobile than men in cross-border mobility from Finland to Russia (Figure 7.3).

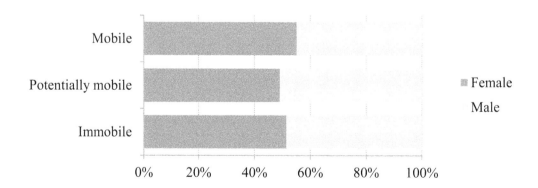

Figure 7.3   Gender division as a descriptor of cross-border mobility ($N = 1097$)

Source: calculation by the authors.

This is somewhat counter-intuitive and would suggest that other thresholds may be more important than gender. The travelling behaviour of men and women may be different due to the fact that cars are currently the only mode of transportation. This suggests that women are less mobile than men, since in general, men are likely to have a driving licence and a car for their personal use, and women are more likely to use public transport (Dargay 2002). In terms of issues of safety it can be expected that women are more cautious in travelling to Russia. This is partly because of the risk of robbery across the border or other forms of hazards that may happen in the Republic of Karelia, since shopping tourism is mostly motivated by the availability of cheap gasoline, alcohol and cigarettes, referred to in the Finnish media as 'vodka tourism' (Izotov and Laine 2013). This may, in turn, result in higher cross-border mobility among men (see Davydova and Pöllänen 2010). In this sense, gender differences and relations may result in greater geographical variations in Karelian's cross-border mobility context, where imaginative geographies of gender are vividly present and articulated.

As to the age groups of the respondents, the data show clear differences in mobility patterns, not only between age groups but also from previous research about mobility. In earlier literature on mobility, daily mobility is seen to vary according to the stage of a person's life: older people tend to be less mobile than the middle-aged ones, and young people are the most mobile (Yago 1983). In the context of cross-border mobility from Finland to Russian Karelia, however, the older age groups seem to be more active whereas the younger groups seem less willing to travel (Figure 7.4).

Mobility patterns based on age also seem to differ according to generational differences. Firstly, many of the older age group had to flee from the former Finnish territory annexed by the Soviet Union during the Second World War (more than 400,000 Karelians, according to Cronberg [2003]). Local travel agencies in Finland and the Joensuu region arrange many forms of 'nostalgia tourism' especially for the older people, so that they can 'take a trip down memory lane' and visit places of their childhood. Secondly, for those aged 30–50 who do not have such relationship across the border, the 'bordering' processes have been based on stories told by the older people about the 'lost Karelia', news reports about the Soviet Union, and random encounters with Soviet visitors. Thirdly, the youngest age groups have been doing their 'bordering' after the border was already open. However, this particular group has neither concrete relationships nor familiarity with the region of the Republic of Karelia. In this respect, age differences, where it concerns Finnish–Russian Karelian cross-border mobility, do create a specific kind of transformation of spatial construction in the border region, particularly as a result of an ageing population.

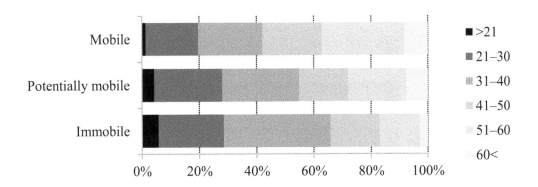

Figure 7.4   Age reflecting the 'bordering' process (*N* = 1097)

Source: calculation by the authors.

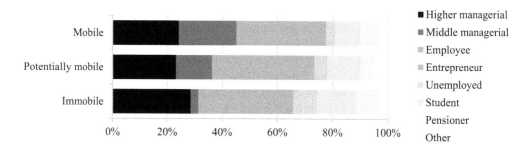

Figure 7.5   Occupational level as a descriptor of cross-border mobility (*N* = 1097)

Source: calculation by the authors.

Based on the data, differences in relation to the respondents' occupations are modest (Figure 7.5). In general, mobility is higher among social groups with higher-status occupations; managers, for instance, are clearly among those who are mobile. However, when the group of managers is divided into two levels, that is, middle and higher levels respectively, the situation is actually reversed as the middle-level managers are more mobile than the higher-level managers. Contrary to popular belief, students are among the most immobile, according to the survey. This finding also contradicts previous studies on general mobility, which show students and young people in general to be more mobile than older people (see, for example, Yago 1983). This may be due to the peculiarity of this particular border region.

As an interpretation of this the functioning of bordering in individual situations is reflected upon and thought over more carefully in the context of bounded cross-border mobility. Although the driving force behind contemporary policies for integration is economic, the interest is not particularly high among the economically powerful social groups. Laine (2013) defines the Finnish–Russian cross-border operation as mostly 'action from below'; motivated by the civil society's aspirations for social integration. Mobility patterns from the survey reveal the same.

With respect to monthly income, the findings of the survey do reflect a similar trend as other mobility research. That is, the people who are mobile tend to be those with high household income (Figure 7.6). Low-income households are more represented in the immobile group but interestingly, they are quite prominent in the potentially mobile groups.

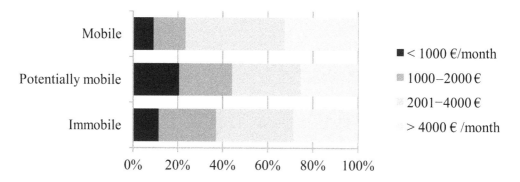

Figure 7.6   Monthly income as a threshold in cross-border mobility (*N* = 1097)

Source: calculation by the authors.

Inclusion of the low-income groups into the 'potentially mobile' group becomes understandable when we include the thresholds to cross-border mobility. Income is a resource of an individual which enables him/her to function as an agent in different settings. If travel costs are high, or travel requires car ownership, low-income groups may be hindered in cross-border mobility.

The Primary Driving Force in Cross-Border Mobility in Karelia

To understand the factors behind individual mobility and what might hinder mobility between the two Karelias, a regression analysis was employed, by including several independent factors. This analysis was to discover which factors are most important in determining travel behaviour Several regression analyses were performed.

Here the result of a stepwise binary logistic regression is presented, performed on the 'mobile'/'immobile' division. For the 'mobile' and 'immobile' groups, questions, such as 'Have you travelled to the Republic of Karelia in the last year?' were used.[2] Gender, age, occupational level, monthly household income, mother tongue, place of residence, preference of transportation mode (privately owned car, public transport, other), and reasons for travelling (work, visiting relatives and friends, holiday, transition, other) served as co-variates. These variables represent some of the features of the thresholds. Putting these features in relation to one another, allows us to obtain insight into some of the driving forces behind Karelian cross-border mobility.

The mother tongue served as a proxy for emigration from Russia to Finland and thus becomes part of the mental threshold that incorporates social networks. The mental threshold also manifests itself in terms of age as personal experiences related to the dramatic history of the ceded Karelia are embedded in age. Furthermore, reasons for travel are part of the mental threshold. The place of residence serves as a partial proxy for the locational threshold, representing the distance to the border. The third threshold, the trajectory threshold, is accounted for by the variable of travelling mode, as it indicates the passage between different places.

The stepwise binary logistic regression procedure created a model in which the reasons for travelling were the clearest indicators for being included in the 'mobile' group of people. Following that, reasons for travel (visiting relatives and friends and work-related) are included in the analysis. The role of the mother tongue in determining mobility is also high. As the regression model extends, it reveals that the individual motives for travelling are quite important, this is indicated in the fourth and fifth cycle of the model as shopping and 'other reasons' for travelling.

The sixth step includes the mode of travel, and although less significant than individual motivations, the significance is still high. As expected, possessing a car is the most important predictor for travel to the Republic of Karelia. The last steps in the model include gender and age divisions. The locational threshold, as far as it is expressed in the place of residence, had no significant importance in predicting the probability of being mobile, so it was not included in this stepwise model. The overall predictability of this model is 53 per cent (Nagelkerke R Square is 0.528). For reasons of clarity, only the results from the last step are presented (Table 7.1).

Upon examination of the final regression model, the two most important factors for cross-border travel were visiting friends/relatives and work. The latter is in line with the joint economic cooperation schemes in the region. The reason to include 'mother tongue' as one of the factors was to assess the extent of mobility among the Russian immigrant population in North Karelia who speak Russian as their mother tongue. The finding here illustrates the interrelation between

---

2   Logistic regression gives more reliable results when the categories to compare are equally sized.

Table 7.1   Predictors and their significance for belonging to the mobile group

		B Probability of 'belonging to mobile' versus 'belonging to immobile'	S.E.	Sig.
Step 8[t]	Reason for travel – visiting (a)	1.896***	0.221	0.000
	Reason for travel – work (b)	2.056***	0.193	0.000
	Mother tongue (c)			0.000
	Finnish	0.181	0.469	0.699
	Other	2.876***	0.556	0.000
	Reason for travel – shopping (d)	1.037***	0.177	0.000
	Reason for travel – other (e)	2.168***	0.496	0.000
	Mode of transportation (f)			0.005
	Car	0.683***	0.249	0.006
	Public transport (train/bus)	0.302	0.344	0.380
	Other mode of transportation	−0.167	0.189	0.377
	Gender (g)			0.023
	Female	0.778	1.292	0.547
	Male	−0.439***	0.167	0.009
	Age (h)			0.037
	<25	−0.461	0.689	0.503
	25–34	−1.138***	0.350	0.001
	35–44	−0.714	0.327	0.029
	45–54	−0.522	0.327	0.111
	>54	−0.514	0.312	0.099

a. Variable(s) entered in step 1: Reason for travel – visiting

b. Variable(s) entered in step 2: Reason for travel – work

c. Variable(s) entered in step 3: Mother tongue

d. Variable(s) entered in step 4: Reason for travel – shopping

e. Variable(s) entered in step 5: Reason for travel – other

f. Variable(s) entered in step 6: Mode of transportation

g. Variable(s) entered in step 7: Gender

h. Variable(s) entered in step 8: Age

Source: calculation by the authors.

ethnicity and mobility, and is in line with the commonly accepted fact that immigrants tend to travel back and forth between their hometown and their adopted place of residence (Pöllänen 2013). Immigrants tend to want to keep up their relations with relatives and friends they have left behind. Hence, their mobility is based on a commonness and familiarity with the host country. It is important to take into account their visits to friends and relatives. The motivation to visit friends and relatives include notions of social networks and the influence on everyday mobility.

The analysis also shows that gender can influence mobility: men seem to be less mobile than women, when other factors, such as age and individual motivations, are kept constant. Also, the age group of 25–34 years old tends to be significantly immobile in Finnish–Russian cross-border contexts. Travelling by car is the prevailing option to cross the border in Karelia. This shows that having a car is an important predictor of mobility.

Overall, based on this model of mobility and immobility in the cross-border context between Finland and Russia, the primary factors behind everyday cross-border mobility are situated in the mental thresholds. Individuals need to be motivated to travel. However, the mental threshold is a combination of personal attitudes and experiences, as well as notions of historically and institutionally built structures within. In other words, it is about the 'politics of memory' (Zhurzhenko 2011).

Conclusions and Discussion

The results of our study demonstrate that the threshold approach, developed by Van der Velde and Van Naerssen (2011), might benefit from a micro perspective. This study indicates that even though macroeconomic and institutional features are important as the context and reference to cross-border mobility, the micro level has its own manifestations in the analysis of this phenomenon. Individual motivations for travelling, as well as cultural and individual closeness or remoteness to the other side of the border, can act as thresholds, as well as resources in facilitating cross-border mobility. On the other hand, variables such as place of residence (whether near or far from the border) do not seem to matter in the light of these data. The actual physical possibilities of travelling and being mobile are more important factors.

It is the duty of public organizations to provide local actors with the possibilities to be mobile. However, in the context of North Karelia, the public interest has not been significant, since the actual levels of mobility have been low. There has been a vicious circle of low mobility, leading to low interest, which does not generate structures for mobility, thus resulting in low levels of mobility. When analysing thresholds in relation to mobility patterns, individual motivations are the driving factor behind cross-border mobility, but the absence of individual and public resources (economic, social, mental and so on) may create thresholds.

The current mobility rests on an ageing population's sense of belonging to the ceded Karelia. If structural limitations for mobility are a hindrance to the cross-border mobility of the younger generation, this space of belonging will change. Furthermore, when space and time are considered in our understanding of this 'bordering' then it is clear that the Karelian border area is created partly by the prevailing social relations. Karelia is a construction of historically built conceptions and contemporarily created social divisions. This study illustrates how the Finnish–Russian border and Karelia can be exploited to both mobilize and fix territory, identities, emotions and memories. In the context of Karelia, the clear differences in age, occupations and income, all affect mobility patterns, thus proving that thresholds and the act of 'bordering' differ between the different social groups.

In conclusion, the integration of the two Karelias should be understood not only in terms of economic endeavours, but also as a space of personal connections. However, while the border is now open, the possibilities to cross it have not really grown. In fact, the actual mobility is still based on images of a Karelian before the era of globalization. It seems that 'instead of bridges, there are doors that are opened selectively' (Paasi 2012, 2306). The micro perspective shows how borders and bordering practices come about, but it is equally vital to critically reflect upon these practices against existing political rationalities and state ideologies.

References

Arosara, T. 2004. "Työvoiman liikkuvuudesta Pohjois-Karjalan ja Karjalan tasavallan välillä" [On labour mobility between North Karelia and the Republic of Karelia]. SPATIA Reports 3/2004. Joensuu: University of Joensuu.

Bachman, K., and E. Stadtmuller, eds. 2012. "The EU's Shifting Borders: Theoretical Approaches and Policy Implications in the New Neighbourhood." London: Routledge.

Cronberg, T. 2003. "Euregio Karelia: In Search of a Relevant Space for Action." In *The NEBI Yearbook 2003*, edited by L. Hedegaard, B. Lindström, P. Joenniemi, H. Eskelinen, K. Peschel, and C.-E. Stalvant, 223–39. Berlin: Springer Verlag.

Dargay, J.M. 2002. "Determinants of Car Ownership in Rural and Urban Areas: A Pseudo-Panel Analysis." *Transportation Research E* 38 (5): 351–66.

Davydova, O., and P. Pöllänen. 2010. "Gender on the Finnish–Russian Border: National, Ethnosexual and Bodily Perspective." In *Ethnosexual Processes: Realities, Stereotypes and Narratives*, edited by J. Virkkunen, P. Uimonen, and O. Davydova, 18–35. Helsinki: Kikimora Publications.

Eskelinen, H. 2011. "Different Neighbours: Interaction and Cooperation at Finland's Western and Eastern Borders." In *The Ashgate Research Companion to Border Studies*, edited by D. Wastl-Walter, 569–83. Farnham: Ashgate.

Eskelinen, H., and D. Zimin. 2004. "Interaction across the EU-Russian Border: Driving Forces at a Local Level." In *Northwest Russia: Current Economic Trends and Future Prospects*, edited by D. Zimin, 70–89. Joensuu: University of Joensuu.

Global Blue Records. 2012. "Global Blue Intelligence." Accessed 18 December. http://localservices. globalblue.com/fi_su/local-news/ (site discontinued).

Hebinck, P., S. Slootweg, and L. Smith, eds. 2008. *Tales of Development: People, Power and Space*. Assen: Van Gorcum.

Hipfl, B., A. Bister, P. Strohmaier, and B. Busch. 2002. "Shifting Borders: Spatial Constructions of Identity in an Austrian/Slovenian Border Region." In *Living (with) Borders: Identity Discourses on East-West Borders in Europe*, edited by U.H. Meinhof, 53–74. Aldershot: Ashgate.

Hurd, M., ed. 2010. *Bordering the Baltic: Scandinavian Boundary-Drawing Processes, 1900–2000*. Berlin: LIT Verlag.

Izotov, A. 2012. "Mental Walls and the Border: Local Identity Construction in Sortavala." *Journal of Border Studies* 27 (2): 167–83.

Izotov, A., and J. Laine. 2013. "Constructing (Un)familiarity: Role of Tourism in Identity and Region Building at the Finnish–Russian Border." *European Planning Studies* 21 (1): 93–111.

Kangaspuro, M. 2011. "Suomi ja Venäjä jatkuvassa sodassa? Talvisodan Venäjä-kuvan muutokset ja muuttumattomuus" [Finland and Russia in constant war? Changes and stability of The Winter War time Russia image]. In *Näin naapurista: Median ja kansalaisten Venäjä-kuvat*

[Consequently about the neighbour: Media and civil image of Russia], edited by L. Lounasmeri, 231–52. Tampere: Vastapaino.

Kiiskinen, K. 2012. "Border/land Sustainability: Communities at the External Border of the European Union." *Anthropological Journal of European Cultures* 21 (1): 22–40.

Kokkonen, J. 2012. "Origins of the Modern Finnish–Russian Border: The Border and Those Crossing It from the Grand Duchy Period until the Winter War." In *Nation Split by the Border: Changes in the Ethnic Identity, Religion and Language of the Karelians from 1809 to 2009*, edited by T. Hämynen and A. Paskov, 24–45. Joensuu: University Press of Eastern Finland.

Korpela, J. 2004. *Viipurin Linnaläänin synty* [Birth of Vyborg Province]. Lappeenranta: Karjalan kirjapaino.

Kosonen, R., M. Paajanen, and N. Reittu. 2005. "Etelä-Suomi venäläisten turistien länsimatkailussa" [Southern Finland in Russian tourists' Western travels]. Helsinki School of Business Publications 59. Helsinki School of Economics.

Laine, J. 2013. "New Civic Neighborhood: Cross-Border Cooperation and Civil Society Engagement at the Finnish–Russian Border." PhD diss. Social Sciences and Business Studies. Joensuu: University of Eastern Finland.

Liikanen, I. 2005. "Bordering Finland, Bordering Europe: Exclusion and Inclusion in the Definitions of Finnish Eastern Border." Paper presented at the 7th World Congress of ICCEES, 25–30 July. Berlin: International Council for Central and East European Studies.

Liikanen, I., and P. Virtanen. 2006. "The New Neighbourhood: A 'Constitution' for Cross-Border Cooperation?" In *EU Enlargement, Region Building and Shifting Borders of Inclusion and Exclusion*, edited by J.W. Scott, 113–30. Aldershot: Ashgate.

Lounasmeri, L. 2011. "Lähellä, mutta niin kaukana? Suomalaisten Venäjä-kuvan äärellä" [So close but so far away? On the strand of Finnish image of Russia]. In *Näin naapurista: Median ja kansalaisten Venäjä-kuvat* [Consequently about the neighbour: Media and civil image of Russia], edited by L. Lounasmeri, 7–15. Tampere: Vastapaino.

Massey, D. 1994. *Space, Place and Gender*. Cambridge: Polity Press.

Massey, D. 2005. *For Space*. London: Sage.

Moisio, S. 2012. *Valtion, alue, politiikka. Suomen tilasuhteiden sääntely toisesta maailmasodasta nykypäivään* [State, region, politics: Finnish spatial relations governance from the Second World War to present]. Tampere: Vastapaino.

O'Dowd, L. 2002. "The Changing Significance of European Borders." *Regional and Federal Studies* 12 (4): 13–36.

Ojajärvi, S., and S. Valtonen. 2011. "Karhun ja kassakoneen naapurissa: Journalismin ja kansalaisten venäjät" [In between the bear and a cash register: 'Russias' journalism and the people]. In *Näin naapurista: Median ja kansalaisten Venäjä-kuvat* [Consequently about the neighbour: Media and civil image of Russia], edited by L. Lounasmeri, 19–67. Tampere: Vastapaino.

OSF (Official Statistics of Finland). 2013. "Foreign Demand for Accommodation Services Grew by One Per Cent in 2013." *Statistics Finland*, 30 May 2014. Accessed 2 October 2014. http://www.stat.fi/til/matk/2013/matk_2013_2014-05-30_tie_001_en.html.

Paasi, A. 1999. "Boundaries as Social Practice and Discourse: The Finnish–Russian Border." *Regional Studies* 33 (7): 669–80.

_____. 2003. "Region and Place: Regional Identity in Question." *Progress in Human Geography* 27 (4): 475–85.

_____. 2012. "Border Studies Reanimated: Going beyond the Territorial/Relational Divide." *Environment and Planning A* 44 (10): 2303–09.

Pöllänen, P. 2013. "Hoivan rajat: Venäläiset maahanmuuttajanaiset ja ylirajainen perhehoiva" [Borders of care: Transnational family care in the lives of Russian immigrant women in Finland]. PhD diss. Helsinki: Väestöliitto.

Raittila, P. 2011. "Venäjä kansalaismielipiteessä" [Russia in civil opinions]. In *Näin naapurista: Median ja kansalaisten Venäjä-kuvat* [Consequently about the neighbour: Media and civil image of Russia], edited by L. Lounasmeri, 125–68. Tampere: Vastapaino.

Rajavartiolaitos. 2012. "Henkilöiden rajatarkastukset ulkorajaliikenteessä rajanylityspaikoittain" [Border checks in Finnish outer border checking points]. Accessed 18 December. http://www. raja.fi/tietoa/rajavartiolaitos_lukuina.

Romanova, O. 2002. "Starye uchastki granits Rossijaskoi Federatsii" [The old parts of the border of the Russian Federation], edited by L. Vardomski and S. Golunov, 326–431. Moscow-Volgograd: NOFMO.

Roudometof, V. 2005. "Transnationalism, Cosmopolitanism and Glocalization." *Current Sociology* 53 (1): 113–35.

Scott, J.W., ed. 2006. *EU Enlargement, Region Building and Shifting Borders of Inclusion and Exclusion.* Aldershot: Ashgate.

Urry, J. 2002. "Mobility and Proximity." *Sociology* 36 (2): 255–74.

Yago, G. 1983. "The Sociology of Transportation." *Annual Review of Sociology* 9: 171–90.

Van Houtum, H., and T. van Naerssen. 2002. "Bordering, Ordering and Othering." *Tijdschrift voor economische en sociale geografie* 93 (2): 125–36.

Van der Velde, M., and T. van Naerssen. 2011. "People, Borders, Trajectories: An Approach to Cross-Border Mobility and Immobility in and to the European Union." *Area* 43 (2): 218–24.

Zhurzhenko, T. 2011. "Borders and Memory." In *The Ashgate Research Companion to Border Studies*, edited by D. Wastl-Walter, 63–84. Farnham: Ashgate.

Chapter 8

Navigating the Thai–Cambodian Border: From Battlefield to a Dynamic Border Space[1]

Pol Fàbrega and Helena Lim

The border between Thailand and Cambodia,[2] like many borders in Asia, is characterized by both conflict and cooperation, a duality that shapes the way governments seek to control and regulate the periphery and the way its inhabitants cope and determine their livelihood activities. The major crossing point between the towns of Poipet in Cambodia and Aranyaprathet in Thailand is no exception and continues to display considerable dynamism and growing mobility where binary tensions between trade and exploitation, opulence and poverty, lawlessness and regulation daily interact to create this intersection's own border rules and particular modus operandi.

This chapter investigates these processes and interactions at the Poipet–Aranyaprathet border area and provides an overview of the migration patterns that operate there. These will be analysed against Van der Velde and Van Naerssen's (2011) 'threshold approach' to migrant decision-making, which is used as a benchmark to gauge migrants' willingness and readiness to cross borders and migrate. This contribution seeks to shed some light on the importance and relevance of border dynamics in shaping migrants' decision-making process. The chapter starts with a background on migration in Cambodia and an overview of the country's historical make-up, before explaining the main migration management policies between Cambodia and Thailand. Then, the case study of the Poipet–Aranyaprathet border area will highlight the importance of cross-border economies and on the ground decision-making. Finally, mobility patterns at this border area will be explored using the threshold approach, revealing some interesting implications for its understanding.

Economic Development and Cambodian Migration

After decades of civil war and political instability, migration in Cambodia has shifted from politically induced forced migration in the 1970s and 1980s, with many refugees going to Europe and the US, to a mobility motivated by economic interests and driven by the voluntary movement of people in search of improved livelihoods. Movement in Cambodia is mostly internal and is predominantly rural-urban and rural-rural migration (IOM 2006, 9). According to the last population census (NIS 2009, 101), rural-rural migration accounts for half of the total internal migration, and rural-urban accounts for a little more than a quarter of internal flows. Although close to 80 per cent of the population works in the agricultural sector, Cambodia's economic growth is generated in urban areas where a booming tertiary sector includes garment factories, construction and services.

---

1 This contribution is derived from findings developed in a paper commissioned by the North East Research Group (NERG), a group of local NGOs and international organizations working in Cambodia on migration, exploitation and human trafficking (see Lim and Fàbrega 2011).

2 The terms Cambodian and Khmer are used interchangeably throughout this chapter.

Despite the internal mobility, the current supply of jobs does not meet the employment demands of a growing population, and the international movement of Cambodian people has intensified since the mid-1990s. A body of migration studies in the region reveals that cross-border migration is mostly conditioned by a sum of push factors, including chronic poverty, landlessness, land grabbing, lack of employment and access to markets, debt, natural disasters and health issues (IOM 2006, 9; Murshid 2007, 197). At the same time, improved communication and infrastructures and an increasing demand for low-skilled workers in more developed Southeast Asian countries have encouraged and facilitated a cross-border labour movement (Tuot 2007, 22). However, a majority of cross-border migrants are irregular, undocumented and employed in non-skilled or low-skilled jobs in construction, agriculture, domestic work, fishing and factory work. Once abroad, migrants find themselves working in so-called 3D jobs (Dirty, Dangerous and Difficult) that rarely improve their economic situation but instead increase migrants' exposure to risk and abuse (IOM 2006, 9).

Except in agriculture where both men and women are employed, a gender-based division of labour strongly marks most sectors and there is an increasing demand for labour in highly feminized sectors, like health care, domestic work, entertainment and textile manufacturing (Huguet and Chamratrithirong 2011). The garment industry attracts a majority of young, single Khmer women to work in urban areas (Chen Chen 2006, 9). In the hospitality industry, Chen Chen estimated that about 80 per cent of workers are women in low-paid positions working as clerks, waitresses and beer promoters. In some instances, these jobs double up as sex or entertainment work, although no reliable figures are available. Domestic work abroad is also greatly gender-driven and female migrants have returned with mixed reviews, some successful, others not. The uncovering of networks of exploitation and abuse has prompted the formation of numerous organizations dealing with counter-trafficking and victim's assistance. Men usually work in agriculture, construction and fishing, which is not exempt from cases of trafficking for commercial exploitation with men forced to stay on fishing boats, sometimes for years at a time (LSCW 2005[3]).

Key destinations for international migration include Thailand, Malaysia, South Korea, Saudi Arabia, Singapore, Japan and Taiwan. Thailand is the largest recipient of Cambodian workers, but no official number of migrants in Thailand has been registered, as a vast majority are irregular. According to the International Organization for Migration (IOM, quoted in Vutha et al. 2011, 10), an estimated 180,000 Cambodians unofficially work in Thailand. Beyond the push factors in Cambodia, the increasing demand for labour as a response to Thailand's growing economy is a key factor for migration. Given Thailand's economic development, the local workforce is reticent to take on unskilled employment that is becoming identified with migrant groups. Thailand continues to act as a magnet for thousands of Cambodian workers who, hard-pressed by economic difficulties at home, decide to take their chances and look for economic opportunities on the other side of the border.

Generally speaking, Cambodian migrants to Thailand fall into two broad categories: (1) long-term and long-range migrants are frequently employed in 3D jobs in fishing, agriculture, factory, and domestic work; and (2) short-term and short-range migrants who live and work in and along the border area. This contribution will primarily focus on the mobility patterns and motivations of the second group of migrants. In this chapter, individuals living and working at the border are referred to as borderlanders.

---

3   For a review of the existing research on human trafficking in Cambodia, see Derks et al. (2006).

Dynamics at the Thai–Cambodian Border

In the last four decades, the border situation between Cambodia and Thailand has undergone tremendous change both politically and economically. During the 1970s Thailand closed and militarized its borders with Cambodia to prevent communist insurgencies spilling over into its territory. When Vietnamese troops overthrew the Khmer Rouge regime in 1979, hundreds of thousands of Cambodian refugees fled to the Thai border to seek refuge. Three Cambodian 'resistance' armies (including the Khmer Rouge) remained at the border with the support of the Thai army in an attempt to overthrow the Vietnamese-led regime (French 2002, 430). Despite the martial law imposed along the entire border area during the civil war, the Thai army posted at the border benefited from the informal border trade by supplying refugee camps with commodities (447).

In the late 1980s, the nature of the border relations changed as Thai Prime Minister Chatichai Choonhaven led a major shift in Thailand's approach to Cambodia and Laos, as branded in his policy slogan of 'turning Indochina's battlefields into market places'. In 1992, this vision for the region was granted an additional boost with the establishment of the Greater Mekong Subregion (GMS) Cooperation Programme, supported by the Asian Development Bank (ADB) and aimed to promote a development model based on market liberalization and on the regional and global integration of previously isolated economies. The GMS countries opened their economies, establishing important cross-border interactions, investing into large-scale infrastructure projects, and leveraging intratrade opportunities and economies of scale.

The creation of regional 'economic corridors' is expected to foster economic growth and social development by facilitating cross-border trade among GMS countries, but also to expand the trade potential by reaching influential neighbours such as China, India and other ASEAN countries (Caouette et al. 2006, 11). The Southern Economic Corridor aims to bolster economic cooperation, trade and investment along the east-west axis between Thailand, Cambodia and Vietnam.

In the midst of these regional developments, the growth of border industries is attractive given the geographical advantages and easy access to abundant and cheap labour force, lower service link costs (lower transport and communication costs), and more reliable and cheaper utility services (Kudo 2009, 20). Border industries in Special Economic Zones (SEZ) have proliferated along the borders of Cambodia, benefiting from government preferential support and relaxed regulations on business activities to attract international investment, foreign technology and expertise. Out of the 21 SEZs approved by the government since 2005, 13 are located along the borders with Thailand and Vietnam (MMN and AMC 2013, 186). Despite being hailed as bringing infrastructures, jobs, skills and enhanced productivity, it is uncertain how labour rights are being monitored there.

Regional infrastructure development and enhanced transportation have facilitated the movement of people and increased labour opportunities in areas previously considered too distant, therefore making cross-border migration a more feasible option. These changing environments and uneven development between countries also exacerbate demographic and socio-economic divides in the region that fuel the expansion of labour migration dynamics across the borders (Cauoette et al. 2006, 12).

Labour Migration Regulatory Framework

Governments in the region have stepped up their efforts to develop policies and international guidelines to better manage and regulate migration-related issues, from labour supply and demand to migrant protection and prevention of human trafficking. Bilateral agreements in the form of Memorandums of Understanding (MoUs) between Thailand and Cambodia (2002), Laos (2002) and Myanmar (2003) were signed 'to establish close collaboration between the contracting states with regard to the sending and receiving of migrant workers, including better matching of supply and demand through quotas for recruitment, enhanced protection of migrant rights, and measures to curb irregular migration, including long-term permanent settlement in Thailand' (ADB 2009, 19). On paper, the MoUs constitute an important milestone in the GMS's efforts to reach a consensus on migration management and curb irregular migration.

In Cambodia, implementation of the MoU began in 2006 but has been slow and costly. High recruitment fees (some up to US $700), long waiting times, complicated procedures, a mismatch between available jobs, skills and salary expectation, and the lack of clear benefits associated with regular migration continue to fuel irregular migration channels that remain cheaper and faster (PATH 2010). According to a study from the Cambodian Development Research Institute (CDRI 2009, 11), it would cost a Cambodian migrant on average US $747 to legally obtain work in Thailand, four times more than opting for irregular migration. For example, up until 2009[4] the issuance of a Cambodian passport (cost: US $150) could only be done at a singular small office in Phnom Penh, making it costly (administration fees and transportation) for prospective migrants living outside the capital. By contrast, smugglers charge an average fee of US $100 to cross the border into Thailand and the process is swift. After a while, many experienced migrants can re-enter Thailand without the services of a smuggler and pay only US $50 in transportation costs (Sophal 2009, 56). The risks of irregular migration, such as arrest, deportation, exploitation and trafficking, are real but most migrants still opt to migrate irregularly as it is more affordable and time-efficient. Furthermore, the monitoring of formal recruitment agencies that process migrants' applications to work abroad is opaque and the benefits of government policies to regular migration are not straightforward. The reputation of training and recruitment agencies has been marred, from basic claims of inefficiency and costly trainings to full-fledged accusations of exploitation leading to being blacklisted.[5]

Regular migration policies are not only costly but are also disconnected from realities on the ground. By seeking to respond to Thailand's private sector demand for cheap labour, such policies fail to address other socio-economic situations at the border (ADB 2009, 16; Chen Chen 2006, 14). In theory, international agreements on migration shape the way borders between Cambodia and Thailand are to be managed and focus on 'immigration control; prevention of irregular migration flows; counter-trafficking, and the sending of workers to countries outside the GMS' (IOM 2006, 17). In practice, these agreements are loosely implemented and do not acknowledge the spontaneous mobility that occurs at the border. Hence borderlanders, who rely on daily border crossings and interaction, are not protected by any legal framework, making them particularly vulnerable to extortion or abuse. All in all, bilateral agreements and MoUs are a step in the right direction, but they do not provide a complete picture of the migration landscape at the border and fail to address the root causes that make irregular migration possible.

---

4   Following the pressure from many migration advocates, the government decided to lower the cost to US $24 for approved migrant workers (Sophal 2009, 56).

5   A ban to send domestic workers to Malaysia was passed in October 2011 as the government of Cambodia no longer deemed Malaysia to be a safe destination for Cambodian citizens. Nonetheless, some Cambodian agencies still work their way around the ban and continue to send domestic workers to Malaysia (LICADHO 2011).

The border area between Poipet in Cambodia and Aranyaprathet in Thailand provides a vivid example of complex mobility patterns at the border and how a border's dynamic characteristics highlight local border management practices.

Poipet–Aranyaprathet: A Dynamic Border Area

Located in Banteay Manchey Province on the main axis between Phnom Penh and Bangkok, the Poipet–Aranyaprathet border checkpoint officially opened as a crossing point in 1993. It now serves as the most important land crossing and transit point between Cambodia and Thailand. Mass transit has promoted the development of the border area and the population of Poipet has grown from 43,300 in 1998 to over 110,000 in 2008 (NCDD 2009), making it more populous than the provincial capital Sisophon. National Road 5 (NR5) that stretches from Sisophon to the border has facilitated passage as well as trade, especially with the neighbouring Rong Kluea market in Aranyaprathet.

Poipet still lacks many basic infrastructures, but is paradoxically driven by the flurry of luxurious casinos that populate the no man's land between Thai and Khmer checkpoints. These casinos generate much of the local employment and border economy in the area. The casinos contrast sharply with the adjacent slum and underserved areas, karaoke bars and massage parlours that double up as brothels (despite government crackdowns) and line NR5. Trade and tourism from Thailand is so important that local businesses in Poipet usually operate in Thai Baht, working as an extension of Thailand into Cambodia.

Poipet's Cross-Border Economies

Formal and informal trade between Cambodia and Thailand has increased in recent years. Official total trade from Thailand to Cambodia increased from US $527 million in 2002 to over US $4 billion in 2012.[6] The Poipet–Aranyaprathet border checkpoint accounts for half of the total cross-border exports from Thailand to Cambodia and two-thirds of total imports (ADB 2010, 18). The major exports from Thailand to Cambodia include cement and other construction materials, fresh and processed foods, cosmetics, consumer goods, car parts and fertilizers. By contrast, Cambodia exports scrap metal, second-hand clothes, handicrafts, soybean grain, tapioca, corn feeds for livestock and fresh and processed fish (ADB 2010, 18; CDRI 2005, 34). In line with Cambodia's policy to promote special economic zones along the Thai border, Poipet was one of the first areas to be developed into a SEZ and to benefit from government subsidies to set up shops.

The larger benefits of Thai-Khmer cross-border economy are concentrated in the hands of a few, with strong ties to high-ranking government officials. These networks influence and manipulate border management to their advantage, circumventing formalities such as licences and permits so that 'even these formal flows (of goods) include an informal component, in terms of under-invoicing, tax evasion and partial payments and partial recording' (CDRI 2005, 12). To a lesser degree, similar rapports are found in the informal trading sector at the international crossing point where petty traders, cart pullers, labourers and artisans interact

---

6   Derived from the ADB Integration Indicators Database (http://aric.adb.org/, accessed 24 September 2014).

with local officials to facilitate their crossing and the flow of goods and commodities and to avoid the formalities of border crossing (CDRI 2005, 11).

Despite being much smaller in terms of volume and value, petty trade between Poipet and the Rong Kluea market in Aranyaprathet is a profitable and stable economic activity attracting many internal migrants throughout Cambodian and fuelling the livelihoods of a vast majority of borderlanders. Hundreds of cart pullers, many of them young migrants with their children, commute along the crossing point, carrying fruit, garbage, clothes, fish and so on into Rong Kluea market and back, often accompanied by their young children. Thousands of Cambodian traders hold stalls in this market, and many more Cambodian petty traders/workers daily cross the border to work in the market. Given the great income disparity at the border, engagement in border trade is 'one of the best possible livelihoods that afford respectability for many people in the commune' (Kusakabe 2004, 56).

Kusakabe (2004, 61) argues that cross-border fish trade in Poipet not only emphasizes different gender roles, attitudes and perceptions of the border but also the importance of networks to maintain trading capacity in times of political tensions in the province: 'When the border was closed because of political conflict between Thailand and Cambodia, Cambodian women traders telephoned their regular customers in Thailand and met at other smaller border checkpoints that were not closed.' Continuous economic opportunities and a general feeling that Thai and Cambodians are similar in many ways constitute a premise for cross-border activities to thrive in Poipet. Rong Kluea market, where Thai and Cambodian traders meet, represents more of an actual border than the physical border established by the state – 'the perceptual border would be extended to the limit of that market' (61). When violent political tensions broke out in February 2011 at the border over Preah Vihear Temple[7] and the security of Koh Kong further south was reinforced, the activity at the Poipet–Aranyaprathet border point continued to operate as usual.

Another remarkable feature of the local economy in Poipet is constituted by the gambling industry. National laws and regulations dictated in Phnom Penh become blurry and flexible at the border to serve the interests of a few joint ventures between Cambodian and Thai businesses, and politicians (Weggel 2007, 143). Close to a dozen luxury casino-hotels have been established in the no man's land in Poipet between the Thai and Cambodian checkpoints, shaping the town's landscape and providing employment to thousands of Cambodian people earning around 250 baht per day (around US $8) (CDRI 2005, 165). Since gambling in Thailand is illegal, Cambodia has developed a niche in this sector, attracting Thai tourists and their money whilst prohibiting their access to Cambodian people (except for workers). This no man's land of a few hundred metres wide offers the possibility for Thai people to gamble outside Thailand without officially crossing into Cambodia, reducing administrative hassles associated with visas, working permits and operating licences for visitors and casino workers (Lintner 2012). Unfortunately, the gambling sector usually creates a deregulated side-industry to meet a demand for sex and entertainment workers. Many non-governmental organizations in Poipet address cases of trafficking and exploitation and work with young women and girls who are exposed to risky working environments.

---

7   Even though the International Court of Justice established in 1962 that the Preah Vihear Temple lies within Cambodian territory, its control remains disputed. In 2008 when UNESCO declared the Preah Vihear temple a Cambodian World Heritage Site, the tension over the border quickly escalated and border clashes have continued ever since.

The strategic location and flexible regulations at the Poipet–Aranyaprathet border foment economic opportunities for large-scale casinos and import/export companies, but also create a risky and uncertain environment for Rong Kluea market's petty traders. These border characteristics determine migrants' decision-making process to travel across the border. The next section will explore the mobility patterns at the Poipet–Aranyaprathet border area through the threshold approach and will test its relevance and application in general to dynamic border areas.

Decision-Making: Migrating to the Poipet–Aranyaprathet Border Area

In Van der Velde and Van Naerssen's (2011) study, migrant decision-making is analysed through a series of thresholds – mental, location and trajectory – all of which influence and determine an individual's choice to migrate. This approach explores an individual's sense of belonging to a place and community, the reasons 'for' and 'against' a move and location, and, finally, the means and itinerary to reach this destination and the various permutations of this journey. In the case of borders, a threshold analysis can be ambiguous and distinctions between 'here' and 'there' are blurred, as the border becomes the destination, and not just a transit on a longer journey. Whilst the border area is home to both Cambodian and Thai people, this contribution will focus primarily on Cambodian migrants travelling from inside Cambodia to the Poipet–Aranyaprathet border area.

Mental Threshold

According to Van der Velde and Van Naerssen (2011), the first step in an individual's decision to migrate requires overcoming the mental distance that is created between the place of origin and the destination on the other side of the border. The economic, social and cultural similarities and differences specific to either side of the border determine prospective migrants' sense of *space of belonging*, of feeling at home. Likewise, the presence of strong networks, a diaspora or transnational communities on the 'other' side may reduce this mental distance and contribute to feeling in a 'home away from home' (221–2).

The dynamic nature of the Poipet–Aranyaprathet border area and its ongoing history of cross-border mobility blurs any distinction between spaces of belonging and spaces of difference. This border area is more an anteroom where strong factors of difference (or otherness) and indifference (or selflessness) are present all at once and are constantly interacting and being (re)created.

On one hand, living and moving across the Cambodia-Thailand border means that the home country – and the space of belonging – is never very far. What is more, in the north-western provinces of Cambodia and Thailand, interaction between Thai and Khmers dates as far back as the Angkorian era. The region was once all part of the Angkor Empire, and to date, Thai and Khmers are known for both their strong linkages and their aversion in times of conflict. Nonetheless, strong networks and long-lasting relationships across the borders now form the basis of solid and enduring cross-border economic interactions. Other commonalities, from religion (Theravada Buddhism) to traditional types of economies (a wet-rice-based economy for example – only recently superseded by manufacturing in Thailand), bring together more than they divide the Thai and Cambodian people (French 2002, 437). For borderlanders, the immediate sides of the border area become familiar. Many people live close to a crossing point as their livelihoods rely on strong and regular interactions with the other side.

On the other hand, the ambivalence of the border area also inherently holds within it the sense of otherness, and with it a sense of difference. Unskilled and irregular Khmer migrants in Thailand are generally perceived negatively and despite many cultural similarities, the differences are regularly highlighted (Cauoette et al. 2006, 56). Prejudice, discrimination and stigmatization remain deeply rooted in dominant public discourses about migration and Cambodian migrants are frequently accused of generating all kinds of social ails, despite their important but unappreciated contribution to the Thai economy by filling a high demand for unskilled jobs. Patterns are perpetuated when Thai employers recruit Cambodian migrants for their disposition to work long hours and undertake hard work at lower pay 'without complaining' (56). In the case of the Poipet–Aranyaprathet border area, cases of exploitation of Cambodian workers by Thai employers are not uncommon. Cambodians often confirm this understanding 'by behaving in self-protective and self-serving ways that Thai read as untrustworthy' (French 2002, 462). This only serves to exacerbate a greater sense of otherness and alienation.

A key feature of the border area is its porosity and accessibility. The various formal and informal channels available to migrants increase their choices and facilitate their border crossing, contributing to lowering the mental barrier of having to overcome a borderline. The international crossing point is the main transit point; hundreds of passers-by cross in and out of Cambodia here every day. Many irregular migrants slip through the checkpoint, either unnoticed or by purchasing a daily border pass and overstaying in Thailand. It is also the main passageway for daily traders commuting to the Rong Kluea market for business or to Thailand

Figure 8.1   Migrants at small irregular crossings around Poipet
Source: Photo by authors.

for short-term small-scale activities such as construction. Hundreds of cart pullers go back and forth along the crossing point, carrying fruit, garbage, clothes or fish into Rong Kluea market. However, the international crossing point is only one gateway amongst many around Poipet town. Many migrants cross into Thailand undocumented through smaller regular checkpoints, village paths or unofficial routes, all along the border. The presence of networks on both sides, including border and local authorities, smugglers, brokers, Thai employers are all part of a system that facilitates the irregular passage into Thailand and back.

Despite the ease of flows at the border, the presence of institutional surveillance mechanisms – in the form of border police, immigration control and so on – is a reminder that the border serves an enforcement or filtering role, creating imbalanced binary distinctions between 'us' and 'them', 'included' and 'excluded', and 'regular' and 'irregular'. Thai people can move freely through the international crossing point, but this mobility is restrained and regulated for Cambodian people. Flows of trucks deport and repatriate migrants back from Thailand (and Malaysia) through the Poipet–Aranyaprathet border checkpoint and are a reminder for would-be migrants of the possible consequences of crossing irregularly and that they do not really have a place in Thailand.

Migrants and borderlanders who daily cross and work in and around the border face a duality of factors that create a sense of belonging and difference. The 'otherness' and 'selflessness' generated from border encounters are mediated by the daily practices of border players that normalize their mobility.

Locational Threshold

Once the mental threshold is overcome, migrants go through a rational process to decide on the location. Known factors for and against the place of origin and place of destination are considered and assessed and would-be migrants have to consider a number of push/pull and repel/keep factors specific to the border area. Cross-border migration is mainly driven by a combination of push factors from the place of origin, including chronic poverty, landlessness, land grabbing, lack of employment and access to markets, debt, natural disasters and health issues (IOM 2006, 9; Murshid 2007, 197). In turn, rapid development of infrastructure and transportation result in greater access to places that were previously seen as remote or distant.

The Poipet–Aranyaprathet border area is located on a direct route between Siem Reap and Bangkok and is a significant springboard for migrants wishing to travel deeper into Thailand. A majority of pull factors are economically driven. Cambodian households are heavily reliant on subsistence agriculture, a sector impeded by seasonal rains and outdated facilities. Bustling Poipet offers year-round employment with a surge in the service sector. Casinos create employment but also generate side businesses related to tourism, entertainment and hospitality. The cross-border economy of Poipet is a continuous hustle and bustle of market traders, cart pullers, taxis and motodops (moto-taxis) and tourists, and continues to attract people from rural Cambodia. Unlike Phnom Penh, migrants arriving in Poipet have a choice either to stay on the Cambodian side or to cross into Thailand for work. Unfortunately, the majority of the available jobs are dangerous and risky, ranging from low-paid jobs to being conducive to more extreme threats such as human trafficking and commercial or sexual exploitation. Despite the presence of luxurious casinos (or perhaps because of their presence) in the no man's land, poverty prevails, local communities continue to have limited access to water and sanitation, and slum dwellings line the main road.

In Thailand irregular migrants are not protected by the law and have very few channels to seek shelter or assistance when faced with cases of abuse or exploitation. The fear of deportation, arrest or retaliation from employers force migrants into a situation where cases of abuse or ill treatment go un- or under-reported. Migrants are also vulnerable to corrupt practices by immigration officials who not only extort migrants upon their irregular entry into Thailand but also take bribes from Thai employers to turn a blind eye to illegal labour practices. Migrants have limited freedom to move or travel outside their workplace. Working and living conditions rarely meet minimum standards and migrants have no right to organize themselves, or to take paid or sick leave. Given their irregular status and low socio-economic condition, migrant women and children are particularly vulnerable to gender-based violence, abuse and trafficking. Men are not exempt from modern-day slavery or abuse, and male trafficking victims are on the rise in Cambodia, many tricked into working aboard on Thai fishing boats and 'forced to work in slave-like conditions without pay' (IOM 2013).

To address these issues, a number of NGOs and advocacy groups are campaigning to raise awareness of the risks associated with migration and to promote safe migration practices through the media, role plays and educational workshops. Despite these preventative measures and migrants knowing the potential risks to health, livelihoods and life, the attraction still outweighs the risks and many are compelled to migrate to Poipet to cross into Thailand irregularly.

Trajectory Threshold

The third and final threshold is linked to the trajectory, that is, which route will be undertaken. The Poipet–Aranyaprathet borderland is a destination, a point of origin and a transit for a wide variety of people on the move. This space reveals how mobility patterns at border areas defy fixed and directional understandings of migration. Internal and international migrations are usually identified and defined as two types of mobility and studied as different and dissociated phenomena. However, King et al. (2008, 3) emphasizes that this analytical distinction is unclear on the ground, where different types of mobility patterns occur at the same time. Internal and international migrations are not mutually exclusive. The actors and processes of one are interlinked with those of the other and one system can lead to the other (and vice versa).

As the Poipet–Aranyaprathet border area is a space of varied trajectories and economic prospects, it is not uncommon to find migrants from different types of mobility such as internal migrants to the border, daily cross-border commuters, short-term migrants in the areas nearby the border, and long-term migrants to Thailand. Broadly speaking, each type of mobility has its own features, trajectories and processes at the border.

For long-term and long-range migrants crossing irregularly into Thailand, the border is a necessary transit to be overcome, marking a clear distinction when crossing from 'here' to 'there'. Long-term migrants look to pass through Poipet quickly, usually through one of the many smaller regular checkpoints, village paths or unofficial routes that dot the borderline. Migrants turn to brokers (or *mekhal*) for a substantial section of their trajectory, including the border crossing. These passers, sometimes former migrants themselves, have strong connections in Thailand and can organize part or the entire end-to-end journey for migrants, from their household in Cambodia to their final destination in Thailand (IOM 2006, 28). 'These individuals know the work situation in Thailand and have good relationships with Thai employers, Thai brokers, and Thai police' (ILO 2008, 26). Broker fees and commissions vary according to the services provided, and include part or all of the following: transportation to the border, smuggling across the border, arrangement of working permits or job placements. This process is costly and risky, and does not usually apply to the most disenfranchised, as initial capital, sometimes in the thousands of dollars, is required to prepare for the journey ahead. Brokers play a decisive role in determining the final outcome of a migration process. Resorting to a broker is a gamble: there is much to gain and even more to lose. In cases where migrants are sold to a Thai employer by brokers, migrants end up on 'on credit' and brokerage fees are deducted from their salaries, creating a debt bond difficult to break.

For borderlanders whose livelihoods depend on the Rong Kluea market or on short-term work in areas nearby the border, the international crossing point is safe and convenient and no broker is needed. Immigration formalities have been adapted to facilitate the mobility of border residents and unskilled workers. Daily border passes can be purchased for 10 baht (US $0.30) and allow commuters to travel no further than the market (CDRI 2005, 34). Weekly passes are dispended to migrants contracted for construction work on longer projects, allowing entry into Thailand but not doubling up as a working permit. Pangsapa and Smith (2008, 496–7) describe the case of Burmese migrants in Thailand and argue that 'the fact that one-day passes can be obtained at official checkpoints emphasizes how the exploitation of workers is facilitated by the state' as 'many migrants use these "passes" as informal work permits into the country.' For Cambodian migrants working irregularly in Thailand, a legal border pass does not protect them from extortion or abuse. The validity of border passes is short-lived (one or seven days), so a migrant's regular status can very quickly shift to an irregular one, thereby increasing the risks of police arrest and detention. In this instance, the border pass extends the physical limits of the official borderline to include the market and neighbouring areas to facilitate the mobility of cheap labour. The border hosts an administrative limit that is different from its economic and trading limit.

Another group of daily commuters and short-term migrants cross the border through smaller regular crossing points to work in plantations (cassava, rice, yam, sugar cane or for weeding) (Lim and Fàbrega 2011, 14). Crossing is relatively safe and supervised by a network of players, including police officials (Thai and Cambodian) and Thai landowners. At one such small checkpoint a few miles from Poipet International crossing point, large groups of people cross en masse in the early hours, paying a symbolic fee to border officials (100 riels, roughly equal to US $0.02). The Thai landowner pays the Thai border police to allow entry and can then employ cheap agricultural labour. The daily commuters, undocumented and otherwise unemployed, sell their labour to the Thai landowner for a higher wage than in Cambodia. One group member, responsible for the group, ensures that everyone returns home every evening, crossing back where they came from to avoid being caught by the authorities. This daily practice is a win-win situation that benefits all parties. At this particular crossing point, a local NGO is present to oversee the process and inform migrants on safe migration practices. Irregular migrants are generally perceived as particularly vulnerable due to high poverty levels, unstable livelihoods and low educational attainment, which also exacerbate their exposure to unstable jobs, police arrest and detention. Despite their unfavourable socio-economic situation, migrants demonstrate a certain resolve and agency in working around the system, rather than being passive bystanders subject to structural forces beyond their control. As active border agents, these daily commuters operate and shape a system in which they benefit from strong local networks of routes and employers. Here, brokers are not needed and borders are perceived not as an obstacle but as a hub for social interaction and opportunity. Table 8.1 summarizes the empirical findings on the several thresholds.

Table 8.1   Thresholds in the Thai–Cambodian border region

MENTAL THRESHOLD	
Indifference factors	**Difference factors**
Historical cross-border interaction	Negative perception/treatment of migrants in Thailand
Strong networks and personal relationships	Exploitation and extortion of migrants
Cultural/religious affinity	Institutional presence and surveillance at the border
'Home' (Cambodia) is never too far for those living/ operating at the border area	
Porosity of border lowers the mental barrier of overcoming a border.	

LOCATIONAL THRESHOLD	
Push/pull factors	**Repel/keep factors**
Push factors at home (poverty, landless and so on)	Migration associated risks (dangerous and low-paid jobs, human trafficking and exploitation)
Economic opportunities at the border area	Awareness raising campaigns on safe migration and human trafficking
Enhanced transportation and infrastructure development	
Strong social networks	
Poipet serves as a springboard into Thailand	

TRAJECTORY THRESHOLD	
Short-term/short-range migrants	**Long-term/long-range migrants**
International crossing point is safe and convenient for daily commuters and short-term migrants	Smaller crossing points, usually with the use of brokers
Smaller crossing points for daily commuters working in agriculture in nearby border areas	

Source: the authors.

Conclusions

The key objectives of this chapter were to explore how Cambodian migrants move across the Thai–Cambodian border and to match their decision-making patterns for mobility with the threshold approach developed by Van der Velde and Van Naerssen (2011). This approach places (im)mobility into the analysis of migration decision-making and provides a useful tool to examine the thought process and motivations of migrants. The investigation of the Poipet–Aranyaprathet border area contributes to this analysis by highlighting the importance of fluctuating border dynamics in shaping a migrant's decisions.

Firstly, at the border between Thailand and Cambodia, distinctions between 'here' and 'there' are unclear. Factors of difference and indifference are present simultaneously and are constantly being renegotiated on the ground. These factors operate in both directions, constantly lowering and increasing the mental threshold to migration. Ongoing social, cultural, historical and economic affinities between Thai and Khmer communities continuously emphasize the 'other'. Greater consideration of the changing configuration of border areas is needed to better understand the daily interactions of borderlanders and the ambiguous sense of belonging that results from these negotiations.

Secondly, the bounded rational process exerted by potential migrants includes conflicting factors that encourage and prevent migration at the same time. The porosity of the border offers an opportunity for irregular migrants to cross easily and consequently can make the prospects of crossing more alluring, but it also inherently fails to provide a protective regulatory framework for undocumented and irregular migrants, increasing their vulnerability to extortion, arrest and abuse. Similarly, the Poipet–Aranyaprathet border area attracts for its year-round range of economic opportunities, but the migration process is not risk-free. Jobs can be dangerous and risky and migrants are subjected to hazardous working conditions, which in some cases can also lead to trafficking and exploitation. Although the government, through various development aid agencies, seeks to raise awareness on the risks of migration, it does not offer any feasible or affordable alternative for regular migration and migrants opt instead for the more precarious but also less costly forms of migration.

Thirdly, the border is not a homogenous border line with a straightforward crossing point. The strategic location of the Poipet–Aranyaprathet border area, the presence of strong networks and the availability of economic opportunities on both sides attract people who constantly negotiate and reshape their mobility patterns and trajectories on a daily basis based on employment opportunities and motivations. Depending on the destination (the market, the Thai agricultural lands near the border or provinces deeper into Thailand), different trajectories are selected, each with its own challenges and risks. Nevertheless, borderlanders display a great deal of agency and resolve when navigating across the border, where the role of networks helps establish safe passage and routes. These 'socio-economic contracts' cultivate solutions for a number of players at the border, including vulnerable migrants.

Finally, this chapter highlights how local players shape the way the border is managed and negotiated. Formal and official agents and border authorities interact with local and regional organizations, civil society and industrial and trade networks. But other, more informal entities such as labour migrants networks, local/indigenous populations and even criminal networks exacerbate the irregular cross-border movement of goods and people and shape the unwritten rules that guide border movements. The volume and frequency of border flows highlight the level of micromanagement and engagement of local players in shaping the outcome of mobility patterns, and the importance of small-scale economies (such as the market) in people's subsistence living. Greater attention for and research of the role of migration management at ground level is needed to capture the motivations and reasons that influence people's decision to migrate as well as their capacity to assess, predict and circumvent the risks and leverage the opportunities at the border.

References

ADB (Asian Development Bank). 2009. *Migration in the Greater Mekong Subregion: A Background Paper for the Fourth Greater Mekong Subregion Development Dialogue*. Mandaluyong City: ADB.

_____ . 2010. *Sharing Growth and Prosperity: Strategy and Action Plan for the Greater Mekong Subregion Southern Economic Corridor*. Mandaluyong City: ADB.

Caouette, T., R. Sciortino, P. Guest, and A. Feinstein. 2006. *Labor Migration in the Greater Mekong Sub-Region*. Bangkok: Rockefeller Foundation.

CDRI (Cambodia Development Research Institute). 2005. *The Cross Border Economies of Cambodia, Laos, Thailand and Vietnam.* Phnom Penh: CDRI.

_____ . 2009. "Costs and Benefits of Cross-Country Labour Migration in the GMS: Synthesis of the Case Studies in Cambodia, Laos, Thailand and Vietnam." CDRI Working Paper Series 5. Phnom Penh: CDRI.

Chen Chen, L. 2006. *Cambodian Women Migrant Workers: Findings from a Migration Mapping Study.* Phnom Penh: United Nations Development Fund for Women.

Derks, A., R. Henke, and L. Vanna. 2006. *Review of a Decade of Research on Trafficking in Persons, Cambodia.* San Francisco: Asia Foundation.

French, L. 2002. "From Politics to Economics at the Thai–Cambodian Border: Plus Ça Change …" *International Journal of Politics, Culture and Society* 15 (3): 427–70.

Huguet, J.W., and A. Chamratrithirong, eds. 2011. *Thailand Migration Report 2011: Migration for Development in Thailand; Overview and Tools for Policymakers.* Bangkok: International Organization for Migration.

ILO (International Labour Organization). 2008. *The Mekong Challenge: An Honest Broker – Improving Cross-Border Recruitment Practices for the Benefit of Government, Workers and Employers.* Bangkok: International Labour Office.

IOM (International Organization for Migration). 2006. *Review of Labor Migration Dynamics in Cambodia.* Phnom Penh: IOM.

_____ . 2013. "Male Trafficking Victims on the Rise in Cambodia." *IOM*, 12 March. Accessed 15 August 2013. https://www.iom.int/cms/en/sites/iom/home/news-and-views/press-briefing-notes/pbn-2013/pbn-listing/male-trafficking-victims-on-the.html.

King, R., R. Skeldon, and J. Vullnetari. 2008. "Internal and International Migration: Bridging the Theoretical Divide." Working Paper 52. Sussex: Sussex Centre for Migration Research.

Kudo, T. 2009. "Border Area Development in the GMS: Turning the Periphery into the Center of Growth." ERIA Discussion Paper Series 15. Jakarta: Economic Research Institute for ASEAN and East Asia.

Kusakabe, K. 2004. "Women's and Men's Perceptions of Borders and States: The Case of Fish Trade on the Thai–Cambodian Border." *Journal of GMS Development Studies* 1 (1): 45–65.

LICADHO. 2011. "Recruitment Agencies Still Sending Maids to Malaysia, Two Days after Prime Minister Signs Ban Order." *LICADHO Press Release*, 17 October. Accessed 7 October 2012. http://www.licadho-cambodia.org/pressrelease.php?perm=260.

Lim, H., and P. Fàbrega. 2011. *Border Assessment on Mobility and Vulnerability in Cambodia.* Phnom Penh: Coalition to Address (Sexual) Exploitation of Children in Cambodia.

Lintner, B. 2012. "Casinos: Betting on the Border." *Asia Pacific Media Services Limited.* Accessed 7 October. http://www.asiapacificms.com/articles/cambodia_casinos/.

LSCW (Legal Support for Children and Women). 2005. *Needs Assessment and Analysis of the Situation of Cambodian Migrant Workers in Klong Yai District, Trad, Thailand.* Phnom Penh: LSCW.

MMN and AMC (Mekong Migration Network and Asian Migrant Centre). 2013. *Migration in the Greater Mekong Subregion Resource Book: In-Depth Study; Border Economic Zones and Migration.* 4th ed. Chiang Mai: MMN and AMC.

Murshid, K.A.S. 2007. "Domestic and Cross-Border Migration from the Tonle Sap." In *'We Are Living with Worry All the Time': A Participatory Poverty Assessment of the Tonle Sap*, edited by B.M. Ballard, 195–216. Phnom Penh: Cambodia Development Resource Institute.

NCDD (National Committee for Sub-National Democratic Development). 2009. *Paoy Paet Krong Data Book 2009.* Phnom Penh: NCDD.

NIS (National Institute of Statistics). 2009. *General Population Census of Cambodia 2008: Provisional Population Totals*. Phnom Penh: NIS.

Pangsapa, P., and M.J. Smith. 2008. "Political Economy of Southeast Asian Borderlands: Migration, Environment, and Developing Country Firms." *Journal of Contemporary Asia* 38 (4): 485–514.

PATH. 2010. *Improving Labor Migration Processes and Protection of Migrant Workers from Cambodia and Thailand: Issues and Recommendations*. Bangkok: PATH.

Sophal, C. 2009. "Review of Labour Migration Management, Policies and Legal Framework in Cambodia." ILO Asia-Pacific Working Paper Series. Bangkok: International Labour Office.

Tuot, S. 2007. "Cross-Border Labor Migration in and out of Cambodia." *Cambodian Economic Review* 3: 21–39.

Van der Velde, M., and T. van Naerssen. 2011. "People, Borders, Trajectories: An Approach to Cross-Border Mobility and Immobility in and to the European Union." *Area* 43 (2): 218–24.

Vutha H., L. Pide, and P. Dalis. 2011. "Irregular Migration from Cambodia: Characteristics, Challenges and Regulatory Approach." CDRI Working Paper Series 58. Phnom Penh: Cambodia Development Research Institute.

Weggel, O. 2007. "Cambodia in 2006: Self-Promotion and Self-Deception." *Asian Survey* 47 (1): 141–7.

Chapter 9

From Spontaneity to Corridors and Gateways: Cross-Border Mobility between the United States and Canada

Victor Konrad

During the first decade of the twenty-first century, cross-border mobility between the United States and Canada changed decidedly and emphatically. Mobility has become restricted, managed, aligned and selectively expedited through security measures imposed at the border and in the borderlands by both countries. The spontaneous travel and relaxed cross-border migration which characterized the twentieth-century border relationship between friends and neighbours has been increasingly replaced by cautious, planned, multipurpose trips and rule-bound migration between business partners and allies in the war on terror. The border now is perceived as more of a barrier by residents in both countries. This perception of the border as an emerging constraint to mobility has been expanded by media representations of a thicker border, and substantiated by the 'security-scapes' constructed in the borderlands.

Crossing the Canada–US border has changed since the events of 9/11. Increased waiting times at border crossings, uncertainty about crossing, profiling of potentially dangerous ethnic minorities, and strident militarization of the border contribute to increasing perceptions that the border is a barrier. Consequently, both Americans and Canadians are obliged to change their border-crossing behaviour from spontaneity to directed and compliant use of the corridors and gateways, fashioned and sustained by the post-9/11 competing pressures of enhancing mobility and assuring security at the border. Yet, the border crossings between Canada and the United States remain among the most frequented and busiest crossings in the world. Also, the number of migrants has increased in the past decade. There are currently more Americans living in Canada and more Canadians living in the United States than at any time since the 1950s (Statistics Canada 2006; US Census Bureau 2007).

The central question guiding this study and others in this book is how, and to what extent, do borders and transnational migrant trajectories impact on human mobility (Van der Velde and Van Naerssen 2011). This chapter explores how cross-border mobility has changed in perception and behaviour, and how perception and behaviour related to borders impact decisions and patterns of border crossing.[1] Specifically, border-crossing perception and behaviour of Americans and Canadians in cross-border regions across the continent are evaluated to assess how people decide to move across the border, for short-term or long-term objectives, the thresholds that either hold them back or convince them to go, and the trajectories that border crossers create to enable their journeys. The study tests the theoretical components of the

1   The interviews of border stakeholders in the Cascade Gateway are part of a larger study funded by the Border Policy Research Institute (BPRI) at Western Washington University, where the author held a research fellowship in 2009.

'people, borders, trajectories' approach and suggests refinements for an emerging theory of cross-border mobility in an era of globalization.

In order to follow the arguments in this chapter, it is necessary to define and clarify terms such as migration, mobility and commuting, and variations of these terms. Mobility has several connotations: changeableness expressed as variegation, fluidity, inequality and transientness; motion defined as the successive change of place, and articulated as transit or passage, progression or recession, and oscillation; and moral sensibility displayed as sensitivity or sensitiveness, and vigour or vitality. Contemporary cross-border mobility encompasses all of these dimensions with perhaps a greater emphasis attributed to the successive change of place with its characteristics of passage and transit rather than the progression once more widely associated with mobility and particularly migration. Migration, in turn, has evolved in meaning from distinct aspects of either departure (escape, egress and ascent) or wandering (passage), to more specific aspects of emigration from and immigration to states, and even more recently to transmigration, remigration and intermigration in a globalizing world. The ascent to a better place once commonly associated with migration may be ceding to more of a descent and wandering with recent forced migrations and diasporas or dispersals of peoples. Commuters, on the other hand, are rooted in both place and the periodicity of their movements. To commute is to be periodic, often with a regularity of occurrence to the point of habit, to substitute, change one place for another, transfer and be metonymic, to interchange, double or engage in a mutual change and transpose, and to compromise in mutual concession or give and take.

In order to understand contemporary cross-border mobility between the United States and Canada it is necessary to review mobility characteristics of the twentieth century. The chapter begins with this assessment to set the stage for the changes in cross-border movements after the events of 9/11. The twentieth-century border between Canada and the United States is both real in its historically documented manifestations and operation, and imagined as a set of places, landscapes and procedures that are now idealized as components of a golden era in US-Canada border relations. The next part of the chapter characterizes cross-border mobility in the twenty-first century and draws on the recent studies that document a new border-crossing geography of corridors and gateways, pronounced regional differences, scheduled and ordered movements, expanded infrastructures and extended uncertainty (Konrad and Nicol 2004, 2008a). This is a geography of motion and transition in which the border is both a manifestation of the dynamic Canada–US relationship and a process that is constantly evolving (Konrad, 2015). Several case studies, primarily from the north-west coast region, but drawing as well on other regional contexts, are offered to explore the geography of cross-border mobility. The final part of the chapter evaluates the findings from the analysis of movement across the Canada–US border in the context of the 'people, borders, trajectories' approach to building a theory of global cross-border mobility.

The Twentieth-Century Border between Canada and the United States

The border between Canada and the United States has often been referred to as the 'longest undefended border in the world'. This view was held and expressed by both Americans and Canadians throughout the twentieth century, and it was referred to constantly by leaders in both countries. The widely maintained perspective, on the one hand, helped to smooth the inevitable differences between two countries meeting at many border regions along an immense and virtually undefendable border. On the other hand, the view had substantial credence and was constructed on the realities of daily interaction along the border as well as the common goals of growing transnational corporations, a massive trading relationship, interlocked consumers, a continental media and an immense set of cross-border agreements extending from resource and environmental concerns to military pacts. The twentieth-century border worked as an intricate set of scaled relationships encompassing communities, regions and countries. Because it worked effectively and quietly underlying the relationship, the border was heralded as symbolic of integration and coordination between the US and Canada.

This border was crossed constantly by people who would return to their homes and also those who would stay in the neighbouring country. Although many of the large group migrations across the border characteristic of the nineteenth century waned after the beginning of the twentieth century, movements such as French Canadian migration to New England and American farm migration to the Prairie Provinces continued well into the 1920s and beyond. As Widdis (1997, 1999, 2011) has documented for English Canada, these migrations and border crossings were quiet, 'with scarcely a ripple', because people living in the borderlands knew each other, they crossed the border routinely, they had business links and operations that extended across the permeable border, and their language among other aspects of culture was shared.

From this tradition of quiet crossings for work, family interaction and relocation in almost seamless border regions, a sense of entitlement to cross the border emerged. During the second half of the twentieth century, as Canadians and Americans became more affluent and mobile, they purchased properties across the border for seasonal recreation. Americans in the Great Lakes region, for example, bought lakefront properties on the Canadian shores of Lakes Erie, Huron, Ontario and even remote Lake Superior, and in the Ontario 'cottage country' of Muskoka, Haliburton and the Rideau Lakes. Hundreds of thousands of Americans travelled annually to these locations and others across Canada for summer and weekend holidays. They were waved into Canada by customs and immigration officers pleased to see the revenue stream entering the country. Similarly, Canadians travelled to the US in growing and eventually even larger numbers to spend increasingly longer periods in the southern US states. Florida and California attracted the largest numbers of seasonal migrants at first, but by the end of the century most of the southern US was impacted by the 'snow bird' migrations that numbered in the millions (Tremblay 2006; Simpson 2000). Some 'snow birds' eventually relocated to the US permanently, but most continued their annual migrations in the assurance of continued Canadian medical coverage if they spent the obligatory six months back in Canada. Also, some Americans decided to stay in Canada and sought Canadian citizenship.

Business across the border has also engendered an entitlement to cross. Many US branch plants in Canada were established by and directed by American managers who lived, and then sometimes stayed in Canada. Canadians flocked to engage in the burgeoning American media industry in New York and primarily in California. They were followed in the 1990s by Canadians seeking success and advancement in high technology industries. Cross-border integration of resource extraction, manufacturing, media and services advanced mobility of managers and technical experts (Simpson 2000). The border was if anything a formality. The question was not if you could cross, but when you would cross.

During the late twentieth century all border crossings between Canada and the US were essentially the same. True, some were smaller and less frequented, and some were busy with substantial commercial traffic, but the border crosser was afforded the same expedient crossing anywhere across the continent. Clearly, it was most expedient to take the border crossing nearby because the crossing was familiar to the person crossing and the officials were familiar with the person crossing. Border crossing then was essentially a straight line from decision to action, and the trajectory was based on the ease of linkage and the almost absolute guarantee of success in spontaneity.

Figure 9.1 illustrates the segmented linear process of spontaneous border crossing that developed and prevailed in the twentieth century. Mental, locational and trajectory thresholds are all aligned and barely recognized or treated as thresholds. The mental border threshold in this spontaneous process is low or almost non-existent because little or no thought goes into the decision to cross the border. For both Canadians and Americans, the mental border threshold is virtually no threshold at all. Crossing for low-order goods and services is routine. In the Okanagan Valley, extending from British Columbia South into Washington State, this is referred to as the 'six pack' border because Canadian residents routinely cross the border for lower-priced gas, milk, and a 'six pack' of beer which may be purchased readily in any convenience store rather than in a regulated and high-priced government liquor store in British Columbia (Konrad 2010). Border officials usually wink at this practice. In the lower mainland of British Columbia and Washington State, the border is the 'milk and cheese' border, and Canadians are identified colloquially as 'cheese heads'. As a result, the locational threshold is also easily and readily crossed: one goes to convenient and close destinations near the border. This 'comfort crossing' is accomplished in familiar cross-border communities. Consequently, the trajectory threshold is low or non-existent because little time or thought has gone into the design and the plan of the crossing. The crossing is spontaneous. You take the nearest and most convenient crossing.

mental border threshold is low or almost gone

little thought about if and when to cross

cross for low order goods and services: gas and 'six pack' border

locational threshold is easily and readily crossed

go to convenient, close destinations near the border

'comfort crossing' in familiar cross-border communities

trajectory threshold is low or non-existent

little time or thought to plan and design crossing

take the nearest and most convenient crossing

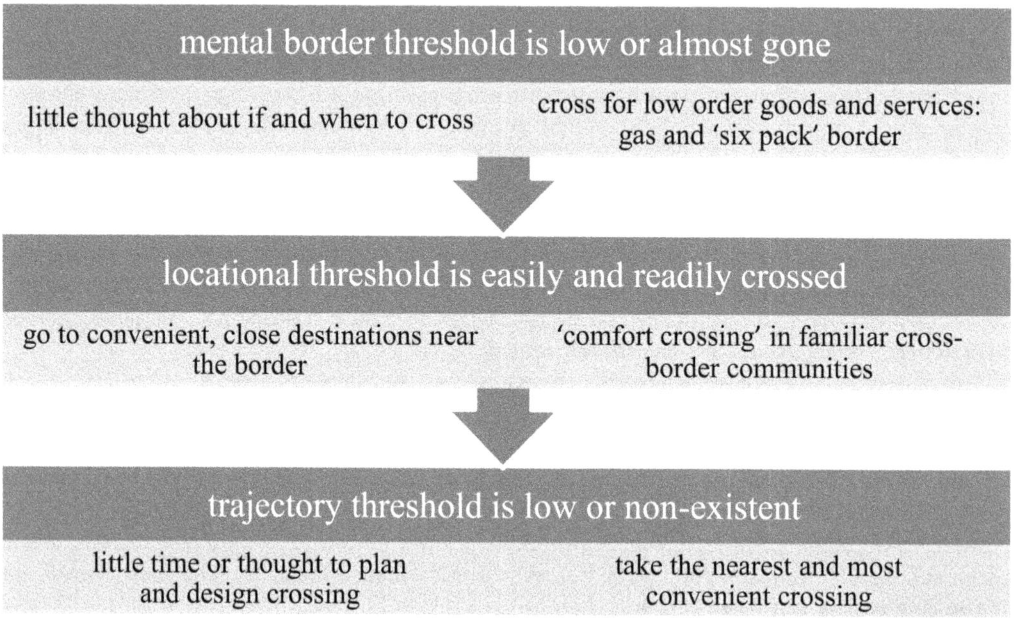

Figure 9.1   The segmented linear process of spontaneous border crossing in the twentieth century

Source: the author.

Whereas cross-border shoppers and casual visitors, and even business travellers, would conform to this spontaneous approach to crossing the border, migrants from the United States to Canada, and from Canada to the United States would devote more thought and preparation to the process. The process, however, remained linear. Migrants drew a straight line between their point of origin to their point of destination and they crossed the border at the most convenient border crossing. This was true of seasonal migrant labour, such as that originating in the US tobacco belt in Virginia and the Carolinas bound for southern Ontario's tobacco harvest, and this straight line approach also appears to have prevailed for migrants relocating between the two countries.

Little research has documented the immigration of Canadians to the United States and likewise the immigration of Americans to Canada in the second half of the twentieth century because few of these migrations have been discernible in the generic flows of people across the border. This contrasts with the growing literature on cross-border migration during the nineteenth century (Widdis 2010). One notable exception is the renewed interest and research on US draft resisters who moved to Canada during the Vietnam War. They remain the singularly most evident and significant group of migrants across the border because they underline political and ideological differences between Americans and Canadians. Recent books by Dickerson (1999) and Hagan (2001), and a wealth of journal literature, explore the relocation of more than 100,000 Americans, mainly to major urban centres in Canada. These immigrants were characteristically well-educated 'mainstream' Americans not associated with minority groups in the United States. They blended into the populations of Montreal, Toronto and Vancouver, and to a lesser extent other, smaller Canadian cities and towns. The evidence suggests that the draft resisters or 'dodgers' aimed to blend into Canadian cities that were large – places where they could remain anonymous more easily. These were also places where liberal, anti-war groups in Canada rallied to support the American resisters and organized efforts to help them settle, find jobs and integrate into the communities that they selected.

If we reconstruct the migration process of these resisters, it is apparent that for many the border crossing was rapid and precipitous if not spontaneous. They received their draft notice from the US military, made a rapid decision to leave the United States, and entered easily across a porous border by simply indicating that they were bound to visit friends or family. Although the gravity of their exit from the US was apparent to them, the resisters crossed thresholds in decision-making that were relatively low and encountered thresholds of obstruction and difficulty at the border that were also relatively low. They were able to take advantage of an ambient border in order to cross easily the mental, locational and trajectory thresholds before them. They could be viewed as overcoming the indifference threshold that existed before the Vietnam War when Canada was mentally non-existent as a place to migrate to. Then during the War it suddenly was revealed as an option.

Canada–US Cross-Border Mobility in the Twenty-First Century

It is fascinating to reflect on how cross-border mobility has changed so markedly from one century to another in the history of movement between the United States and Canada. Until the end of the eighteenth century, the fledgling United States was consumed with consolidating its own territory, and formerly French and then British North America was doing the same. Early in the nineteenth century, American and British interests collided and the border between Canada and the US became a reality and a line to be defended. Yet, by the time the nineteenth century came to an end, both Canada and the United States traded people as well as goods across a border that was porous across the continent. This soft, permeable and thin border would prevail throughout the twentieth century, although the opposing outlooks of American and Canadian authorities on issues such as prohibition, selective military service, civil rights and immigration created a border that could be invoked and enhanced if required.

The border of the early twenty-first century is decidedly different, keyed emphatically by the events of 11 September 2001 (Ackleson 2009). Yet, the processes of change at the border were in motion for at least a decade before as the North American Free Trade Agreement (NAFTA) was put into effect, enhanced technologies were applied to monitoring cross-border movement of people and goods, and undetermined globalization forces required bordering responses by countries (Konrad and Nicol 2008a). The events of 9/11 punctuated an immense process of change already in motion by the mid-1990s. One component of this change was to expedite the flow of goods, and to differentiate and restrict the flow of people.

Canadian and American migration flows across the border in the early twenty-first century fall into three distinct and more differentiated categories than before: short-term crossings in the borderlands, seasonal migrations beyond the borderlands and migration related to resettlement across the border both in the borderlands and beyond. Although these flows are made more discrete by a 'thicker' border and its enhanced regulation, technologically assisted sorting and generally increased scrutiny, the flows have in fact become parts or components of a reinstated and smoothed movement of people across the Canada–US border. In this sense, the flows may be conceptualized as subsets of a general process where consistent rules apply across the categories of cross-border movement.

None of these contemporary flows has been examined extensively by scholars of migration. A few studies focus on French Canadian seasonal migration to Florida (Tremblay 2003, 2006; Tremblay and Chicoine 2011) and American migration to western Canada (Hardwick 2010; Hardwick and Mansfield 2009). There is a Migration Policy Institute study on Americans who moved to Canada subsequent to the re-election of George W. Bush (Kobayashi and Ray 2005). There is also a collection of essays about the migrant experiences across the newly enhanced borders in North America, and some of the papers deal with indigenous and visible minority migrations (Sadowski-Smith 2002, 2008).

Whereas most Canadians who moved to the United States in recent years moved there for better jobs and better careers with the easing of skilled-worker restrictions under NAFTA, many American immigrants in Canada have moved there for political and social reasons. The political reasons pushed them out of the US because, again, they resist American wars abroad, and they feel that politics in the US is overly conservative. The social reasons have pulled them toward Canada because Canada has a better social safety net and health care, there is better public safety and security, there are fewer guns and there is greater tolerance for alternative lifestyles (Hardwick 2009, 2010; Hardwick and Mansfield 2009).

The result of these migrations is a substantial concentration of the 2007 Canadian-born population in Florida and California (200,000 or more), and concentrations approaching these in Texas and Arizona. However, the concentrations of Canadian-born populations are just as high in the border-adjacent states of Washington, Michigan and New York, and in Massachusetts. Other East Coast and West Coast states and Colorado have moderate concentrations under 75,000 people. The mid-western and lower Mississippi Valley states have small Canadian-born populations under 10,000 people. The percentage change in the Canadian-born population between 1970 and 2007 is highest in the sunbelt where increases between 250 and 500 per cent are evident in all of the western mountain states, Texas, Oklahoma, Arkansas, Tennessee, Kentucky, Virginia, the Carolinas, Mississippi, Alabama, Georgia and Florida. Meanwhile, the northern border states and California have characteristically seen moderate declines of 25 to 75 per cent (Hardwick and Smith

2012). Hardwick and Smith go on to emphasize that from the mid-1990s on to the present, the increases in migration have been due mainly to the increasing availability and affordability of real estate in the sunbelt states during the global recession and the relative ease of obtaining work visas in the United States.

Most of the Americans who left the United States for Canada in recent years are white, well-educated and sufficiently wealthy to establish themselves readily in Canada. Like the wave of more than 100,000 Vietnam War resisters who preceded them in the 1970s, many of these immigrants came to Canada and in some instances returned to Canada for political reasons. Others are pursuing economic opportunities in Canada because the recession has not impacted Canada to the same extent as the United States. A host of social pull factors include universal health care, tolerance for alternative lifestyle communities such as gays and lesbians, retirement and strict gun control (Jedwab 2008; Hardwick and Mansfield 2009). Most Americans migrating to Canada have selected to live in major cities including Toronto, Vancouver, Calgary and Halifax but small towns along the border and in picturesque areas of the country such as the Canadian Shield and the ranges of western mountains also draw American immigrants (Hardwick and Mansfield 2009).

Hardwick and Smith (2012) document the American-born population in Canada based on the 2006 Census. Concentrations of over 100,000 people are found in Ontario and British Columbia, and concentrations of between 12,000 and 60,000 are characteristic across the rest of the provinces except the northern territories where only a few thousand Americans reside. Yet, the percentage increase of Americans in the north rivals that of Ontario and the Maritime provinces where increases between 1970 and 2006 range between 11 and 50 per cent. Quebec and the Prairie provinces of Alberta, Saskatchewan and Manitoba all have seen decreases in excess of 50 per cent during this time. The inflow of American immigrants to British Columbia is maintaining the numbers in the province with only a 2 per cent decrease recorded between 1970 and 2006.

The overall contemporary immigration pattern of Americans in Canada indicates a sustained concentration in Ontario where the greatest number of economic and other benefits prevail. Both coasts also draw Americans but the interior provinces do not appeal and Quebec poses a language barrier to immigration from the United States. Hardwick and Mansfield (2009) suggest that the recent influx of Americans to British Columbia originates largely in the western US and primarily California. It is an extension of the movement of Californians up the West Coast of the US. In this sense the trajectory is evident. A similar pattern appears to be evident in the relationship between Americans in the Northeast, and particularly New England, moving to the Maritimes. Ontario draws more widely from across the United States. Americans now living in the Perth, Ontario area, interviewed by the author, originate in Pennsylvania, New York, Michigan and Illinois. Their trajectories were all guided by prior seasonal destinations in Canada. One person actually returned to a family hometown where he was able to reacquire the ancestral home occupied by his grandparents.

The pattern of origins and destinations for Canadians in the United States is not as clear, and it requires more research. Evident, however, is that westerners in Canada characteristically are concentrated in the mountain and desert states and in Texas. This large state also draws immigrants from Ontario, and to a lesser degree eastern Canada. Most Ontarians as well as virtually all the Quebecers reside in Florida, and in recent years increasingly find homes in the other sunbelt states up the Atlantic coast and across the Gulf coast. Traditional Canada–US borderland concentrations remain but these are giving way to the greater number of Canadians deciding to reside in the sunbelt. In this pattern, we can see the linkage between seasonal migration and immigration to the United States. Seasonal visitors from Canada to the sunbelt have eventually purchased reasonably priced properties and in some instances decided to stay in the US. These Canadian residents of the

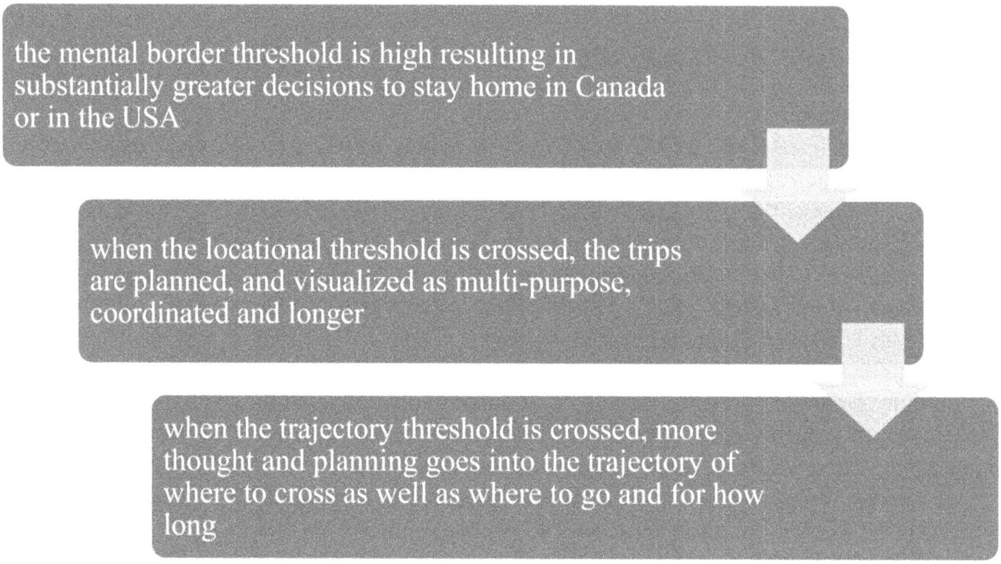

the mental border threshold is high resulting in substantially greater decisions to stay home in Canada or in the USA

when the locational threshold is crossed, the trips are planned, and visualized as multi-purpose, coordinated and longer

when the trajectory threshold is crossed, more thought and planning goes into the trajectory of where to cross as well as where to go and for how long

Figure 9.2   The staggered process of cross-border mobility in the twenty-first century
Source: the author.

US, however, remain relatively few compared with the millions of the earlier mentioned 'snow birds' who temporarily reside in the US sunbelt.

Although these aggregate, spatial data provide a general assessment of the geographical patterns of migration between the United States and Canada, they provide only a context for evaluation of the process of cross-border mobility. In the rest of this chapter, the focus is on short-term crossings in the borderlands. This form of migration is an integral component of the overall spectrum of migration and cross-border mobility. Examination of short-term crossings should reveal similar conceptual dividends if the forms of migration are indeed theoretically linked. Also, the information and data regarding short-term crossings in the borderlands is more readily available and more relevant to the analysis of the imposition of stricter border controls. It may be argued that immigrants to the US from Canada, and similarly immigrants to Canada from the US, are not impacted by border thickening and thinning as much as by the push and pull factors that cause them to emigrate and immigrate.

Figure 9.2 describes the 'staggered' process of cross-border mobility in the twenty-first century. Each step in the process is expedited if conditions are met in order to cross thresholds that are now higher due to the imposition of a thicker, more regulated border. There may be situations, however, where thresholds are not necessarily passed consecutively in the cross-border mobility process. Seasoned and frequent cross-border travellers, for example, collapse these decisions into an almost seamless process. Others may dwell on the mental threshold or focus on the trajectory.

Currently, the mental border threshold is high due to the perception among both Canadians and Americans, situated near and living a great distance from the border, that the border is simply more difficult to cross. This perception is supported and reinforced by the media in constant stories about people who have been halted and arrested because of some past misdemeanour, or instances where border crossers have had vehicles and property confiscated by overzealous officials, and recently by US calls for fencing the border (*Seattle PI*, 4 October 2011). The documented result is that many Americans and Canadians now simply do not cross the border (Konrad 2010; Konrad and Nicol 2008a, 2008b). Also, now that identity documentation at the Canada–US border requires passports or equivalent proof of citizenship, many Americans (about 60 per cent) and a substantial number of Canadians (about 40 per cent) do not comply with identity confirmation requirements (Konrad and Nicol 2008b). These people are now effectively 'country locked' whether they have any inclination to cross the border or not. The overall result, through greater regulation, more effective enforcement, media confirmation and personal experience, is to heighten the mental threshold, and for many Americans and Canadians, to make this threshold insurmountable.

When this mental threshold is crossed, in most instances, the decision ahead is to decide where to go and where to cross the border. This locational threshold is more daunting simply because the place of destination now lies on the other side of a more imposing border, and the previously simple act of moving across the border at the most convenient crossing between point A and point B may no longer be assumed. Informants have substantiated that trips need to be planned in order to pick the right time to cross so as to avoid delays at peak crossing hours. They explain as well that the trips across the border are becoming multipurpose trips in which shopping is combined with a business purpose, a visit to family across the border, a sporting event or some other form of entertainment. Consequently, the trips become longer and require overnight stays in some instances. These trips also require more coordination (Konrad 2010).

Usually, when the previous thresholds are crossed, more thought and planning goes into the trajectory of where to cross as well as where to go. The trajectory then becomes a significant part of the border-crossing process whereas formerly, in spontaneous crossing, the trajectory was relatively unimportant. In the Niagara Gateway of the Great Lakes border region, trajectory planning has become a part of many border-crossing decisions for both Americans and Canadians travelling to Ontario, and Canadians and Americans travelling to New York State. With four land border crossings to choose from, and rapid highway access to at least three of these crossings, motorists approaching the border from either Canada or the US have options to cross if they are informed about traffic congestion and waiting times at the respective crossings. This information is increasingly readily available through designated radio stations and electronic signs at a distance from the border, and the waiting time information combined with accumulated knowledge about preferred border crossings, aids the motorist with trajectory planning.

Table 9.1 outlines how thresholds are altered and impacted with the advent of border enhancement. For short-term crossings the immediate impact is to cause many previous spontaneous crossers to not bother, and to stay home. Whereas risks of being stopped, detained and fined or even arrested were minimal with permeable border conditions, these risks have increased substantially at the enhanced border. As outlined above, those people who do cross tend to prepare more effectively and take longer, multipurpose trips. The trajectory now becomes important, and it is visualized and planned. Seasonal migrants are impacted as well but since they have a well-developed pattern and experience of at least annual crossings, and since they often own property in the other country, they have too much to lose if they are not compliant. Their locational and trajectory thresholds are sustained but with greater attention to details of compliance with regulations for crossing the border. Migrants who move to the other country also need to be compliant with the demanding US and Canadian immigration statutes and regulations. With an enhanced border, regulations are enforced more effectively, and those potential migrants with criminal records and other barriers to crossing tend to be dissuaded by the mental border threshold. Locational thresholds see little or no change with a thicker border but trajectories become more uncertain.

Table 9.1   Cross-border mobility with enhanced borders

	Short-term crossing	Seasonal migration	Emigration/Immigration
Mental border threshold	Stay home or risk setbacks	Sustained with enhanced compliance	Sorted, stratified, more compliant
Locational threshold	Trips planned, multi-purpose, longer, coordinated	Enhanced planning	Little or no change
Trajectory threshold	Planned trajectories	Planned trajectories	Increased uncertainty

Source: the author.

After the events of 9/11, the border between the United States and Canada was reconstructed rapidly into a greater barrier to mobility than had ever existed between the two countries. Yet, at the same time, mobility was enhanced substantially through the expedited construction of corridors and gateways at selected major crossing points along the border (Konrad and Nicol 2004). These corridors and gateways institutionalized a geographical logic of a few larger crossing points, and several orders of smaller crossings along the border, and they developed a distinct hierarchy of crossing points (Konrad and Nicol 2008a). In each cross-border region – Pacific, Plains, Great Lakes, St. Lawrence and Atlantic – the largest crossings became even larger as they were singled out for enhancements in infrastructure, staffing, and the application of digital technologies to enable the crossings to process more people and goods. The corridors, already evident in the NAFTA trade, transportation and migration channels, now channelled more human and goods traffic through the enhanced larger crossings. These enhanced crossings at the upper end of the crossing hierarchy became identified as gateways with specific names. The gateways – Champlain, Niagara, Windsor-Detroit, Cascade – characteristically offer two or more crossings, and often differentiate the crossings for different modes of traffic. The Champlain Gateway offers two crossings from the Montreal area to the US via Vermont and New York State. Niagara Gateway provides three crossings in the Niagara Falls vicinity, and another close by in Buffalo/Fort Erie. Whereas the Buffalo and Queenston crossings are both linked directly to superhighways in both the US and Canada, the crossings in Niagara Falls serve local traffic in the case of the Whirlpool Bridge and international visitors from around the world in the case of the Rainbow Bridge. The Windsor-Detroit crossing, the busiest in the world, currently offers a tunnel restricted to passenger cars, and the massive Ambassador Bridge. Another link is planned. The Sarnia-Port Huron crossing is close by to relieve some of the traffic pressure in this region (the Cascade Gateway is profiled in the next section of this contribution). All of these gateways are associated with international airports located nearby. Second-order crossings are found in all of the cross-border regions and they have grown during the past decade substantially as well. In the Prairies and in the Atlantic region, these second-order crossings actually serve as the primary crossing points and are linked to the main corridors in these regions.

Changing Canada–US Border Mobility in the Cascade Gateway

The Cascade Gateway is the border-crossing system of the Pacific corridor of road, rail, sea and air connections along the narrow coastal zone of settlement between the Cascade Mountains/ Coast range and the waters of Puget Sound and the Strait of Georgia. It is the primary border gateway between the US and Canada on the Pacific coast. At the core of this crossing system are four separate yet linked border crossings: the Peace Arch crossing which links US Interstate 5 to Highway 99 in Canada, the Pacific Highway crossing several blocks to the east, the small Aldergrove crossing, and the Sumas crossing at the base of the Cascades. This set of crossings forms a functional hierarchy that provides a set of alternative crossings for both Americans and Canadians, particularly if they have knowledge of the respective crossings and where they lead.

In the twentieth century, the four crossings were distinct ports of entry between the US and Canada serving the interactive needs of cross-border communities as well as the growing transnational movement of goods and people from the region and beyond. With the substantial growth of Vancouver and lower mainland communities, and the northward expansion of Seattle, as well as expansion of cross-border shopping opportunities, integrated regional tourism and the growth of trade under the NAFTA, the immense increase in traffic required the greater rationalization of the border-crossing system. During the 1990s, substantial enhancements were made at the Pacific Highway crossing in order to expedite trade in goods whereas the other crossings utilized to a greater extent by passenger vehicles saw only incremental changes in infrastructure and technology. After 9/11, and the focus on identifying and interdicting terrorists, the people crossings of the gateway received substantially more attention. The scrutiny of human and goods traffic across the border was elevated significantly and the crossing system became severely strained resulting in massive queues and waiting times exceeding an hour and more during peak crossing periods.

Fortunately, a coalition of border stakeholders known as the International Mobility and Trade Corridor Program (IMTC), developed to expedite mobility in the 1990s, was able to expand its efforts after 9/11 to 'troubleshoot' problems with the system, convene the stakeholders in monthly forums, recommend mobility enhancements and monitor the efficacy of changes made by federal border agencies, state and provincial government bodies responsible for infrastructure, and other stakeholders from private and public sectors (IMTC 2009). In 2009, a large scale survey conducted among 160 border stakeholders as well as detailed interviews with a coincident group of 98 stakeholders provides an extensive data set for analysis of perceptions, opinions and attitudes of stakeholders on a range of border issues and policies (Konrad 2010). The interviews are particularly useful in analysing the mobility characteristics of people who live near the Cascade Gateway.

These interviews combined with IMTC reports and documentation from the Border Policy Research Institute enable a conceptualization of the hierarchical crossing structure of the Cascade Gateway and a detailed assessment of border-crossing perception and behaviour in the Gateway. A visualization of how the Cascade Gateway operates entails articulating the four land border crossings from west to east: Peace Arch/Douglas, Pacific Highway, Lynden/Aldergrove and Sumas/Huntingdon. Rail traffic runs along the coast at Peace Arch but inspection is carried out at origin and en route. Watercrafts are inspected at terminals in Washington State and in British Columbia. Primary airports are in Vancouver and Seattle, both outside the gateway, whereas new international facilities at both Bellingham Airport in Washington and Abbotsford Airport in the lower Fraser Valley, BC, are approached by international passengers crossing the border through the gateway by auto to access these convenient air connections into the US or Canada. The Cascade Gateway has become a key link between the growing settlements of the lower Fraser valley and the adjacent borderlands of northern Washington State as well as the urban growth poles and world cities of Vancouver and Seattle. All four land crossings accommodate passenger vehicles whereas the Peace Arch facility is the only land crossing in the gateway restricted to cars. It is the primary crossing point in the gateway for passenger vehicle traffic between the two countries. When this crossing becomes very busy, vehicles are directed to adjacent Pacific Highway which also serves as a crossing for bus traffic and the primary crossing for truck traffic. Truck traffic may also use the Sumas crossing which is particularly convenient for eastbound truck traffic which does not need to go through the Vancouver area. During peak traffic periods, some trucks are also processed at the small Aldergrove crossing. This port of entry is only open until midnight daily whereas all other gateway crossings are available 24 hours a day. During the first decade of the twenty-first century, this crossing structure has evolved slowly to deal with the increasing demands for expedited crossing in an era of enhanced security. The short-term crossings in the Cascade Gateway borderlands, according to informants interviewed in 2009, had changed significantly in less than a decade. The main reason for this change was the imposition of security primacy, where security trumps all other border concerns (Konrad 2010). The immediate impact was to reduce the number of crossings and this altered the nature of short-term crossings. This change is evaluated in the context of the 'people, borders, trajectories' approach based on the analysis of thresholds of spatial migratory behaviour.

The mental border threshold has emerged from relative insignificant in the migrant decision-making process, to loom as a primary consideration and substantial initial barrier in short-term crossing. Informants from both countries refer to the 'scary border experience' which develops fears that range from being mistreated to simply being late for a commitment. There is a common sense among informants interviewed (Konrad 2010) that the augmented border system is in fact 'breaking down', that uncertainty is the rule, that 'people cannot live with uncertainty,' that 'there are no quick trips anymore' and that 'spontaneous travel is not an option.' One informant indicated that older people are not crossing like they used to. Another felt that people were no longer being regarded and 'treated as customers'. One resident of Lynden, WA, actually conceptualized the heightened mental threshold with the statement that 'Canada is further away; Lynden and Langley are no longer connected effectively.' An informant from a Canadian border community Chamber of Commerce pointed out that 'gateway is a misnomer.' Virtually everyone who crosses the border, including privileged crossers who hold NEXUS[2]-trusted traveller cards, need to cross a mental threshold before they cross the border. Uncertainty is a major consideration for all border crossers, and this uncertainty is the main component of the heightened mental threshold.

The locational threshold is crossed by fewer migrants in the contemporary Cascade Gateway than in the pre-9/11 regime of spontaneous travel and cross-border community interaction. Even those who cross the mental threshold in an effort to consider trips across the border may not engage in the act of crossing but rather are content to contemplate the possibility or resigned to nostalgic reminiscence of how it used to be or where one could go. Informants in the Lynden area spoke fondly of the 'church rush' on Sundays when Canadians lined up at the border to attend church on the other side, and to bring back a load of groceries as well. Others make plans to cross, and these plans are often complex and closely scheduled efforts to engage with as many tasks and destinations as possible in a day, or even to extend the trip to several days in order to make the trip and the wait at the border worthwhile. Some informants spoke of making elaborate plans but ultimately 'people don't want to be bothered' by uncertainty and a long wait at the border.

---

2   NEXUS is the trusted traveller programne instituted along the US-Canada border. Upon completion of successful security checks and an interview, a trusted traveller may be provided with a NEXUS card to enable rapid entry at land, sea and air crossings.

Increasingly, those who do cross for short terms are commuters, business travellers, cross-border property owners, frequent shoppers, and destination-oriented travellers such as gamblers and recreationists (skiing, hiking, cycling and so on). Others do cross as well to visit relatives and friends, engage in cultural group and religious events and to tour in the borderlands. These visits, however, are multipurpose, combined ventures which include at least two or three core purposes. Informants consistently described trips that combined an *incentive* purpose such as a visit to see an ailing family member or friend, with one or more *auxiliary* purposes such as stopping to retrieve mail from a US or Canadian mailbox, shop for bargain price goods, maintain a boat at a cross-border marina or attend a sporting event. An incentive purpose often involved a destination deeper in the borderlands and might require an overnight stay. A longer stay of course provided for a larger customs exemption on return. The trajectories for short-term crossers, even those with trusted traveller status, are often a combination of several discrete destinations at different distances from the border. In this process of ranging farther afield across the border, the short-term crossers are actually extending their borderlands experience and expanding the borderlands. Their trajectories define a broader borderlands with enhanced security rather than the narrower cross-border community borderlands of the spontaneous border-crossing era. Consequently, it is not surprising that informants describe a new geography of the borderlands in the Cascade Gateway region, a geography where cross-border cultural communities such as the Sikh community are experiencing a downturn in the number of binational marriages, and the Dutch Reform community straddling the border is documenting a reduction in cross-border participation in religious services and community interaction. Moreover, the border towns, particularly the smaller towns of Blaine, Lynden and Sumas on the American side of the border, are experiencing business decline, tax base erosion and real estate slumps. The trajectories of short-term crossing in an era of enhanced security delineate a borderlands geography where the cross-border community is being hollowed out as the borderlands are being extended.

People, Borders, Trajectories and the Emerging Theory of Cross-Border Mobility

In their thought-provoking paper outlining an approach to cross-border migration, Van der Velde and Van Naerssen (2011) emphasize that the mobility apparent in global migration patterns is actually overshadowed by substantial immobility, even in active immigration theatres like the European Union. They attribute this to the substantial 'spaces of indifference' where mental and perceived thresholds to cross borders are not broached if even considered. This immobility exists as well in encountering migration thresholds to relocate from Canada to the United States, and particularly from the US to Canada. Overall, relatively few Canadians move to the US and proportionally even fewer Americans come to Canada to stay. Nevertheless, this chapter offers the proposition that even these relatively small numbers of migrants matter because they are part of a migratory time/space system that exhibits centenary changes of considerable importance and significance. Also, the border crossings are but one component of a continuum of migration that extends from permanent relocations, to seasonal and cyclical migrations, and short-term crossings in the borderlands. This spectrum of mobility can offer an instructive framework for examining mobility and immobility in a Canada–US and perhaps North American context. Its wider application needs to be tested.

Cross-border mobility and immobility between the US and Canada does confirm that migrants do make strategic decisions to either stay or go across the border whether they are crossing for an afternoon or to relocate from their home country to their new host country. Most Canadian and American migrants by far who cross the borders are among the 300,000 and more short-term travellers who traverse the border daily for business and for pleasure. This massive wave is followed by a seasonal accretion of millions of Canadians and a smaller but still significant number of Americans who are seasonal and cyclical migrants. Their geographical impact on regions such as the sunbelt and destination places in both countries has long been felt across the continent and it is expanding. Recently, migrant relocation from the US to Canada has increased more substantially than at any time since the Vietnam War. Canada has to a certain extent traded places with the US as a land of opportunity or certainly the country of greater human well-being and benefit.

The border too has changed, and it has changed significantly. This calls for a dynamic conceptualization of borders in any approach to cross-border mobility and immobility, and this chapter has offered such an approach by assessing border crossing in the context of the kind or type of border which prevails. For the United States-Canada border, the discontinuity is clear, and the substantial change in border security enforcement reveals two borders for comparing mobility and immobility: the relaxed, permeable or spontaneous border prevailing throughout much of the twentieth century, and the enhanced, thickened and militarized border of the last decade and more. The comparison of these two border contexts provides a rich and detailed illustration and analysis of how the spatial routes of migration or trajectories change with changes at the border, whether the lens is focused on short-term crossings, cyclical migration or migrant relocation. Clearly, there is more research required, particularly on the typical quiet migrations that occur between friends and neighbours like Canada and the United States. Also, the field of short-term migration study between Canada and the US requires more comparison with other cross-border regions outside the Pacific Corridor and Cascade Gateway. Yet, even these modest examples of cross-border migration afford some broader insights.

One of these insights is that whereas the relaxed border allowed and promoted spontaneous crossing for short-term and even cyclical migration, relocation between Canada and the US has characteristically been planned and carefully articulated since early in the twentieth century, and it continues to be so under conditions of an enhanced border. All crossings of the enhanced border are indeed considered carefully and planned.

Transnational migrant trajectories are a growing consideration in globalization. If we are to understand borders in globalization, as well as comprehend cross-border mobility, we need to develop a clearer sense of the intersection of borders and migration. This chapter has illustrated that one does not simply cross the other. Rather, the intersection exhibits multiple layers of crossing, each layer imbued with unique characteristics of course, but also alignments of agency, structure and meaning.

References

Ackleson, J. 2009. "From 'Thin' to 'Thick' (and Back Again?): The Politics and Policies of the Contemporary US-Canada Border." *American Review of Canadian Studies* 39 (4): 336–51.

Dickerson, J. 1999. *North to Canada: Men and Women against the Vietnam War.* Westport, CT: Praeger.

Hagan, J. 2001. *Northern Passage: American Vietnam War Resisters in Canada.* Cambridge, MA: Harvard University Press.

Hardwick, S.W. 2009. "Borders and Boundaries: Geographies and Identities in Canada and the United States." *Canadian Journal for Social Research* 3: 19–26.

_____. 2010. "Fuzzy Transnationals? American Migration, Settlement, and Belonging in Canada." *American Review of Canadian Studies* 40 (1): 86–103.

Hardwick, S.W., and G. Mansfield. 2009. "Discourse, Identity, and 'Homeland as Other' at the Borderlands." *Annals of the Association of American Geographers* 99 (2): 383–405.

Hardwick, S.W., and H. Smith. 2012. "Crossing the 49th Parallel: American Immigrants in Canada and Canadians in the US." In *Immigrant Geographies in North American Cities*, edited by C. Teixeira, W. Li, and A. Kobayashi, 288–311. Toronto: Oxford University Press.

IMTC (International Mobility and Trade Corridor Program). 2009. *2009 IMTC Resource Manual.* Bellingham, WA: Whatcom Council of Governments.

Jedwab, J. 2008. *Home is Where the Heart Is: Census Reveals Growing Numbers of Americans in Canada.* Montreal: Association for Canadian Studies.

Kobayashi, A., and B. Ray. 2005. "Placing American Emigration to Canada in Context." *Migration Information Source: Feature*, 1 January. Accessed 11 September 2014. http://www. migrationpolicy.org/article/placing-american-emigration-canada-context/.

Konrad, V. 2010. "'Breaking Points,' but No 'Broken' Border: Stakeholders Evaluate Border Issues in the Pacific Northwest Region." BPRI Research Report 10. Washington, DC: Western Washington University.

_____. 2015. "Toward a Theory of Borders in Motion." *Journal of Borderlands Studies* 30(1): 1–17.

Konrad, V., and H.N. Nicol. 2004. "Boundaries and Corridors: Rethinking the Canada-United States Borderlands in the Post-9/11 Era." *Canadian–American Public Policy* 60: 1–28.

_____. 2008a. *Beyond Walls: Re-Inventing the Canada-United States Borderlands.* Aldershot: Ashgate.

_____. 2008b. "Passports for All." *Canadian–American Public Policy* 74: 1–64.

Sadowski-Smith, C., ed. 2002. *Globalization on the Line: Culture, Capital and Citizenship at U.S. Borders.* New York: Palgrave.

_____. 2008. "Unskilled Labor Migration and the Illegality Spiral: Chinese, European and Mexican Indocumentados in the United States, 1882–2007." *American Quarterly* 60 (3): 779–804.

Simpson, J. 2000. *Star-Spangled Canadians: Canadians Living the American Dream.* Toronto: Harper Collins.

Statistics Canada. 2006. "Census of Canada: Foreign Born by Province." Accessed 10 July 2011. http://www12.stacan.gc.ca/census-recensement/2006/as=sa/97-577/p3-eng.cfm (site discontinued).

Tremblay, R. 2003. "Le déclin de Floribec." *Teoros: Revue international de recherché en tourisme* 22 (2): 63–6.

_____. 2006. *Floribec: Espace et communauté.* Ottawa: Presses de l'Université d' Ottawa.

Tremblay, R., and H. Chicoine. 2011. "Floribec: The Life and Death of a Tourism-Based Transnational Community." *Norsk geografisk tidsskrift* 65 (1): 54–9.

US Census Bureau. 2007. "Census of the United States: Foreign Born by State." Accessed 10 July 2011. https://www.census.gov/population/www/scdemo/foreign/cps2007.html (site discontinued).

Van der Velde, M., and T. van Naerssen. 2011. "People, Borders, Trajectories: An Approach to Cross-Border Mobility and Immobility in and to the European Union." *Area* 43 (2): 218–24.

Widdis, R.W. 1997. "Borders, Borderlands, and Canadian Identity: A Canadian Perspective." *International Journal of Canadian Studies* 15: 49–66.

_____. 1999. *With Scarcely a Ripple: Anglo-Canadian Migration into the United States and Western Canada, 1880–1920.* Montreal: McGill-Queen's University Press.

_____ . 2010. "Crossing an Intellectual and Geographical Border: The Importance of Migration in Shaping the Canadian–American Borderlands at the Turn of the Twentieth Century." *Social Science History* 34 (4): 445–97.

_____ . 2011. "'Across the Boundary in a Hundred Torrents': The Changing Geography of Marine Trade within the Great Lakes Borderland Region during the Nineteenth and Early Twentieth Centuries." *Annals of the Association of American Geographers* 101 (2): 356–379.

Chapter 10

When Fencing Is Not Protecting:
The Case of Israel–Gaza

Doaa' Elnakhala

Border barriers are built to stop perceived or real threats usually defined as undocumented immigration, trafficking or cross-border violence (Andreas 2000). Just like migrants who cross borders in search for a better life, militants[1] are involved in a decision-making process before crossing borders to perpetrate their attacks. This implies crossing a line identified as the border of a state's territory and employing different tactics and routes to cross this line. This chapter is especially interested in hardened borders represented by physical barriers. It employs the approach towards mobility of Van der Velde and Van Naerssen (2011), henceforth called 'the threshold approach'.

The chapter at hand demonstrates how barriers, built in the context of a conflict, under certain conditions escalate violence instead of preventing it. It argues that barriers without cross-border security cooperation may enable militants to continue attacking despite the barrier and even increase the frequency of their attacks. Barriers, however, affect militants'

Figure 10.1   The Gaza Strip fence

Source: Wikimedia Commons 2005.

---

1   Militants and militant groups are used interchangeably in this chapter. They include militant groups that attack a state. Others may refer to these same groups as terrorist in nature. Here we refrain from doing so based on the preference to avoid value-laden expressions and use more neutral terms.

choice of attack tactics, which depends on available military resources. The argument that a continuous barrier that seeks to achieve security by stopping cross-border attacks may instead promote violence is both counter-intuitive and theoretically new. Since an increasing number of states are building border barriers to fend off perceived threats, such as violence, undocumented immigration and drug trafficking, this argument potentially has important empirical implications for other barriers.

Although the threshold approach is mainly meant to explain (labour) migration, it is applicable to the Israeli border barriers as well. The approach includes three analytical pillars for understanding cross-border movement: people, borders and trajectories. People in this chapter are militant border crossers, who are bounded rational actors making decisions based on available but incomplete information. Borders are both real and imagined; in this chapter they specifically concern physical borders that act as barriers. Finally, trajectories followed by border crossers are flexible and change for example according to updated information. This contribution focuses on tactics followed by militants as border crossers and how these tactics change.

While the threshold approach helps in developing the argument of this chapter, for our purposes this approach can be enhanced. It focuses quite strongly on undocumented migrants, especially where it concerns trajectories. However, the latter are not the only border crossers involved in a decision-making process. Learning about what has been written on other border crossers may help to broaden the scope of the threshold approach. Furthermore, due to its specific focus on undocumented migration, the approach relies on mechanisms usually employed by migrants, that is, trajectories or geographical routes. A more general concept, such as border-crossing tactics, may be relevant and have the potential of further expanding the applicability of the approach.

Therefore, the current chapter attempts to fill these gaps by adding to the threshold approach in a number of ways. First, the argument of this chapter adds the element of fortified borders and their effect on decision-making by militants as border crossers. Second, the chapter highlights the role of cross-border cooperation by introducing the fortification–cooperation dynamic, which sheds light on the effect of the interaction between fortification and cooperation on militant border crossings. Third, it develops the concept of border-crossing tactics selected in the decision-making process by militants. It does so by explaining how border barriers lead to another decision-making process by militants when they decide to use new tactics. Fourth, the threshold approach considers trajectories as geographical routes followed by border crossers but this chapter posits that routes form one example of border-crossing tactics; militant border crossers for instance may choose to switch the type of their attacks in order to achieve their goals. This expands the applicability of the threshold approach by viewing trajectories as part of a more general concept, namely the tactics of border crossings. Fifth, this chapter applies the threshold approach to other kinds of border crossers, that is, cross-border militants. By doing so, this chapter uses the literature of political violence. This helps in developing the threshold approach by learning from this literature.

By studying the case of the Gaza barrier, this contribution addresses questions of relevance to several other cases. Do physical barriers on borders stop cross-border attacks? What are the actual outcomes of barriers on borders? What causes militants to decide to attack and change their attack tactics? This research examines the outcomes of barriers in light of the frequency of attacks and their nature.[2] While other variables, such as lethality of these attacks are important,

---

2   Nature of attacks means type of attack or attack tactic.

the number and nature of the attacks are more telling about militants' decision-making and are important to the security of the concerned states.

This chapter employs a single case study as a plausibility probe to determine the theory's validity and applicability. Plausibility probes are 'preliminary studies on relatively untested theories and hypotheses to determine whether more intensive and laborious testing is warranted' (George and Bennett 2005, 75–6). Although it is not the only Palestinian area involved in a conflict with Israel, in the past years, Gaza attracted increasing political, public and media attention partly because of its intense involvement in the conflict. This case is selected for two main reasons. First, Gaza is an extreme case, because it has one of Israel's strictest border controls with barriers as intensified forms of border policing. A study of the dynamics between barriers and borders, the (lack of) local security cooperation, presence of militants' institutions and escalation of the conflict should thus offer clear, discernible results (for similar studies see Van Evera 1984). Second, the Gaza-Israeli barrier was completed nearly twenty years ago, which offers sufficient time to examine its effects. Thus, the Israeli conflict with the Palestinians in the Gaza area is multifaceted because it offers many observation points although it is a single case. Because this study is an initial probe, into a newly identified causal mechanism, however, and because it involves only one case, the possibilities to generalize from its findings should be considered carefully.

The first section of this chapter examines present contributions on border security and barriers. The second and third sections define barriers and borders and explain how the threshold approach applies to the interaction between barriers and militants as border crossers. The fourth section establishes the relationship between presence or lack of local security cooperation and militant activities and explicates the effect of barriers on the strength and weakness of militant groups' supply institutions. Then, it examines the effect of these institutions on militants. The fifth section considers the Gaza barrier case as a preliminary probe of the plausibility of the chapter's theory.

Border Security

The number of contemporary states that built barriers by fencing or walling off borders has significantly increased in the past two decades (Jones 2012; Rosière 2011). To illustrate, the US is erecting fences on its southern borders to stop undocumented immigration, drug trafficking and 'terrorism' (Andreas 2000), Greece started a wall on its borders with Turkey to stop undocumented immigrants (*Telegraph*, 3 January 2011), Spain fenced off its enclaves of Ceuta and Melilla to stop undocumented migrants, drug trafficking and 'terrorism' (Saddiki 2010; Ferrer Gallardo and Espiñeira in this volume), and India launched a barrier project on the borders with Bangladesh to stop cross-border violence (Jones 2012). States who build barriers assume these will protect their sovereign territory by making it easy to defend (Jervis 1978, 194) and deter external threats. Barriers are often supplied with complex systems of surveillance and weapon technologies, and towers that may extend beyond the border into a web of control mechanisms on one or both sides of the border.

Between 2000 and 2010, 25 barriers were initiated and substantially fortified by different states, including Botswana, Greece, the US, India and Saudi Arabia. The number of barriers particularly increased after 9/11 and wall-building states justified fortification of their borders in the context of the 'global war on terror'. Unlike classical state enemies, the new enemy, presented by militant groups or non-state actors, is repeatedly depicted as amorphous, evil and uncivilized. Thus, in the face of an enemy that can be anywhere, states can take extreme measures to protect their population and territorial sovereignty, including barrier building (Jones 2012).

Among the 25 physical border barriers built in the past decade, three were unilaterally built by Israel on its borders with Lebanon (2000), the West Bank (2002), and Egypt (2010). In the early 1990s, before the 'global war on terror', Israel signed the Oslo Accords with the Palestinian Liberation Organization (PLO) and created the Palestinian Authority (PA).[3] Two Palestinian militant groups, Hamas and Palestinian Islamic Jihad (PIJ), launched a wave of suicide bombings in objection to the accords. In a speech in January 1995, the late Israeli Prime Minister Yitshak Rabin proposed building a barrier to ensure peace and create separate Palestinian and Israeli territories, adding that the barrier would stop Palestinian attacks in Israel. A few months later, Israel finished its high-tech and militarized fence on the Gaza Strip[4] borders (*Jerusalem Post*, 24 March 1995; *Guardian*, 24 April 2007; Jones 2012, 87).

The academic literature devotes little attention to theories on the role of physical barriers in cross-border security. The earliest discussion of the issue briefly appeared in the defence–offence debate when Jervis (1978, 194) posited that barriers enable states to diminish attacks by raising the costs of infiltrations because of the need for extra information, logistics and technology. They also lower the efficiency of attackers' actions and give the defenders time to prepare.

With the surge in militant groups' attacks and the development of literature discussing such attacks, new contributions emerged that built on the above understanding. These works based their arguments on one of two logics. The first is identified as the 'punishment logic', according to which barriers facilitate punishment of militants, which leads the latter to stop their attacks. The second is the 'logic of denial' as barriers deny the attackers' access to their targets. By this logic barriers have the best prospects for reducing the frequency and intensity of militant groups' attacks (Kaufman 1996, 1998; Makovsky 2004; Pape 2003; Dutter and Seliktar 2007, 438).

A few scholars disagree with these views. Andreas (2000, 9–15) contends that rather than punishing or denying militants, barriers stimulate innovative behaviour since infiltrators may shift their border-crossing routes, change the nature of their activities and methods of transportation, and morph their organizations. Jones (2012) also demonstrates that rather than providing security, barriers institutionalize legal, political and economic distinctions. This also fits in with the threshold approach as it explains the flexible nature of border-crossers' routes and their changeable destinations. This flexibility leads to a new phase in the process of decision-making by border crossers (Van der Velde and Van Naerssen 2011, 221).

While these arguments are important contributions to the understanding of barriers on borders, they do suffer from several shortcomings. First, the offense–defence idea is simplistic and has been subjected to lengthy criticisms, particularly that offence and defence are relative to strategy and technology (Van Evera 1984). Also, these claims have state-centric perceptions of world politics and should not be extended to state-militant groups' relations without deeper exploration. Unlike states, militant groups are tactically flexible in the sense of the ease of changing operational fighting techniques usually due to their non-hierarchical structures (Horowitz 2010, 36). This contribution builds on the above literature by offering an empirical study on militants' response to barriers.

Scholars agree that militant groups often are motivated by a long-term fundamental goal, such as self-determination, independence and policy changes, as well as short-term instrumental and

---

3   The PA is an interim Palestinian self-government institution created by the Oslo Accords.

4   The Gaza Strip (or Gaza in other places in the text) is a sandy 62 sq. km area on the Mediterranean, to north-west of the Negev. It is bordered by Israel from all sides except in the south, where it borders Egypt (CIA 2011). The biggest city is Gaza; other population centres are Beit Hanun, Beit Lahya, Jabalya, Deir Balah, Khan Yunis and Rafah.

operational goals such as bombings, rocketing and tunnelling (Dutter and Seliktar 2007, 431; Kydd and Walter 2002; Bloom 2004) that pave the way to the long-term ones. The persistence of the long-term goals keeps militants motivated to attack despite all obstacles and deterrence policies even if one specific attack does not return much gain.

Yet, while this motivation to attack is important, two additional conditions are necessary to incentivize militants to attack and to determine what tactics to adopt. The first is absence of local security cooperation with the barrier builder, while the second is retention of military capabilities. If militants keep a continued flow of military supplies and know-how after the construction of the barrier, they will be able to continue attacking and to diversify their attacks. Since these two conditions vary across time and place, they can help us understand how the security outcomes of barriers are not guaranteed and vary as well.

Barriers and Borders

Barriers are a subset of borders. They serve the same positional and administrative functions as any border. Positional functions include actual demarcation and delimitation of the borders. The administrative tasks include prevention of border jumping, suppression of smugglers and contraband, extraction of customs duties, and the regulation and facilitation of legal crossings (Prescott 1987; Gavrilis 2008). Borders both exist and are imaginary. While there is much imagination about what a state is, its identity, and its demarcations, there is something tangible about the idea of a border. It is tangible in the sense of wars fought over lines separating states, though they were possibly fought for an imagined idea. It is tangible in the sense of the border patrols, checks and laws and regulations.

In this sense, a barrier visualizes the imagined. Barriers are distinguished from borders by being hardened lines. They make state sovereignty more visible and more distinguishable. Nevertheless, a barrier also implies the failure of a 'normal' border to fulfil its functions. In the post-9/11 era, a barrier is an attempt by a state to clearly define its own territory and defend its population against 'global terrorist networks', the 'ungoverned spaces' that nurture terrorism and the 'uncivilised' (Jones 2012, 12 and 32). Barrier-building states highlight the distinction between the self and the other as a distinction between civilization versus barbarianism, good versus bad, order versus disorder and the known versus the unknown. Barriers signify the prevention of movement of people rather than just serving the function of controlling it (32).

Borders have served different purposes through history. Traditionally, borders were markers of military defensive lines of states as the vast majority of interstate conflicts were about territorial defence and conquest (Keegan 1993). Gilpin (1981, 23) argues that states had incentives to pursue the 'conquest of territory in order to advance economic, security, and other interests'. Under this conception, borders are strategic lines to be militarily defended or breached. State survival is based on the deterrent function of borders against military incursions by other states (Waltz 1979). In fact, according to many scholars, states themselves are formed through these wars that constantly shape and reshape their borders (Tilly 1990). Consequently, the rulers of a territory would not allow the army of a neighbouring state to cross their borders. Some of these defensive borders are still present in our world today and are exemplified by, for example, the demilitarized zone on the Korean peninsula.

After World War II, interstate wars decreased and borders were rarely contested. This is partly the result of growing international respect for what Zacher (2001) calls the 'territorial integrity norm', under which borders became markers of sovereignty. Under this norm of fixed borders, and within the UN system, states agreed that borders separate their sovereign regimes. A border has become a line where a legal and administrative system ends and another begins (Elden 2009; Atzili 2006; Murphy 1996). Even under this system, however, some states are weaker than others, which results in cross-border economic inequalities that motivate people to move across the borders in search for better conditions. The purpose of borders became prevention of unauthorized movement of people to consolidate the state authority. The 'global war on terror' intensified this process and consequently, many states built barriers.

To the states that build them, barriers are defence structures. As unmovable structures, walls and fences may be presented as the most defensive systems (Sterling 2009, 3). However, this perception of barriers is not universal. Particularly if these barriers are unilaterally constructed, people on the other side may consider them offensive and/or restricting access to resources and opportunities.

Militant Groups as Border Crossers and Their Tactics

A tactic is the method, route or strategy border crossers decide to employ to reach their goal or destination. A tactic could mean using a certain type of transportation, a certain road, disguising or forging documents or even employing a certain attack type. While tactics can be chosen at an early phase of the decision-making process, they can be modified as militants obtain new information. In this case, over time one can trace changes in the patterns of those militants' attacks.

Like other border crossers, militant groups go through a decision-making process at least at two levels. First, they decide whether or not to attack across the border, and second, they decide on what tactics to employ. In both decision processes, they weigh their options and make a calculus to achieve a goal or a set of goals. Yet they are not rational in the (neo)classical sense. Rather, they are boundedly rational because they do not have access to complete information about all options and they have to choose among available alternatives.

When militants decide to cross the borders, they do not do so in search for jobs or better welfare opportunities, as other border crossers do. Rather, they cross primarily in order to perpetrate attacks. They weigh their options (for instance to attack or not to attack) and try to behave instrumentally to maximize the attainment of their goals. The second decision takes place when they examine what tactics they can follow to fulfil their goal(s). These tactical decisions are defined by their military capabilities, knowledge and skills. One of the tactics is choosing specific trajectories.

These trajectories are another pillar in the threshold approach and are introduced to shed light on the flexibility of the routes, destinations and settlement of migrants. The approach posits that these trajectories are not fixed but choosing them is rather another phase in the decision-making process undertaken by border crossers. Often, migrants, while en route to their initially chosen destination, acquire new information about the route of their trip. Based on this new information border crossers may modify their routes and perhaps even destinations to achieve their goal (Van der Velde and Van Naerssen 2011, 221). Routes and trajectories, seen as a subset of tactics of border crossing, are examples of the tactics employed by border crossers to reach their target.

Conventional wisdom suggests that barriers on borders dissuade militants from crossing the borders to perpetrate attacks on the other side. The following sections, however, discuss the importance of the context in which each barrier is built and the concerned militants operate. In doing so, some light will be shed on the decision-making process carried out by militants on the other side and how barriers might cause increased cross-border violence in areas where militants can locally act freely and are resourceful.

Barriers, Security Cooperation, Militants' Supply Institutions and Cross-Border Attacks

Barriers, Local Cooperation and Militant Activity

In his seminal work on violence in civil war, Kalyvas (2006) indicates that availability of information is a central determinant for armed forces that engage in selective or indiscriminate violence. In areas where the enemy has full territorial control, the opposing actor resorts to indiscriminate violence. Yet, if local information is available, this actor can resort to selective violence. For instance, Israel resorted to an indiscriminate military campaign in Gaza late 2008–early 2009 because Hamas had a full territorial control over the strip and Israel did not have enough information about violent activism in the area. In other instances, Israel had so much information from Palestinian informants that it resorted to selective punishment and the assassinations of militants, such as the assassination of Yahya Ayyash of Hamas in 1996. While this research has a different focus and a different interest, the logic and the implications behind Kalyvas's argument are very instructive.

Of relevance here is Kalyvas's independent variable: information. Local residents providing information to local or external governments is one sort of security cooperation. However, this kind of information is not the only nexus of cooperation. Another example of cooperation is having a local security apparatus that shares security information and cooperates with the state that built the barrier by cracking down on militants within their defined area of control. Local punishment of militant activity is vital for a barrier to be able to suppress militant activities in general and attacks in particular.

Local forces that cooperate with the barrier builder on security issues can assist in creating security for the neighbouring states by detecting strongholds of militants in their towns. They can arrest and crack down on these militants. If a barrier is accompanied by such cooperation, it can be successful at stopping attacks. According to analysts, having local security forces cooperating with the barrier builder is the most effective way to paralyze militants (see Zeidan and Rahman 2000). In this context, rationally bounded militants will likely decide not to attack and not to further develop their capabilities. A barrier without such cooperation, however, leaves the militants active and is doomed to fail in stopping attacks. If militant activities are not suppressed locally, militants can enhance their capabilities and perpetrate attacks.

Thus, the efficiency of a barrier ultimately depends on the local context in which militants operate. If militants are trying to operate in an area where the local authorities have an agreement with the state that built the wall to stop militant activities, militants usually find their activities to be too costly and will likely decide not to attack. Alternatively, where there is no security cooperation of this kind, attacks are expected.

Institutions and the Possibility to Act Freely

Once militants assess that they are free to act, they look for supplies for their attacks, such as weapons, raw materials, funding and training. With a barrier that limits their access to such resources, militants resort to local networks. If these networks are available, barriers indirectly perpetuate and strengthen them to become institutions that can provide continued supplies through linkages with other militant groups that guarantee continued attacks and the introduction of new tactics (Desouza and Hensgen 2007, 598–600).

Militants' supply networks often establish linkages among different militant groups. Horowitz (2010) refers to the importance of these linkages by demonstrating that knowledge acquired through connections with other groups is a key factor to the introduction of new tactics by militant groups.

These supply networks can even form institutions that often emerge from smuggling networks that initially existed for criminal or other reasons. When access to sources is blocked by a barrier, militants resort to these institutions for supplies. These institutions help militants cope, survive, sustain and diversify their activities despite the circumstances (Andreas 2008, 18). Barriers on borders do not necessarily create these institutions but they do strengthen the existing ones.

These supply networks are usually dubbed criminal by states and as a result take on a clandestine nature. These networks appeared in the history of border areas when the state carried out law-enforcement campaigns against illegal trafficking (Andreas and Nadelman 2006, 4–5). When militants decide to use these economic networks for military purposes, politics gets mixed with the economies of the existing networks as cooperation becomes mutually beneficial for militants and illegal economic entrepreneurs. Because they provide necessary resources for militants and revenues for illegal entrepreneurs, illegal economic and violent webs intertwine (197).

Over time, as the demand for these networks rises, they may become organized to represent institutions in the legal world (see Gavrilis 2008, 1523; Tilly 1990; Anderson 1996, 1). For instance, these networks may become staffed with people, such as service and goods providers and consumers embedded in networks of selling, buying and transferring knowledge and materials. In some cases, use of these networks becomes an exchange for mutually established fees, paid to members who own the networks. At a later phase, even network owners may have to pay taxes to a higher authority to sustain their business in another attempt to regulate the striving interaction.

To summarize, building barriers as a deterrent state policy surely affects attacks. However, this effect is neither always direct nor actually deterrent. A deterrent policy can influence a militant group's choice of operations by either affecting their resources or the relative costliness of different kinds of attacks (Enders and Sandler 2002). Yet, militants with a drive to attack find themselves fenced off and unable to sustain their attacks due to the obstacles created by the barrier. If they are not repressed locally, they will likely decide to use new tactics depending on available materials and skills. If available, militants resort to their networks to continue their attacks and shift to new tactics if old ones are not successful anymore. Cooperation among different militant groups through their institutions contributes to an increase in militant capabilities (Desouza and Hensgen 2007, 593). Thus, there is a continuation of attacks despite the barrier. Furthermore, if militants access new military skills and materials, they may even introduce new attack tactics. In both cases, rather than stopping attacks, militants may even intensify their attacks.

Militants will be able to increase their attacks and diversify their tactics when they have abundant resources and if there is a lack of local security cooperation. Militants in areas with no security cooperation between local forces and the barrier builder can operate freely and obtain military resources to launch attacks. Once motivated, militants are free to act locally, but they will look for new venues for resources when they are fenced off. If they find supply networks, these networks get institutionalized over time and will provide supplies that allow a shift to new attack tactics that can be implemented despite the barrier.

Thus, four claims regarding the role of barriers on borders in motivating violence sum up the above. These claims shed light on the effect of local punishment and militant institutions as intervening variables. *Claim 1*: Barriers in areas with no security cooperation between local authorities and the wall builder leaves militant activity unpunished and so, militants can act freely. *Claim 2*: When militant activity is possible, barriers on borders can perpetuate and strengthen militants' supply institutions. *Claim 3*: Militants' supply institutions can create conditions that enable militants to continue attacking despite the barrier and even increase the frequency of their attacks. *Claim 4*: Barriers on borders, accompanied by the existence of militants' supply institutions, could cause a shift to new tactics.

The Gaza Case

The Gaza case illustrates the mechanism through which building barriers on borders can increase attacks. This section gives a background on the security cooperation between the Palestinian Authority (PA) and Israel. It shows how the Israeli barrier perpetuated and indirectly strengthened militants' supply institutions when Israel and the PA did not cooperate. It also demonstrates how the combination of absence of security cooperation and presence of supply institutions triggered mechanisms to escalate the violence. The observations are based on descriptive statistics of the Gaza attacks between 1990 and 2010. These statistics are derived from a larger dataset created by the author in 2010–11 covering all Gaza attacks.[5]

Prior to 2000, several militant groups operated from Gaza. They included the armed wings of Hamas, Fatah, Palestinian Islamic Jihad (PIJ), and the Popular Front for the Liberation of Palestine (PFLP). After 2000 and particularly after 2005, new (and smaller) groups, such as the Popular Resistance Committees (PRC), Abu Rish Brigades, and the Army of Islam, proliferated. These militants have different ideological frameworks but they agree on the larger goal of ending the Israeli occupation and establishing Palestinian national independence (Al Watan Voice 2012; Al-Qassam 2007).

Gaza went through different phases between 1990 and 2010 during which militant activities were suppressed at times and permitted at others.[6] When militants were not punished locally, they opted to attack Israel and improve their military capabilities. Over time, militants gained experience and acquired more information about the situation on the border. As information was updated and their skills improved, militants decided to change their attack tactics. Thus, after Israel built its barrier, militants made choices about what attack tactics to use based on their resources and skills and on whether they were free to act.

Between 1990 and 2010, the relationship between the Gaza barrier and attacks went through different phases characterized by intensification of policing strategies on the barrier on the one hand and the availability of militant supplies and presence or absence of security cooperation with Israel on the other. The following sections will explain how the interplay between the internal context of militants and the barrier affected the outcomes, in the form of attacks from Gaza.

1990–2000: From Direct Military Occupation to Self-Rule

In 1993, Israel and the PLO signed the Declaration of Principles, indicating a new era after thirty-six years of direct Israeli military occupation of the Gaza Strip. A year before the agreement, Israel deported members of military groups to southern Lebanon, where Hezbollah was operating. Many scholars of political violence trace the beginnings of linkages between Palestinian armed groups, particularly Hamas, to that time (Horowitz 2010, 37). Through this encounter, Hamas and other militants learned about suicide bombings.

---

5 Attacks data cover suicide bombings, rockets, tunnel attacks and roadside and settlement attacks, collected from world newspapers on *LexisNexis* (English), militant groups' websites (Arabic) and the Israeli Foreign Ministry website (English). The dataset also includes barrier data collected from world newspapers on LexisNexis, OCHA websites (http://www.unocha.org/ocha-websites, accessed 17 October 2014) and the Applied Research Institute-Jerusalem/Society (http://www.arij.org, accessed 17 October 2014).

6 Yasser Arafat, head of the PA until 2004, cracked down on Hamas and PIJ in the 1990s but was believed to encourage military operations against Israel after the failure of the Camp David talks in 2000 (BBC News 2003).

The effect of the Hamas–Hezbollah link was seen in the Israeli streets. In 1993, as peace negotiations were proceeding between the Israeli and Palestinian representatives, PIJ and Hamas militants decided to launch their first suicide bombing originating from the Gaza Strip. Other militants, however, preferred political gains through peace and decided not to attack. Some studies explain the attacks by Hamas and PIJ as protests against the peace process and an attempt to spoil the agreement (see Kydd and Walter 2002). Others like Pape (2005) argue that foreign occupation and religious differences between the concerned groups and the occupying state drove the suicide bombing.

When the Oslo Agreement was signed, internal administration was transferred to the PA headed by the late Yasser Arafat (Efrat 2006, 42). By mid-1995, Israel withdrew its armed forces from the Gaza Strip and redeployed them on its borders. Israel maintained control over the borders, but coordinated crossings with the newly established Palestinian security apparatus.

The agreement established a Palestinian police force to guarantee law and order in the territories and created a joint Israeli-Palestinian Liaison Committee to deal with issues requiring coordination and other matters of common interest and disputes (BBC News 2001). After settling in Gaza, Arafat's forces demonstrated commitment to the agreement with Israel. After each suicide bombing in Israel, Palestinian security forces rounded up militants. On 11 November 1994 for example, the Palestinian police raided homes of suspected militants and clashed with them after an attack on an Israeli settlement. According to media reports, within a few days of the attack, more than a hundred militants were arrested. The Palestinian security forces continued to punish militant activities until 2000 (BBC Summary of World Broadcast 1994).

Security cooperation between Israel and the PA in the 1990s was a major blow to the activities of Hamas and PIJ militants. Arafat's security apparatus harshly suppressed these groups, and Israel made it more difficult for them by building a barrier around the Gaza Strip. Given the cooperation of the PA with Israel, Hamas and PIJ could not train freely or even import supplies. Thus, Gaza militants at the time had very limited access to external military resources and minimal freedom of movement and as a result, militants decided not to attack during that period.

At that time, the tunnel network already existed but was not used militarily. These tunnels dated back to 1982 when Israel fortified its border with Egypt, separating relatives from Rafah. Families started digging at the closest points on both sides (*New York Times*, 24 January 2006). In the 1980s and 1990s, the tunnels were used to smuggle people and illegal commodities. As fees were charged, profit became a new factor that enhanced the sustainability of tunnels. Gaza underground traffic prospered, but mainly for economic reasons (*Boston Globe*, 25 January 1995).

At that time Israel launched its fence at the borders with Gaza in response to a wave of Palestinian suicide attacks. By mid-1995, the Israeli fence completely sealed off the Gaza Strip from Israel. It consisted of a 300 m buffer zone, fixed surveillance cameras and touch sensors and electric wires. Israeli officials claimed that this barrier would protect Israel and its population from the Gaza attacks. If these claims were true, attacks should stop or at least decline after building the barrier.

Figure 10.2 depicts the number of Gaza attacks in relation to the construction of the barrier from 1990 to 2010. It also shows the attack tactics militants decided to use. The data indicate that they relied on two types of attacks: suicide bombings and settlement and roadside attacks. Around mid-1995, the Gaza militants stopped their attacks after the construction of the barrier. Attacks from Gaza declined from about 25 in late 1993 to zero in late 1997. Arguments for the efficacy of barriers preventing militant attacks often use this period as a key example. However, this cessation was short-lived and verification of this claim would depend on this effect continuing into later phases of the barrier's existence.

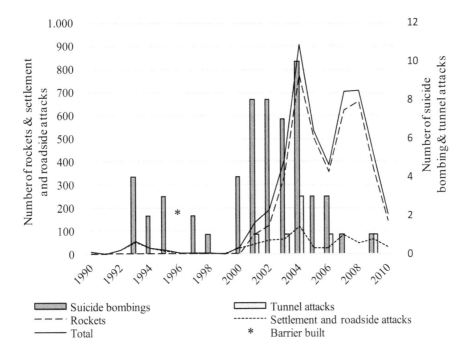

Figure 10.2   Attack frequencies and tactics, 1990–2010

Note: Suicide and tunnel attacks are presented on the secondary axes in bar charts to distinguish them despite their relative low frequency.

Source: the author.

2000–2005: Rising Tensions

In 2000, the Palestinian Al-Aqsa Intifada broke out[7] and Israel experienced many attacks from Gaza, not only by Hamas and the PIJ but also by the PRC and the PFLP, and the Aqsa Martyrs Brigades, the armed wing of Fatah, who was party to the negotiations with Israel. This phase is characterized by two interesting and relevant dynamics. First, while the Israeli barrier remained a constant, the security cooperation between the PA and Israel waned as a result of failure of negotiations in Camp David in 2000. As the levels of repression by the Palestinian security forces decreased dramatically, Gaza militants were left free to operate locally, launch attacks against Israel and receive supplies from abroad. In the same year, Israel extended its barrier system by increasing the buffer zone to one kilometre and adding new technologies including continuous videotaping and surveillance equipment. In addition to new cameras, Israel constructed high-tech overlapping observation posts that enabled soldiers to monitor about six km day and night. It also added improved high-tech sensors to the fence.

---

7   Clashes between the Palestinians and the Israeli escalated after a visit by Ariel Sharon, who later became Israel's prime minister, to the yards of Aqsa Mosque, which is one of the holiest sites for Muslims.

Second, Gaza militants began to use tunnels to obtain raw materials and training from Hezbollah and Iran because they found themselves in an escalating conflict with Israel (*Weekend Australian*, 7 April 2007). However, the barrier continued to impose many obstacles to the execution of attacks. Seeing an opportunity in using the smuggling tunnels to import military materials and bring foreign trainers to Gaza or having militants travel abroad through the tunnels for training, militants decided to use these networks.

Thus, with the outbreak of the Aqsa Intifada in 2000, Gaza tunnels took on a political nature as they supplied militant groups with weapons, other military materials and training that then were used to attack Israel. The smuggled materials were used for rocket making. Sometimes tunnels were also used to attack Israeli military bases just outside Gaza. Although these tunnel attacks were few, they were unique as they required more preparations and planning than the earlier attacks.

The pattern in the Gaza attacks between 2000 and 2005 reveals a counter-intuitive picture for supporters of the idea of barriers being successful deterrent strategies. With security cooperation between Israel and the PA frozen, and militant materials flooding through the Gaza tunnels, militants increased attacks from 30 in 2000 to about 950 in 2004 (see Figure 10.2). Gaza militants intensively used suicide bombings, rocketing, tunnel attacks and settlement and roadside attacks. All in all, this period witnessed one of the most diverse patterns in cross-border attack tactics.

2005–2010: The Siege

In late 2005, Israel had unilaterally withdrawn from the strip and intensified its barrier policing by even adding more technologies, such as aerial surveillance, ground sensors and distantly controlled weapons. It also expanded the buffer zone to between two and nine kilometres. A year later, Hamas won the majority of seats in the Palestinian elections. In objection to the results of the elections, Israel and the Western countries imposed a tight siege on Gaza.

During the Gaza blockade the tunnel system flourished. Different commodities and weapons were smuggled into Gaza. Tunnels provided jobs for many unemployed who worked in Israel before. By 2009, about 12,000 Gazans worked in the tunnels and tunnel-related business. Furthermore, militant groups funded themselves through local taxes imposed on the use of the tunnels for commercial purposes. Hamas for example, started imposing taxes after taking over the Gaza Strip in mid-2007. They also provided protection for the diggers and tunnel business in general (*Washington Post*, 18 June 2008). With the new realities on the ground, tunnels have become the Gaza lifeline and its militants' main venue for weapons and trainings. With Hamas and other militant groups acting freely in Gaza, and with more military supplies flooding through the tunnels, militants continued to launch attacks (Figure 10.2) in defiance of the barrier that got more sophisticated over the years. In this period, attacks continued without any periods of cessation, albeit fluctuating in frequency.

Between 2005 and 2010 militants relied on rockets and settlement and roadside attacks. Gaza militants claim that the shift in tactics largely to rockets was a decision they made independently from the Israeli barrier-building policies. On its official website for example, Hamas describes the development of its arms over time and claims that it decided to rely on rockets after 2005 and give up suicide bombings, because it managed to develop the range and accuracy of their rockets enough to inflict harm on Israel, striking Israeli towns and settlements beyond the border area and crossing points, without having to cross the border (The Media Center 2003).

The fluctuation in the number of attacks between 1990 and 2010 correspond closely to the presence or absence of security cooperation between Israel and the local Palestinian authorities. This highlights the importance of this variable for a barrier to be successful as a security tool. Thus, it seems that the argument that barriers alone are responsible for stopping attacks has no strong empirical support from the Gaza case.

The data also reveal that the number of attacks fluctuated after 2000. For example, when there was a major Israeli military operation, attacks increased. In 2004, Israel launched several ground operations, especially in Rafah (BBC Monitoring 2004). Later in the same year, Israel executed several ground operations in the strip to fight rockets (*Daily Telegraph*, 18 October 2004). In 2006, Israel launched a wide-scale military operation after militants kidnapped the Israeli soldier Gilad Shalit (*Guardian*, 6 July 2006). In late 2007, Israel executed a big incursion accompanied by tank and airborne strikes (*Montreal Gazette*, 12 December 2007). In late 2008–early 2009, Israeli launched its operation 'Cast lead'. During each of these operations, attacks increased.

The declines followed a similar pattern. In late 2005 Israel unilaterally disengaged and evacuated its Gaza Strip settlements (BBC News 2005). Fawzi Barhoum, Hamas spokesman, commented: 'This is the victory of Hamas against the occupation' (Bengali 2009) and the group's attacks went down. In mid-2008, Israel offered a ceasefire deal with Hamas (*Jerusalem Post*, 13 May 2008). In June, Israel and Hamas started a six-month truce (*Guardian*, 19 June 2008). Right after 'Cast Lead', Israel and Hamas announced a ceasefire that explains the decline in attacks after January 2009 (BBC Monitoring 2009). In all of these cases, attacks across the Gaza border declined.

From these data, we can conclude that the construction of the barrier in 1995 did not stop militant attacks, and in fact, with the eruption of the Aqsa Intifada in 2000, increased attacks. The key explanation for this is that the security cooperation between the PA and Israel stopped and tunnels started to be intensively used by Gaza militants. Additionally, new attack tactics were introduced: tunnel attacks and rocket attacks. Importantly, both the skills and materials needed for those tactics, which the Gaza militants until then had lacked, were provided by using the tunnels.

Conclusion

This chapter focused on hardened borders, which are meant to stop cross-border mobility. It addressed the effect of these barriers on cross-border movement and questioned whether they really achieved security. In the case of Gaza, the construction of the Israeli barrier at first seemed successful witnessing a significant decline in the number of attacks. Yet, after 2000 the number increased again, as security cooperation between the PA and Israel stopped and militants became free to act. Additionally, militants decided to use Gaza tunnels to supply their activity. After 2005, militants shifted to new tactics. Because the Gaza barrier completely encircled the Gaza Strip since the day of its construction, attacks fluctuated in a manner that cannot be explained by the mere construction of the barrier. This finding supports the argument that barriers alone fail to deter attacks.

Instead, the above case demonstrates that hard barriers achieve security only when there is security cooperation between the barrier builder and local authorities. If not, militants will be able to look for resources to supply their activities and introduce new tactics. Barriers do of course create constraints on the militants attempting to attack the state that built them. However, if these barriers are not supported by cooperation of authorities on both sides of the border, they may lead to the unintended consequences of aiding militants in developing and using networks to support the flow of materials and knowledge needed for executing attacks. In the Gaza Strip it pushed the militants to develop the tunnels networks in such a way that they became institutions.

While much can be learned from this case, caution is urged when attempting to generalize from a single case study. A carefully designed study that compares the Gaza case with other past or present cases is needed in order to confirm the conclusions presented here. However, the link that is drawn between the Gaza barrier, security cooperation, the tunnel networks and attacks does aid in understanding the increased intensity of Gaza attacks at certain times as well as the shift to new tactics. It is also instructive when trying to understand the changing tactics or trajectories. What stands out here is how resourceful border crossers can be in choosing their tactics and trajectories.

With barriers becoming a common contemporary phenomenon due to the threat posed by militant groups, this chapter offers an empirical study of the effect of one of the most intensively policed barriers in the world. By employing the threshold approach of Van der Velde and Van Naerssen (2011) this chapter attempted to answer important questions related to the security of modern states. The approach offers a framework that assists in understanding complex phenomena, such as cross-border violence. In their efforts to cross the border, militants decide on, among other things, the destination of their attacks, and the employed tactics and, just like migrants, militants switch their border-crossing tactics and may change their destinations.

This chapter contributes to the threshold approach by adding the element of border barriers and their effect on decision-making militants. Moreover, it explains how attack tactic selection by militants is part of a decision-making process and how changing these tactics is, in fact, another decision-making process. Finally, this chapter expands the scope of the threshold approach by presenting trajectories as one category of overarching border-crossing tactics, which makes the approach applicable to a wider range of cross-border dynamics.

References

Agnew, J. 2009. *Globalization and Sovereignty.* Lanham, MD: Rowman & Littlefield.

Al-Qassam. 2007. "Hamdan: We Will Persist in Resistance till the End of Occupation, Restoring the Palestinian Rights and the Return of the Palestinian Refugees." *Al-Qassam*, 15 April. Accessed 30 April 2011. http://www.qassam.ps/news-297-Hamdan__We_will_persist_in__resistance_till_the_end_of_occupation_restoring___the_Palestinian_rights_and_the_return_of_the_Palestinian_refugees.html.

Al Watan Voice. 2012. "Arouri: Hamas is a Resistance Movement against Occupation and Talking about Prisons for the Movement in the West Bank is Nonsense." *Al Watan Voice*, 23 September. Accessed 30 September 2012. http://www.alwatanvoice.com/arabic/news/2012/09/23/319170.html.

Anderson, M. 1996. *Frontiers: Territory and State Formation in the Modern World.* Cambridge: Polity Press.

Andreas, P. 2000. *Border Games: Policing the U.S.-Mexico Divide.* New York: Cornell University Press.

_____ . 2008. *Blue Helmets and Black Markets: The Business of Survival in the Siege of Sarajevo.* New York: Cornell University Press.

Andreas, P., and E. Nadelman. 2006. *Policing the Globe: Criminalization and Crime Control in International Relations.* Oxford: Oxford University Press.

Atzili, B. 2006. "When Good Fences Make Bad Neighbors: Fixed Borders, State Weakness, and International Conflict." *International Security* 31 (3): 139–73.

Baldwin, R.E., P. Martin, and G.I.P. Ottaviano. 2001. "Global Income Divergence, Trade, and Industrialization: The Geography of Growth Take-Offs." *Journal of Economic Growth* 6 (1): 5–37.

BBC Monitoring Middle East. 2004. "Israeli Army Completes Rafah Operation, Houses Razed."
12 August.

_____ . 2009. "Hamas Government to stop Rocket Firing at Israeli Towns." 22 May.

BBC News. 2001. "Text: 1993 Declaration of Principles." *BBC News*, 29 November. Accessed
12 September 2012. http://news.bbc.co.uk/2/hi/in_depth/middle_east/israel_and_the_
palestinians/key_documents/1682727.stm.

_____ . 2003. "Palestinian Authority Funds Go to Militants." *BBC News*, 7 November. Accessed 13
June 2012. http://news.bbc.co.uk/2/hi/middle_east/3243071.stm.

BBC News 2005. "Israel Completes Gaza Withdrawal." *BBC News*, 12 September. Accessed 13
October 2014. http://news.bbc.co.uk/2/hi/4235768.stm.

BBC Summary of World Broadcast. 1994. "PNA Rounds up more than 100 Islamic Jihad activists
in Gaza crackdown." 14 November.

Bengali, S. 2009. "After Israeli Withdrawal, Hamas Asserts Victory in Gaza." *Christian Science
Monitor*, 22 January. Accessed 10 September 2012. http://www.csmonitor.com/World/Middle-
East/2009/0122/p04s01-wome.html.

Bloom, M.M. 2004. "Palestinian Suicide Bombing: Public Support, Market Share, and Outbidding."
Political Science Quarterly 119 (1): 61–88.

CIA (Central Intelligence Agency). 2011. "The World Factbook: Gaza Strip." Accessed 30 April.
https://www.cia.gov/library/publications/the-world-factbook/geos/gz.html.

Desouza, K., and T. Hensgen. 2007. "Connectivity among Terrorist Groups: A Two Models
Business Maturity Approach." *Studies in Conflict & Terrorism* 30 (7): 593–613.

Dutter, L., and O. Seliktar. 2007. "To Martyr or Not to Martyr: Jihad is the Question, What Policy
is the Answer?" *Studies in Conflict & Terrorism* 30 (5): 429–43.

Efrat, E. 2006. *The West Bank and Gaza Strip: A Geography of Occupation and Disengagement.*
London: Routledge.

Elden, S. 2009. *Terror and Territory: The Spatial Extent of Sovereignty*. Minneapolis: University
of Minnesota Press.

Enders, W., and T. Sandler. 2002. "What Do We Know about the Substitution Effect in
Transnational Terrorism?" Accessed 28 March 2010. http://www.utdallas.edu/~tms063000/
website/substitution2ms.pdf.

Gavrilis, G. 2008. *The Dynamics of Interstate Boundaries*. Cambridge, MA: Cambridge University
Press.

George, A.L., and A. Bennett. 2005. *Case Studies and Theory Development in the Social Sciences*.
Cambridge, MA: MIT Press.

Gilpin, R. 1981. *War and Change in World Politics*. Cambridge, MA: Cambridge University Press.

Horowitz, M.C. 2010. "Nonstate Actors and the Diffusion of Innovations." *International
Organization* 64 (1): 33–64.

Jervis, R. 1978. "Cooperation under the Security Dilemma." *World Politics* 30 (2): 167–214.

Jones, R. 2012. *Border Walls: Security and the War on Terror in the United States, India and Israel*.
London: Zed Books.

Kalyvas, S.N. 2006. *The Logic of Violence in Civil War*. Cambridge, MA: Cambridge University
Press.

Kaufmann, C. 1996. "Possible and Impossible Solutions to Ethnic Civil Wars." *International
Security* 20 (4): 136–75.

_____ . 1998. "When All Else Fails: Ethnic Population Transfers and Partitions in the Twentieth
Century." *International Security* 23 (2): 120–56.

Keegan, J. 1993. *A History of Warfare*. New York: Alfred A. Knopf.

Kydd, A., and B.F. Walter. 2002. "Sabotaging the Peace: The Politics of Extremist Violence." *International Organization* 56 (2): 263–96.

Makovsky, D. 2004. "How to Build a Fence." *Foreign Affairs* 83 (2): 50–64.

Media Center of the Islamic Resistance Movement, Hamas. 2003. "Development in The Quality of the Resistance Weapons." Accessed 15 June 2011. http://www.palestine-info.com/arabic/hamas/glory/tatwer.htm (site discontinued).

Murphy, A.B. 1996. "The Sovereign State System as Political-Territorial Ideal: Historical and Contemporary Considerations." In *State Sovereignty as Social Construct*, edited by T.J. Biersteker and C. Weber, 81–120. Cambridge, MA: Cambridge University Press.

Pape, R.A. 2003. "The Strategic Logic of Suicide Terrorism." *American Political Science Review* 97 (3): 343–61.

———. 2005. *Dying to Win: The Strategic Logic of Suicide Terrorism*. New York: Random House.

Prescott, J.R.V. 1987. *Political Frontiers and Boundaries*. London: Allen & Unwin.

Rosière, S. 2011. "Teichopolitics: The Politics of Border Closure." *Si somos Americanos: Revista de estudios transfronterizos* 11 (1): 151–63.

Saddiki, S. 2010. "Ceuta and Melilla Fences: a EU Multidimensional Border?" Paper presented at the 82nd CPSA Annual Conference, 1–3 June. Ottawa: Canadian Political Science Association.

Sterling, B.L. 2009. *Do Good Fences Make Good Neighbors? What History Teaches Us about Strategic Barriers and International Security*. Washington, DC: Georgetown University Press.

Tilly, C. 1990. *Coercion, Capital and European States, AD 900–1990*. Cambridge: Basil Blackwell.

Van der Velde, M., and T. van Naerssen. 2011. "People, Borders, Trajectories: An Approach to Cross-Border Mobility and Immobility in and to the European Union." *Area* 43 (2): 218–24.

Van Evera, S. 1984. "The Cult of the Offensive and the Origins of the First World War." *International Security* 9 (1): 58–107.

Waltz, K.N. 1979. *Theory of International Politics*. Boston, MA: Addison-Wesley.

Wikimedia Commons. 2005. "Gaza Barrier." Accessed 13 July. http://commons.wikimedia.org/wiki/File:GazaBarrier.jpg.

Zacher, M.W. 2001. "The Territorial Integrity Norm: International Boundaries and the Use of Force." *International Organization* 55 (2): 215–50.

Zeidan, F., and A.A. Rahman. 2000. Interview by F. Al-Qassem. *Al Jazeera*. Accessed 30 September 2011. http://www.aljazeera.net/programs/pages/c08d5217-c725-4c35-a5f8-806b87b7c40c.

Chapter 11

Homeland Security? The Effects of Border Enforcement in Guatemala

Ninna Nyberg Sørensen

On 12 May 2008, the Department of Homeland Security (DHS) raided the Agriprocessors kosher meat plant in Postville, Iowa. The raid was hitherto the largest border enforcement operation carried out at a workplace by immigration authorities and was meant to serve as a pilot model for future raids. A total of 390 migrant workers, 314 men and 76 women, were detained on charges of use of false work permits and illegal residence in the United States. The majority were Kaq'chikel Mayans from Guatemala. Like disposable commodities, the apprehended workers were brought to a cattle fairground in nearby Waterloo that first served as a temporary detention facility, and, in the aftermath of the raid, as a makeshift courthouse. A few workers, the majority of whom were mothers of unattended children, were temporarily released on humanitarian grounds, after being attached to electronic homing devices monitoring their whereabouts. Within a few weeks, and in what was considered unusually swift proceedings, 297 migrant workers pleaded guilty to having taken jobs without proper documentation. Threatened by prosecutors to plead guilty – so as to avoid being tried on identity theft charges carrying a mandatory two-year minimum jail sentence – most agreed to immediate deportation after serving detention time in prison. This agreement included signing away their right to go to an immigration court. According to court documents, only five of the 390 initially arrested migrants had criminal records. When the court proceedings ended later in May, 270 undocumented Guatemalans were sent to a private prison where they spent approximately five months before being deported. Shortly after, Somali migrant workers were taking over the jobs left by the apprehended, detained and deported Guatemalans.[1]

Late January 2009, three Postville deportees related their experience during an International Forum on Migration and Peace in Guatemala. 'Why did they treat us like criminals, when all we were doing was to provide a better future for our children?', one deportee asked. Another compared the DHS raid to the violence sweeping the Guatemalan countryside during the armed conflict: 'Back then surviving family members were scattered all over when they fled the military. Today families are divided due to deportation.' Outside – and out of earshot of the state, church, and NGOs – a deportee explained that he was arranging to get back to the United States within the next couple of months, the only problem being that he didn't know how to gather the money needed for his next attempt, including paying for the necessary services to overcome obstacles along the trail. The probability that he actually attempted to remigrate is high. Local migrant organizations estimate that 70 per cent of deported migrants embark on new journeys.

---

1   On the Postville immigration raid, see for example Camayd-Freixas 2009, Herbert 2009, Juby and Kaplan 2011 and McCarthy 2010. See also the documentary *abUSed* by Luís Argueta (http://www. abusedthepostvilleraid.com, accessed 18 September 2014), or the multimedia series *Life after Deportation* by Peter O'Dowd and Nadine Arroyo Rodriguez, downloadable from http://www.fronterasdesk.org/news/ Guatemala (accessed 18 September 2014).

The majority of those deported during the Postville raid originated in scattered rural communities around the municipality of San Andrés Iztapa in the department of Chimaltenango, some 60 km west of Guatemala City. Four years after the raid, approximately 80 of the original 270 deportees had returned to the United States, the majority as *mojados* or undocumented. The rest lived in extreme poverty: the seriously indebted ones under threats from loan sharks, those slightly better off under threats from racketeering. Others had been tricked to pay 10,000 quetzal (US $1,260) for fake temporary migration permits to Canada, the newest sales item offered by the unscrupulous local migration industry. Yet others were still recovering from incidences of abuse, violence and robbery by criminal gangs and corrupted migration authorities during the original journey through Mexico, as well as from violations experienced during their detention in the US. Motivated by the opportunity to pay off debts and remit money to relatives – and further disillusioned by a shaky local environment characterized by state neglect and lack of rule of law – they all had their hopes vested in getting out of Guatemala as soon as possible.

For the last thirty years outmigration has provided the Guatemalan state with a safety valve against poverty and unemployment, as well as with remittances sent by migrants abroad, contributing significantly to the survival of remaining family members and financial stability. In 2012, remittances to Guatemala reached a record high of US $4.7 billion, an increase of US $404 million on the previous year. The growth occurred despite the fact that 40,635 people were deported from the United States during the year, some 10,000 more than those deported in 2011. Realizing that the amount of Guatemalan citizens abroad – without proper migration documentation and hence high risks of being arrested and deported – has increased exponentially during the last three decades, foreign policy state discourses in support of legalizing Guatemalan migrants abroad have begun to accompany domestic migration-development policy parlour. The government has attempted to negotiate Temporal Protection Status (TPS)[2] for undocumented migrants in the US for the last couple of years. A National Council for Migrant Attention (CONAMIGUA in the original Spanish abbreviation) has been formed with the intention to establish an official policy on migration, emphasizing mechanisms of counsel, assistance and legal defence for Guatemalan migrants abroad. Despite high deportation levels, CONAMIGUA programmes have stayed focused on skills training aimed for the North American, not the Guatemalan, labour market, and few resources have been available for other activities. For the Postville deportees, discursive protection policy rhetoric has included legal counselling aimed at applying for asylum in the US and Canada. Less attention has been given to providing social protection and security to the deportees' home communities.

The Postville deportations are instructive on multiple levels. Apart from those stemming from efforts to create state linkages with migrant populations and 'tap' into migrant resources for national development purposes (Popkin 2003; Baker Cristales 2008), they provide insight to the geopolitics of exclusion (Herbert 2009, 826) and to current aims at externalizing migration and border control (Kron 2013). They remind us how mobility patterns and border dynamics tend to change in periods marked by crisis, and how crisis-rhetoric – be it related to the global financial crisis, unmanaged migration or border security – appears to justify harsher mobility policy discourses and

---

2   The US secretary of Homeland Security may designate a foreign country's nationals for TPS due to conditions in the country of origin that temporarily prevent migrants from returning safely or, in certain circumstances, where the home country is unable to handle the return of its nationals adequately, such as armed conflict, environmental disaster, and other extraordinary and temporary conditions. Salvadoran migrants have benefited from this programme since its inception in 1990, Honduran migrants since 1999. The Guatemalan state did not apply for TPS during and after the armed conflict, nor after the devastating effects of Hurricane Stan in 2005.

stricter border control measures. Migration management has targeted undocumented migrants and instituted new forms of governing movement, people and borders. Hereby, migration decisions, routes and border conceptualizations have changed, as has the conglomerate of actors linked both to the facilitation and to the control of migration (Sørensen and Gammeltoft-Hansen 2013). In this process the United States has gained political currency by demonstrating effective border management. Private companies involved in detention and deportation have made huge profits, the Guatemalan state has gained hard currency from remittances, while economic and social gains for the individuals, families and migrant communities involved have increasingly been substituted by exposure to risk, violence and insecurity as well as abysmal debt.

Mobility Discourses and Approach

This chapter locates Guatemalan migration in a global terrain marked by unequal access and opportunities. It directs our gaze south of the border and examines the effects of border enforcement on local indigenous migrant communities. In this terrain mobility and immobility has historically been distinguished by the degree of force involved. When referring to the flight from Guatemala's violent conflict (1960–96), 'displacement' has been the term invoked, whereas 'mobility' has been used to indicate a 'voluntary' nature of post-conflict migration. 'Emplacement' may better capture the experience of the Postville deportees, as the term indicates 'being placed by others' and often involves a redisplacement (Jansen and Löfving 2009). This view parallels Ballinger (2012) who argues that it is not unusual to see displacement and emplacement as opposite or mirror phenomena, separated by borders, with emplacement denoting the act of 'remaining in place'. In her perspective 'returning to the place from which one was displaced is usually not about an effort to go home', but is rather about re-emplacing oneself in places undergoing change, partly through migration (397). The case of current mass deportations to Guatemala points to larger theoretical dilemmas about how to understand migration trajectories, and underlines the need to analyse mobilities and enclosures as interrelated processes; processes whose dynamics are facilitated and constrained by state and non-state actors, whose workings often become most apparent at and around borders.

From another vantage point Chalfin (2012) sees border security as a 'late capitalist fix' that places the burden of an untrammelled quest for border security on border zones across the United States and allied states worldwide. This 'fix' involves the privatization and outsourcing of migration management, and is most pronounced where neo-liberalization is most advanced. Chalfin's approach resonates with conceptual work currently undertaken by Gammeltoft-Hansen and this chapter's author. The latter argue that the growing commercialization of migration takes on significance through a host of business opportunities that capitalize on migrants' desire to move as well as on states' attempts to control migration. The markets for the facilitation, regulation and control of international migration comprise a wide variety of actors, ranging from small migrant entrepreneurs facilitating the transportation of people, to multinational companies carrying out migrant detention and deportation. A somewhat middle position is occupied by NGOs, humanitarian organizations and intergovernmental actors, who facilitate migration through providing information and offering shelter along the route (Sørensen and Gammeltoft-Hansen 2013; Sørensen 2012). At the same time, they also constrain it by carrying out anti-trafficking campaigns or promoting technocratic northern standards of migration control to migrant-sending and transit countries in the Global South (Andrijasevic and Walters 2010).

In line with the constitutive parts of the threshold approach – people, borders and trajectories – suggested by Van der Velde and Van Naerssen (2011), this chapter addresses the effects of current border enforcement on migrant (im)mobilities. The analysis builds on several years of engagement with Central American migration and, in the particular case of deportation, a shorter ethnographic study conducted during June 2012 in the City of Guatemala and the Department of Chimaltenango.[3] For contextualization, a brief overview of past and present Guatemalan mobilities is given. The second part of the chapter focuses on the social economy of contemporary migration related to deportations, debts and dangers. The discussion underscores the need to include the effects of enhanced border control in the threshold approach. It also highlights the inherent conflict between migration management and safeguarding migrants' lives and rights.

Guatemalan Mobilities

The movement of people, goods and money in Central America has a long history and dates back at least to the twelfth century, when maritime Mayan traders transferred goods and slaves along the Gulf coast of Mexico. During the sixteenth century, the Spanish colonizers transformed the Central American region into a key link in a colonial empire stretching from Argentina to California (Wainer 2012, 2). Migration for work took off in the nineteenth century, when colonial demand for labour led to the establishment of a draft system of *repartamiento* (coerced labour) and debt patronage (Hamilton and Stolz Chinchilla 1991). However, until the 1970s, Guatemalan migration was generally characterized by internal or regional movements of a transborder, binational and temporal character, with the purpose of sustaining local livelihoods (Morales Gamboa et al. 2011).

The armed conflict (1960–96) changed this pattern, first and foremost by causing massive forced displacements during the second half of the 1970s and throughout the 1980s. During this period state-sponsored campaigns of terror and mass killings left a death toll of approximately 200,000 Guatemalans, and an additional 40,000 forced disappearances. Some 440 villages were razed to the ground, 80,000 women became widowed and 200,000 children orphaned (Sanford 2003). The conflict uprooted as many as one million people, of whom up to 400,000 fled to Mexico or other Central American countries, fewer to the United States or Canada, while the rest remained internally displaced within Guatemala's borders. When the Peace Accords were finally reached, many refugees returned to Guatemala. In 1999 the last-assisted repatriation of Guatemalans from Mexico took place, bringing the number of repatriated refugees to around 43,000 (Stepputat 1999). Others returned on their own behalf. The bulk of the approximately 120,000 asylum applications handed in by Guatemalans in the United States were routinely rejected without necessarily leading to return. Of those remaining without documentation in the US, many were later able to legalize

---

3 During this short and targeted fieldwork interviews were conducted with Guatemalan state representatives, Mexican and US embassy officials, central and regional representatives of CONAMIGUA, the Human Rights Defender of the Guatemalan Migrant Population, and representatives of the Catholic Pastoral de la Movilidad Humana and the Casa del Migrante in Guatemala City, both involved in advocacy and relief work. The fieldwork also included talks with activists and researchers involved in ongoing migration projects on deportation and disappearances at the Instituto Centroamericano de Estudios Sociales y Desarrollo (INCEDES), the Facultad Latinoamericana de Sciencias Sociales (FLACSO), the Mesa Nacional para las Migraciones en Guatemala (MENAMIG), and the Fundacion de Antropologia Forense de Guatemala (FAFG) in Guatemala City. Finally, interviews were carried out in the reception area for deportees in Guatemala City's military airport and in one of the Chimaltenango rural communities affected by the Postville raid and subsequent deportations. Here the author talked with community members and male and female members of the local association of deportees, Asociación Pro Mejoramiento de los Deportados de Guatemala (APRODE).

their status. They provided information and social networks for family members, whom later, due to poverty, a lack of public policies and increasing local insecurity, decided to invest their hopes for a better future up north (Hagan 1994). Soon transnational communities connecting families and villages in Guatemala to migrant destinations in the United States began to form.

Estimates of current Guatemalan international migration vary slightly between 1.2 and 1.5 million, including undocumented flows. The United States remain the primary destination, where up to 60 per cent live and work without proper documentation (Sørensen 2011). A majority (around 70 per cent) are male and primarily employed in the service sector, manufacturing and/ or agriculture. Young women (including single mothers) have recently joined the migration flows. Other women are left behind in Guatemala, where they have become known as *viudas blancas* (white widows), often suffering from long-distance or in-laws social control and, occasionally, abandonment. The movement has an important ethnic dimension. It involves all ethnic groups and social classes, but as the civil war primarily targeted indigenous Maya communities, distinct indigenous migration patterns have developed, for example of Chujes, Quichés and Kanjobales in Los Angeles, and, relevant to this case study, Kaq'chikeles in Postville, Iowa. Most migrate in order to provide for family members remaining in Guatemala. Hard-earned remittances comprise around 10 per cent of GDP and dwarf both foreign direct investment and overseas development assistance. They help households to reduce poverty, but also facilitate the government's passing of the cost of public services (Wainer 2012).

Although overwhelmingly undocumented, a few opportunities for authorized migration exist. Since 2003, a temporary foreign workers programme has granted workers from rural areas of Guatemala access to temporary working opportunities in Canada. Since its establishment, the International Organization for Migration Guatemala has been the main player and placed some 6,000 Guatemalan migrants with Canadian businesses (IOM 2012). Both Canada and IOM have lately been criticized for supporting worker exploitation, especially after the 2010 protests that led Guatemalan workers to get laid off and repatriated. The Canadian government has, moreover, been criticized for not providing any legal protection for maltreated workers, the Guatemalan government for allowing foreign recruiters to select workers without regulation and IOM for acting on behalf of private interest and for violating fundamental workers' rights by assisting a programme without oversight and regulation. Rumours of corruption abound, in particular among NGOs involved in migration and anti-corruption efforts.

It is difficult to ignore the role of US policies in encouraging Guatemalan migration. As highlighted by Juby and Kaplan (2011), US policy has reduced the marketability of Guatemalan farmers and, moreover, has a history of supporting repressive military regimes. Due to the supportive stance of the US towards the Guatemalan government during the civil war, less than 3 per cent of asylum seekers received asylum (Brabeck et al. 2011), whereas the rest were readily absorbed into the unskilled labour market. Since the 1990s, however, the United States has been pursuing a simultaneous strategy for managing migration from Mexico and Central America, pressuring their southern neighbours to stem migration, and thereby taking an active role in stopping migrants long before they set eyes on the border (Flyn 2002). The border is now present at multiple locations: 'It travels into Mexico where US officials try to engage in operations that discourage undocumented migrants from even attempting the trek north, and it migrates into small-town America, to places like Postville, where ordinary police officers are now formally deputized to enforce federal law' (Herbert 2009). It reaches Guatemala and Guatemalan NGOs in the form of donor funding for particular projects, for instance migration-development projects when these were in fashion; presently, anti-trafficking and other campaigns warning about the dangers along the migration trail seem to be *la moda*.

Burgeoning budgets, growing numbers of control personnel, increasing amounts of border control equipment and intensified outsourcing of migration control functions to private actors define current US border enforcement efforts. Undocumented Guatemalan migrants live in constant fear of deportation from the United States, as any encounter with US authorities can lead to incarceration and removal proceedings. A mother can be left to provide for the family by herself if the father is deported, children can end up in foster care if both parents are deported. If deported parents take children with US citizenship back to Guatemala, these children have no right to schooling and health care. But even deportees with Guatemalan citizenship are likely to be exposed to high levels of insecurity when emplaced to the rural communities they once left behind.

Deportations

Beginning in the mid-1990s, a hardening of US immigration policies elevated the power of the federal government to arrest, detain and ultimately deport undocumented migrants. This led to increased deportation, now strategically referred to as removals (Hagan et al. 2008). During the fiscal year 2012, the US Immigration and Customs Office of Enforcement and Removal Operations removed the largest number of undocumented migrants (409,849) from the United States in the agency's modern history. Guatemala was the destination of the highest number of air repatriations to any country in the world (Wainer 2012). According to the Homeland Security Office of Immigration Statistics, a total of 215,846 Guatemalans were 'removed' between 2003 and 2012. Behind the mere numbers, the deportees consist of a diverse group of migrants, spanning settled migrants who have lived and worked for ten, twenty or thirty years in the US, to new arrivals apprehended during a first attempt to unauthorized entry. In addition to the problems faced by individual migrants, US deportation policy undermines long-standing family reunification principles and poses dire social, economic and psychological costs for deportees and their families in the US and in the country of origin (Hagan et al. 2008). The threat of deportation is particularly poignant for families of mixed status (Brabeck et al. 2011), who, in the incidence of deportation of one or more family members, become subjected to the 'disruption of family ties' that for a long time has been seen as an undesirable outcome of the initial migration.

Deported Guatemalan migrants come back to a country of origin lacking an effective state programme for reception and integration. To the deported migrant, deportation represents a catastrophe. To the receiving state, the deportee represents a burden. Stripped of his or her economic capacity, the 'migrant hero' of the remittance-dependent country instantly becomes 'deportee trash' (Sørensen 2011). The disposability of deportees is apparent during their reception in Guatemala City's International Airport, where seven to eight weekly deportation flights have landed during the last couple of years. Each aircraft carries around 135 passengers, usually only Guatemalans. Their arrival is not displayed on the electronic arrival/departure monitors in the new modern arrival hall. Nor are the deportees entering Guatemala through the official arrival and immigration facility. They are kept invisible to the dominant social order, entering through a 'secret door', exiting through a 'backdoor' that takes them right out to a noisy, busy and somewhat dangerous street far away from the exit area for other international travellers and the public gaze. As soon as the planes touch ground, the human cargo departs the public landing strips and sets course for the military part of the airport. Here the deportees are released from the plastic flexi handcuffs used during the journey and then escorted to the deportation reception facility.

Inside the building the deportees are assigned a white plastic chair, where small sandwiches, juice cartons and biscuit packages are distributed along with information material (for instance from the Casa del Migrante and IOM). They are welcomed by Foreign Ministry posters saying, 'You are finally in your own country and with your own people – Welcome to Guatemala,' and given a pep talk by a government representative assuring everyone that 'if you were abused abroad, this will now come to an end as you have arrived home to a country that cares about you.' Despite this welcome, the deportees are subject to state authority, enacted by not allowing them to go to the toilet or make phone calls before they hours later have been through first the ordinary migration control procedures, then a second check by the national police (for felonies committed in Guatemala). Apart from the modest meal, the service offered consists – depending on available funding – of a voluntary health check, a free phone call to a relative in Guatemala, and a bus ride to the central bus terminal. Those arriving on afternoon planes and whose home communities are located long bus drives away from Guatemala City are dropped off at the Casa del Migrante, the migrant shelter operated by the Catholic church. Ironically, the only existing reception programme – the Guatemalan Repatriates Project – is funded by USAID and operated by the IOM. Since the project was launched in June 2011, it has provided basic communication and health services, transportation and shelter to about 23,000 returnees. Apart from a few basic and immediate services, there is no long-term reintegration plan for deportees. Until 2012, only four deportees have found jobs through the programme (Wainer 2012).

State engagement stops the moment the deportees leave the reception area. Outside the gate, money exchangers, coyotes and loan sharks line up with family members to 'welcome' deported compatriots. In a city with vast numbers of police and private security personnel guarding both public and private buildings, their absence in this particular place is conspicuously striking.

Debts

While working abroad, remittances have contributed economically to the improvement of family conditions, and in some instances to more gender equality and other social changes (Ugalde 2010). With deportation, remittances are abruptly interrupted as are women's experiences of working alongside and earning equal wages with men. In addition to the immediate loss of livelihood opportunities, most of those who have left Guatemala during the last ten years have incurred substantial debt to finance their migration. Consequently they return to poorer economic conditions than those they left behind (Dingeman and Rumbaut 2010). In the case of the Postville deportees, the majority had spent only one to three years in the United States. A young female migrant explained:

> We worked for twelve hours, the work the men did we did as well. It is hard work. And they didn't pay us much, some US $6.25 that rose to US $7, and from there we didn't get more. And that is the reason why we didn't manage to pay off the debt, it is impossible to come up with that amount of money in a year's time. And now [deported], we don't have any means to pay, with the 10 to 15 per cent of interests, the debt has only risen.

A male migrant expanded:

> Life around here is quite difficult because of the lack of employment opportunities. We don't have our own land, we don't have any place from where to drag subsistence for our families … You see, many of us that came from there, we came back with a bit of debts, we didn't succeed in paying the

travel to that place, and that's what really preoccupies us. Because there are some persons who are ... that consider this ... well who continue to collect interests. Persons who verily don't have any human sense, they continue to demand interests to the effect that our debts keep raising instead of falling.

The costs involved and debts incurred to finance cross-border mobility are seldom taken into account in analyses of migration's effect on home country economies. Nor are the consequences of not being able to repay debts.

According to Stoll (2010), migration is nevertheless a process that runs on debt, with migrants and families indebting themselves in ways that many are unable to repay, often resulting in the loss of home and productive assets. During a study of Nebaj, a highland community in northern Guatemala, he found that 75 per cent of the migrants households surveyed had borrowed money on property titles. In a town where many get by on US $1,500 a year, the average debt reported by migrant households amounted to 126,500 quetzal or US $16,000. Of the 270 Postville deportees in Chimaltenango, more than half had returned to debts ranging from 5,000 to 100,000 quetzal (US $650–$12,500). In the Nebaj study, migrants were being pulled by the promises of higher wages and pushed by easy access to microcredit that became invested in undocumented journeys to the US. In the Chimaltenango case, landless migrants took loans with local loan sharks, some of whom became involved with larger organized criminal networks, capable and willing to threaten those unable to pay in order to get the rest under their control. In May 2012, three members of a migrant family no longer capable of paying the debt quotas were brutally killed, 'by loan sharks', most people believed, 'to set an example'. Criticizing the migration-development parlour of international institutions and the Guatemalan government, Stoll asks if migration in reality sucks more value from the sending communities than it returns. But value is not the only issue here. The Chimaltenango case suggests that human life is at stake, not only on the increasingly dangerous routes migrants use to avoid detection, but also in their home communities. From a threshold perspective we should thus explore changes in the local contexts people are returning to, for instance what prevents former villagers from staying safely in (what was once) their local community.

Dangers

From the displacements of the 1970–80s to today's labour mobility, human insecurity is probably the main factor generating current undocumented migration from Guatemala. The state's lack of territorial control in Guatemala (as well as the Mexican state's lack of territorial control in Mexico) increases the risks that migrants face prior to (in origin), during (in transit) and after forced return migration (in origin upon deportation). In addition to deportation and debt discussed above, the migration trajectory is paved with danger. A recent analysis conducted by the Mexican INEDIM (Instituto de Estudios y Divulgación sobre Migración) and the Guatemalan INCEDES reveals that restrictive control policies are key factors in increasing the risks migrants face during transit, for instance by throwing them into the arms of criminal gangs operating along the trail. The risks include extortion, sexual violence, kidnappings and murder by gang members and organized crime actors, but also abuse of authority, detention and extortion by authorities as well as private agents (security companies, transportation companies, police and border control agents). The ransom demanded for releasing kidnapped migrants range from US $1,000 to $5,000 (Córdova Alcaraz 2011). Organized crime has in several instances been found to act in complicity with government agencies. In Guatemala, migration authorities are believed to be among the most corrupt state actors making huge profits on migrant extortion and smuggling.

Corrupt migration officials are allegedly playing an integral part by, in the words of Rosales Sandoval (2013, 2015), 'greasing the wheels of the migration industry through corruption'.

When asking the Postville deportees whether increasing risks might prevent them from remigrating, a male deported migrant answered: 'I know that I might die along the trail, but here my children are dying of hunger, and how will I ever be able to sleep peacefully if I do not take the risk.' Another elaborated:

> The thing is that early on, yes there were rumours of the migration (authorities), you also heard stories about someone being robbed along the way, but it wasn't much. But now ... Well, the migration and the US police have become an authority, not dedicated towards [solving] the national problems but only against the immigration of the persons ... But it is the mere necessity that makes you take the risk, even the risk of travelling through Mexico, yes ... People decide to leave their families behind and take the risk of dying along the journey. Why? Because of our situation, of not having any jobs and being able to provide for our families.

Until recently, Guatemala did not keep records of migrant kidnappings, disappearances and deaths along the migratory trails. However, the 2010 Tamaulipas Massacre, where 72 undocumented Central American migrants were killed (among them 14 Guatemalans), and the second mass murder of 193 migrants in 2011 in San Fernando, have raised national consciousness about the problem. During 2011, approximately a hundred Guatemalan nationals reported the kidnapping or the disappearance of a family member to the Foreign Ministry or the Human Rights Ombudsman's Office. According to the Foreign Ministry's statistics for 2011, 280 dead bodies were repatriated from the US and Mexico, the majority victims of road accidents, labour accidents, criminal attacks or just dying in the desert while attempting to cross (MENAMIG 2011, 2012). Other reports, in particular from El Salvador or Honduras, indicate that Mexican security forces are involved in extortion and kidnapping of Central American migrants (Sørensen 2013).

The dangers, disappearances and death tolls referred to above are direct consequences of hardened efforts to control and manage international migration. While the actors involved in facilitating or constraining migration earn huge profits, the outcome for Guatemalan migrants is increasingly one of economic and human loss.

Conclusion

This contribution has explored the risks undocumented migrants face in their attempts to cross the US border. Many succeed in overcoming locational and trajectory thresholds. Increasing numbers nevertheless encounter problems. While recent publicity and scholarly writing have begun to give attention to the dangers involved in the overland trajectory, less attention has been paid to the fact that the locational threshold changes over time, and, for example, may change from being a poor but secure home community to becoming an even poorer and insecure location.

Additionally, US border enforcement has targeted undocumented migrants and instituted new forms of governing movement, people and borders. Hereby migration decisions, routes and border conceptualizations have changed, as have the conglomerate of actors linked both to the facilitation and control of migration. Unlike labour migration to the United States, which for the Kaq'chiquels of Chimaltenango to a certain extent was actively decided upon and planned, deportation is externally imposed, suddenly, and often unexpectedly (Brabeck et al. 2011, 284). It is 'profoundly disruptive and debasing for those affected' (Peutz and De Genova 2010, 2), not least due to the resemblance to previous experiences with the forced displacements and disappearances many indigenous Guatemalans experienced during the armed conflict in the 1970s and 80s. Deportation becomes a disciplinary practice that in the Guatemalan case is foreshadowed by historical labour subordination and racial discrimination practices, intimately related to former experiences of coercive movements such as forced dispersal, forced displacement, extradition and rendition (Peutz and De Genova 2010, 6; see also Kanstroom 2012 and King et al. 2012). The well-known notion and embodied experience of the 'disappeared', traditionally referring to the people disappeared by the military regime and local death squads, takes on new meaning as unscrupulous smugglers and organized crime take on the business of disappearing migrants journeying through Mexico (Caminero-Santangelo 2010).

Ironically, as pointed out by Dingeman and Rumbaut (2010), the United States deports people to countries that the state department declares too dangerous for travel. As migration is perceived as one of the few strategies that could make a real difference, poor people nevertheless continue to migrate despite the risks involved. Expected high gains legitimize high risks (Alpes 2011, 50), experienced local vulnerabilities and insecurities explain the continuous flows. As stated by Burrell (2010, 91–92), 'Migrants find themselves moving in and out of entitlements to rights as they cross various types of borders. In these crossings – transnational, national and local – at each of these different scales, the anxiety around issues of security, taken to mean freedom from crime and victimization, animates the often violent and dangerous spaces negotiated.' By centring the analysis on the lives and experiences of Guatemalan migrants and deportees, this chapter has argued in favour of extending the threshold approach to take the effects of border enforcement into account. To the extent that not only the territories traversed to reach the United States (the migration trajectories through Mexico), but also the places of origin have become dangerous, the notion of borderland has expanded way into peoples' lives, even when considered 'at home'.

High gains also explain the deportation logic up north. As the Postville case indicates, stepped-up border enforcement not only yields growing numbers of undocumented migrants bound for detention and deportation, but also vast business opportunities. The business of incarcerating detainees or providing charter flights to remove undocumented migrants has become a lucrative commercial enterprise. From 2005 to 2012, the budget for detention operations more than doubled, soaring from US $864 million to more than US $2 billion, with major contracts outsourced to the two largest private prison companies, Corrections Corporation of America and the GEO Group Inc. According to its own cost estimates, the DHS spent approximately one billion dollar on apprehending, detaining and deporting 76,000 Central Americans in 2010, including US $132.36 million on chartered removal flights. In Washington, DC, the industry's lobbyists have influenced policy to secure growing numbers of immigrant detainees in its facilities; in small rural and economically depressed towns, support for the industry has been gained by offering an influx of public money and jobs (*Huffington Post*, 6 July 2012).

Attempts to control and manage migration through enhanced border controls, stricter border policing and high-tech surveillance facilities have not stopped the human traffic. Rather it has forced migrants to take more dangerous routes and raised the price journeying migrants have to pay. The movement of undocumented migrants through Mexican territory has multiplied since the 1990s and is now worth two to three billion US dollar in yearly revenues, divided amongst criminal cartels and corrupt police forces on both sides of the border (Sørensen 2013). In the eyes of the receiving state, migration management and border security refers to protection from undocumented immigration. In the eyes of the sending state, undocumented migration is fundamental to the national economy, whereas having to deal with the structural inequalities that make migration necessary in the first place would take far more effort in providing distributive public policies than passing round sandwiches to deported nationals. From the undocumented migrants' point of view, immigration enforcement means violence and the risk of becoming trade items between the cartels, a business opportunity for loan sharks and local coyotes in Guatemala, and a commodity yielding huge profits for the control and deportation industry up north, as well as for the remittance-dependent state back home.

The social economy of deportation seriously questions conventional notions within migration theory of departure, settlement and return. Such notions conveniently overlook the complex dilemmas posed by various forms of mobility, immobility and deportability, and the thresholds that prevent people from carving out a living in particular places, especially upon forced returns to communities plagued by violence and insecurity. Deportations have personal, societal, economic and political consequences. Apart from the social economy of deportation, the chapter has provided arguments for an overall analysis also including the state agencies charged with the apprehension, detention, deportation and reception of migrants, the private corporation or corporate mercenaries that benefit from these practices, the licit and illicit groups offering services to beat enhanced border controls, the money earned and the debts incurred and the thriving criminal networks making travel and homecoming a dangerous affair.

This chapter has sought to connect the effects of neo-liberal migration policies to an analysis of the social economy of deportation. The analysis suggests that US border enforcement has targeted undocumented migrants and in the process instituted new forms of governing movement, people and borders. Hereby migration decisions, routes and border conceptualizations have changed, as has the conglomerate of actors linked both to the facilitation and control of migration. A joint outcome for the individuals, families and communities affected is increased exposure to violence and insecurity.

Acknowledgement

The author owes great thanks to the deportees – whether from the Postville raids or from the deportation flights – for giving their precious time to share difficult and traumatizing experiences. The author would also like to express appreciation for the guidance and information exchange provided by Guatemalan activists, NGOs and researchers. Finally, the author is grateful to Programme Secretary María Sofía Villatoro, the Danish Office for Regional Cooperation with Central America, for her assistance in setting up the tight interview schedule and accompanying her during several appointments.

References

Alpes, M.J. 2011. "Bushfalling: How Young Cameroonians Dare to Migrate." PhD diss. Amsterdam: University of Amsterdam.

Andrijasevic, R., and W. Walters. 2010. "The International Organization for Migration and the International Government of Borders." *Environment and Planning D: Society and Space* 28 (6): 977–99.

Baker-Cristales, B. 2008. "Magical Pursuits: Legitimacy and Representation in a Transnational Political Field." *American Anthropologist* 110 (3): 349–59.

Ballinger, P. 2012. "Borders and the Rhythms of Displacement, Emplacement and Mobility." In *A Companion to Border Studies*, edited by T.M. Wilson and H. Donnan, 389–404. Sussex: Wiley-Blackwell.

Brabeck, K.M., M.B. Lykes, and R. Hershberg. 2011. "Framing Immigration to and Deportation from the United States: Guatemalan and Salvadoran Families Make Meaning of their Experiences." *Community, Work and Family* 14 (3): 275–96.

Burrell, J. 2010. "In and Out of Rights: Security, Migration, and Human Rights Talk in Postwar Guatemala." *Journal of Latin American and Caribbean Anthropology* 15 (1): 90–115.

Camayd-Freixas, E. 2009. "Postville: La criminalización de los migrantes." Cuadernos del presente imperfecto 8. Guatemala: F & G Editores.

Caminero-Santangelo, M. 2010. "The Lost Ones: Post-Gatekeeper Border Fictions and the Construction of Cultural Trauma." *Latino Studies* 8 (3): 304–27.

Chalfin, B. 2012. "Border Security as Late-Capitalist 'Fix'." In *A Companion to Border Studies*, edited by T.M. Wilson and H. Donnan, 283–300. Sussex: Wiley-Blackwell.

Córdova Alcaraz, R. 2011. "Seguridad para el migrante: Una agenda por construir/Security for Migrants: Building a Policy and Advocacy Agenda." INEDIM Working Paper Series 2. Mexico: Instituto de Estudios y Divulgación sobre Migración.

Dingeman, M.K., and R.G. Rumbaut. 2010. "The Immigration–Crime Nexus and Post-Deportation Experiences: En/Countering Stereotypes in Southern California and El Salvador." *University of La Verne Law Review* 31 (2): 363–402.

Fitz, M., G. Martinez, and M. Wijewardena. 2010. *The Cost of Mass Deportation: Impractical, Expensive and Ineffective.* Washington, DC: Center for American Progress.

Flyn, M. 2002. "U.S. Anti-Migration Efforts Move South." *Americas Program*, 3 July. Accessed 4 February 2014. http://www.cipamericas.org/archives/1066.

Ford, M., and L. Lyons. 2012. "Labour Migration, Trafficking and Border Controls." In *A Companion to Border Studies*, edited by T.M. Wilson and H. Donnan, 438–54. Sussex: Wiley-Blackwell.

Hagan, J.M. 1994. *Deciding to Be Legal: A Maya Community in Houston.* Philadelphia: Temple University Press.

Hagan, J.M., K. Erschbach, and N. Rodriguez. 2008. "U.S. Deportation Policy, Family Separation, and Circular Migration." *International Migration Review* 42 (1): 64–88.

Hamilton, N., and N. Stoltz Chinchilla. 1991. "Central American Migration: A Framework for Analysis." *Latin American Research Review* 26 (1): 75–110.

Herbert, S. 2009. "Contemporary Geographies of Exclusion II: Lessons from Iowa." *Progress in Human Geography* 33 (6): 825–32.

IOM (International Organization for Migration). 2012. "The Temporary Foreign Workers Programme." Accessed 3 September 2012. http://www.iom.int/jahia/guatemala (site discontinued).

Jansen, S., and S. Löfving. 2009. "Introduction: Towards an Anthropology of Violence, Hope, and the Movement of People." In *Struggles for Home: Violence, Hope, and the Movement of People*, edited by S. Jansen and S. Löfving, 1–23. Oxford: Berghahn Books.

Juby, C., and L.E. Kaplan. 2011. "Postville: The Effects of an Immigration Raid." *Families in Society: The Journal of Contemporary Social Services* 92 (2): 147–53.

Kanstroom, D. 2012. *Aftermath: Deportation Law and the New American Diaspora.* New York: Oxford University Press.

King, R.D., M. Massoglia, and C. Uggen. 2012. "Employment and Exile: U.S. Criminal Deportations 1908–2005." *American Journal of Sociology* 177 (6): 1789–825.

Kron, S. 2013. "Central America: Regional Migration and Border Policies." In *The Encyclopedia of Global Human Migration*, edited by E. Ness, 916–21. New Jersey: Wiley-Blackwell.

McCarthy, A.L. 2010. "The May 12, 2008 Postville, Iowa Immigration Raid: A Human Rights Perspective." *Transnational Law and Contemporary Problems* 19: 293–315.

MENAMIG (Mesa Nacional para las Migraciones en Guatemala). 2011. *Balance migratorio 2011: Entre continuidades y cambios*. Guatemala: Fundación Ford.

_____. 2012. *Tiempos innombrables: Migración, criminalización y violencia; Balance hemerográfico y análisis de tendencias 2010–2011*. Guatemala: Fundación Ford.

Morales Gamboa, A., M. Herradora, and K. Andrade-Eekhoff. 2011. "Movilidad humana en Centroamérica: Un intento de mapeo de los flujos migratorios." In *Migración de relevo: Territorios locales e integración regional en Centroamérica*, edited by A. Morales Gamboa, 15–25. San José: Facultad Latinoamericana de Ciencias Sociales.

Peutz, N., and N. de Genova. 2010. "Introduction." In *The Deportation Regime: Sovereignty, Space and the Freedom of Movement*, edited by N. de Genova and N. Peutz, 1–29. Durham, NC: Duke University Press.

Popkin, E. 2003. "Transnational Migration and Development in Postwar Peripheral States: An Examination of Guatemalan and Salvadoran State Linkages with Their Migrant Populations in Los Angeles." *Current Sociology* 51 (3–4): 347–74.

Rosales Sandoval, I. 2013. "Public Officials and the Migration Industry in Guatemala: Greasing the Wheels of a Corrupt Machine." In *The Migration Industry and the Commercialization of International Migration*, edited by T. Gammeltoft-Hansen and N.N. Sørensen, 215–37. London: Routledge.

Stepputat, F. 1999. "Politics of Displacement in Guatemala." *Journal of Historical Sociology* 12 (1): 54–80.

Stoll, D. 2010. "From Wage Migration to Debt Migration? Easy Credit, Failure in El Norte, and Foreclosure in a Bubble Economy of the Western Guatemalan Highlands." *Latin American Perspectives* 37 (1): 123–42.

Sørensen, N.N. 2011. "The Rise and Fall of the 'Migrant Superhero' and the New 'Deportee Trash': Contemporary Strain on Mobile Livelihoods in the Central American Region." *Border-Lines: Journal of the Latino Research Center* 5: 90–120.

_____. 2012. "Revisiting the Migration–Development Nexus: From Social Networks and Remittances to Markets for Migration Control." *International Migration* 50 (3): 61–76.

_____. 2013. "Migration between Social and Criminal Networks: Jumping the Remains of the Honduran Migration Train." In *The Migration Industry and the Commercialization of International Migration*, edited by T. Gammeltoft-Hansen and N.N. Sørensen, 238–61. London: Routledge.

Sørensen, N.N., and T. Gammeltoft-Hansen. 2013. "Introduction." In *The Migration Industry and the Commercialization of International Migration*, edited by T. Gammeltoft-Hansen and N.N. Sørensen, 1–23. London: Routledge.

Ugalde, M.A. 2010. "Migrant Indigenous Guatemalan Women as Agents of Social Change: A Methodological Essay on Social Remittances." Accessed 16 February 2012. http://www.scribd.com/doc/27862160/Article-Miguel-a-UGALDE.

Van der Velde, M., and T. van Naerssen. 2011. "People, Borders, Trajectories: An Approach to Cross-Border Mobility and Immobility in and to the European Union." *Area* 43 (2): 218–24.

Wainer, A. 2012. "Exchanging People for Money: Remittances and Repatriation in Central America." Bread for the World Briefing Paper 18. Washington, DC: Bread for the World Institute.

Chapter 12

Reflections on EU Border Policies:
Human Mobility and Borders – Ethical Perspectives

Fabio Baggio

This contribution will focus on and propose a wider comprehension of the concept of borders, aiming at adding elements for reflection on this threshold for mobility as well as tracing criteria for an ethical assessment of it. In the era of globalization – while borders seem to be disappearing for ideas, capitals and goods – new walls and fences, either physical or legal, are built to stop migrants and refugees. The issue of human mobility and borders has been hotly debated in the international arena for the past years. The 2005 final report of the Global Commission on International Migration highlighted the human rights dimension, which should be considered when states shape their policies concerning the admission of foreigners (GCIM 2005, 59). In its *World Migration Report 2008*, the International Organization for Migration questioned the assumption that more attention to protection of national interests and the securing of borders would result in more effective migration management (IOM 2008, 2). During the second Global Forum on Migration and Development, held in the Philippines in October 2008, civil society organizations stated in their final report to the governments the need to separate the migration and development agenda from national security issues (GFMD 2008, 4).

The European Union (EU) decided to take a different path. In 2010 the European Commission adopted the EU Internal Security Strategy in Action aiming at strengthening the security of the EU territories through border management. To this purpose a tight European Border Surveillance System (EUROSUR) was created. In the same year the European Council approved the Stockholm Programme. One of the two major components of this programme entails more control on external borders, expanding the role of the border agency Frontex and providing the latter with a budget of 83 million euro. At a national level, in the last three years several European governments (for instance Switzerland, the United Kingdom, Denmark, Italy and Spain) adopted stricter regulations for the entry of migrants, either reducing the quotas or tightening the selection of candidates (IOM 2011, 71–4).

The EU does not represent an exception in the world of migration policy scenarios, however. The global economic crisis and the political transition in the Middle East and North Africa boosted a general drive towards more restrictive migration policies and regulations, which characterized the first decade of the third millennium (IOM 2011, 49–52). Moreover, 'external-territorial' and 'internalized'[1] borders have been added to the traditional national borders. This chapter aims to raise several ethical questions based on a concerned look at the effects of such developments. In many cases human rights are differently applied to border crossers. It is not unusual that discriminative

---

1   For the aim of this contribution, with 'external-territorial borders' the author means the result of transferring national borders (immigration controls, custom controls and so on) offshore. With 'internalized borders' the author means the result of the incorporation of discriminative principles within the receiving societies.

practices on the basis of skills and origin are adopted by receiving countries. The externalization of territorial borders seems to 'deresponsibilize' the governments of the receiving countries. Internalized borders often contradict and/or jeopardize the integration policies enacted in many receiving societies. This chapter intends to ethically assess the present relationship between human mobility and borders, with a special focus on the EU, and proposes a path towards an ethicization of borders.

Human Mobility and Borders in the EU

In their struggle for survival or 'greener pastures', international migrants, refugees and other itinerant people have to deal with different 'borders', whose crossing seems to become every day more complicated. There are three kinds of borders that will be considered in the perspective of migration: national borders, internalized borders and external-territorial borders.

National Borders

Within the present nation-state system, the legal difference between a citizen and a foreigner determines a different approach to admission into the national territory, being a right for the former and a concession for the latter, with few exceptions related to permanent residency and family reunification. This is the main idea of the so-called 'asymmetry of the right to migrate': the universally recognized right to leave one's own country is not guaranteed by the right to enter another country. Every attempt to amend such asymmetry has been understood as a threat to national sovereignty and invited the opposition of many countries.

In the last decades several receiving countries decided to invest a huge amount of capital on border controls and immigration checks. This move was mainly intended to respond to the worrisome increase of unauthorized foreigners residing in the national territory. The trend was confirmed and strengthened in the aftermath of the terrorist attack on the US on 11 September 2001. The development of new technologies applied to personal and document screening led to the introduction of hyper-sophisticated controls and electronic passports with recorded biometric data.

The most evident example of the reinforcement of borders is their 'fortification' through the construction of walls and barriers. The 700-mile long fence between the US and Mexico, the wall built to mark the border between Thailand and Malaysia, the barrier dividing Bangladesh and India and the third fence around Ceuta and Melilla (Spain) respond to new standards of national security (Battistella 2008, 222). To protect the south-eastern border of Europe from the entry of unauthorized migrants from Turkey, the Greek government has built a 12.5-kilometre long and 4-metre high barbed wire fence along the Evros River. The estimated costs of the wall are 5.5 million euro. This will be added to huge funds invested by the EU in the construction of the Border Surveillance Operational Centre, equipped with thermal cameras. This centre has been operational since February 2012 (Abdallah 2012).

The fortification of borders is just one aspect of the general EU drive towards stricter immigration policies in recent years. As was mentioned above, since 2008 the severe economic crisis and the North African political turmoil boosted the fear for an imminent 'invasion' of people coming from the 'Global South'. Control mechanisms like the EU Internal Security Strategy in Action, EUROSUR and Frontex were activated and/or reinforced to prevent unauthorized entries and expel irregular foreigners. Many national immigration regulations became stricter as well as several quota systems and bilateral agreements were revised rewarding the 'virtuous' sending countries – whose nationals usually migrate to the EU through legal channels – and punishing the 'vicious' ones. Nonetheless, beside a regional protectionism, European governments are far from sharing a common stand in terms of immigration policies. In 2011, the chaotic and indecent reception of thousands of asylum seekers fleeing from the burning south coasts of the Mediterranean Sea revealed the lack of coherent EU policies and solidarity among countries. Italy had to deal alone with 60,000 'irregular migrants', who arrived by boat from North Africa in a few months time. While formally upholding the rights and dignity of all the people involved in such dramatic exodus, the Italian government characterized its response to the 'emergency' with demeaning and discriminative practices (Boca 2012). According to official data, from 1 January 2013 to 14 October 2013 over 35,000 migrants reached the Italian coasts with no entry permit. Of these, 9,805 were coming from Syria, 8,443 from Eritrea, 3,140 from Somalia, 1,058 from Mali and 879 from Afghanistan. Considering the countries of origin, it was estimated that 73 per cent of them (around 24,000 people) needed international protection, but their right to access the refugee recognition process was often violated (CIR 2013).

Despite the enormous publicity given to improvements in immigration screening worldwide, migration experts seem to agree on the fact that tougher migration controls and stricter migration policies have not been effective in curbing unauthorized migration (Cornelius et al. 2004; Papademetriou 2005). On the opposite, easier migration processes and more open immigration policies seem to generate positive effects on the regularity of the inflows (Battistella 2007, 211).

The reasons for the ineffectiveness of the migration policies in the EU can be identified as follows: Firstly, restrictive immigration policies are generally shaped unilaterally by the receiving countries, with little consideration to the fact that irregular migration is deeply rooted in the governance constraints of the countries of origin (Baggio 2007). Secondly, migrants are used to rely more on their social networks – which are able to provide instrumental support (housing, job opportunities, access to social services and so on) to clandestine migrants and overstayers in the countries of destination – than on government institutions (Asis 2004, 29). Thirdly, the demand for unauthorized migrant workers as well as the resolution of many people to migrate at any cost fuelled the flourishing of illegal migration channels managed by illegal recruiters, professional people smugglers and able brokers for the underground labour market (Baggio 2010, 53–4). However, the main reason for the persistence of unauthorized migration may be found in the fact that irregular migrants seem to respond to a particular 'economic need' of many countries of destination. The contribution of unauthorized migrants is critical to fuel the informal labour market, which has become a normal trend in the globalized economies (Vogel 2006, 6–8).

The complex relationship between human mobility and national borders in the EU as well as in other regions of the world poses the first of some ethical questions. Groups of foreigners are often singled out for special checks due to their passport or their ethnic origins. Discrimination by nationality at the immigration controls is a generally accepted practice. The same discrimination is used as criterion for admission by many receiving countries, which justify the selection of specific ethnic groups to fill up the immigration quotas with virtuous migration behaviour, colonial bonds, shared cultural backgrounds, common languages and other affinities. All the restrictions and discrimination mentioned above should be ethically reconciled with the free circulation of workers, which has been a reality for the citizens of the EU member states since 1992. They can decide to reside and work in any European country of their preference, being entitled to equal treatment with the nationals.

Moreover, with the object of improving the EU ability to attract highly qualified workers from third countries, the Council Directive 2009/50/EC was adopted in 2009. This directive determined the issuance of a special work permit, called the 'Blue Card', to highly qualified third-country nationals – and their family members – admitted to the territory of a EU country for the purpose of employment. The Blue Card grants freedom of movement within the EU and other privileges to the holders.[2] While the regulation and implementation of the directive is still under discussion in many EU countries, two ethical questions can be raised here. On the one hand, the EU directive appears to increase the existing discrimination between highly skilled and less-skilled immigrants. On the other hand, the declared object seems to justify proactive brain drain practices by EU member states, which formally committed to reverse the brain drain phenomenon and mitigate its negative effects in the sending countries.[3]

Internalized Borders

When geographical borders are legally crossed, other borders may appear to keep foreigners at the margins of the local society. They are patterns of behaviour, written and unwritten rules, privileges, prejudices and attitudes, which are incorporated as conscious or subconscious guiding principles in the locals' relationship with aliens. These internalized borders may jeopardize migrants' efforts towards their positive inclusion in the receiving community, resulting in the establishment of ghettos and emergence of deviances (Pécoud and de Guchteneire 2007, 18).

The first internalized border is the 'prudential distance' that generally regulates the first approach of locals to foreigners. While prudence is normally caused by previous negative experiences, in this case it is essentially rooted in the widespread fear of the unknown. As a paradox, the unknown will remain such as long as one does not know it and, in the case of human beings, it is only the encounter and engagement with the 'other' that can provide some factual basis. And when the prudential distance is kept particularly with specific foreign groups because of their race, colour, descent or national or ethnic origin, one has to call it bluntly racism, whose expressions are unanimously condemned by 175 countries that are party to the International Convention on the Elimination of All Forms of Racial Discrimination (1966, Article 2).[4] Fairly recent studies have identified widespread ethnic and racial discrimination against migrants in Europe (Wrench et al. 1999). There are serious claims that in the last years racist police raids targeting specific groups of migrants are carried out in the Netherlands and the United Kingdom (Webber 2009).[5]

Within the receiving societies migrants, refugees and other itinerant people are often discriminated because they are not citizens. The assignment of privileges to nationals and restrictions to aliens is a common practice in many countries and constitutes an internalized border that marks the exclusion of foreigners from minor or major sectors of social life. While some of these privileges/restrictions are stated by law in the name of nationalism or protection of the local culture and traditions, others are just part of the daily practice in many receiving societies. This social discrimination is particularly felt by migrants residing in countries that are explicitly upholding ethnic and cultural nationalism, but even in so-called multicultural

---

2   See Directive 2009/50/EC of the Council of the European Union on 25 May 2009 on 'the conditions of entry and residence of third-country nationals for the purposes of highly qualified employment', available at http://eur-lex.europa.eu/LexUriServ/LexUriServ.do?uri=CELEX:32009L0050:EN:NOT (accessed 6 October 2012).

3   See the Joint Africa-EU Declaration on Migration and Development, signed in Tripoli, 22–23 November 2006: http://www.refworld.org/docid/47fdfb010.html (accessed 6 October 2012).

4   See http://www.ohchr.org/EN/ProfessionalInterest/Pages/CERD.aspx (accessed 9 September 2014).

5   See also the open letter, of 30 August 2007, from the Campaign against Racist Police Raids, available at http://no-racism.net/article/2247/ (accessed 6 October 2014).

societies one can find elements of exclusion (Tibe-Bonifacio 2005; Shields et al. 2006). The widespread assumption that there is some kind of 'ontological' difference between nationals and foreigners feeds sentiments of xenophobia within many local communities. Xenophobic attitudes in European countries are also justified with the citizens' fear to see their 'welfare state' undermined by the massive arrival of foreigners (Fireside 2002). Xenophobic manifestations are rampant in the EU in this time of economic crisis, to such an extent that the Declaration of the World Conference against Racism, Racial Discrimination, Xenophobia and Related Intolerance (2001, Article 3 and 7)[6] clearly states:

> Xenophobia, in its different manifestations, is one of the main contemporary sources and forms of discrimination and conflict, combating which requires urgent attention and prompt action by States, as well as by the international community ... Xenophobia against non-nationals, particularly migrants, refugees and asylum-seekers, constitutes one of the main sources of contemporary racism and that human rights violations against members of such groups occur widely in the context of discriminatory, xenophobic and racist practices.

Discrimination against foreigners by locals in the workplace represents the third internalized border. According to a study published by the International Labour Organization (ILO), migrants and ethnic minorities are discriminated at the point of access to employment in Belgium, Germany, the Netherlands and Spain (Zegers de Beijl 2000). However, even in EU countries where they are protected by a regime of equality in wages, foreign workers are frequently less paid than locals for the same job. Moreover, immigrants' professional titles are generally not recognized and therefore they suffer chronic dequalification, as was documented in the case of Italy (Galossi and Mora 2012).

Religious belief is also used as a criterion of discrimination. In Saudi Arabia, Christian and Hindu migrants are not allowed to publically display symbols of their religion. According to a report of the ILO (2007, 37), in Senegal and Sudan Christian job applicants, including foreign workers, had been obliged to convert to Islam to get employed. EU member states are not exempted from such discrimination. By the end of the 1990s the issue was object of a European public discussion to include a special chapter in the formulation of the EU anti-discrimination policy.

There has been an increasing awareness in the last decade that the problem of racial discrimination extends beyond unjustified differential treatment on the basis of skin colour. In particular, it has been argued that there is a growing phenomenon of 'Islamophobia', or 'anti-Muslim' discrimination, and that this has rapidly risen since the late 1980s. Furthermore, it is arguable that most religious discrimination amounts to indirect racial discrimination (Directorate-General for Research 1997).

Despite the efforts of all the EU member states, recent reports (ENAR 2007; Amnesty International 2012) reveal that religious discrimination practices have not been eradicated. Migration inflows contributed to further diversify the religious scenario in Europe. The challenging coexistence of different beliefs unveiled ancient prejudices and unsolved antagonisms.

Internalized borders and exclusion of migrants, refugees and other itinerant people often result in a series of negative consequences. The 'discrimination' between national and migrant workers, on the basis of demagogic protectionisms, often justify practices of migrants' labour exploitation: illegal extension of working time, abusive conditions at the worksite, contract substitution upon arrival, delayed payment or reduction of wages and other kind of cheating (Grant 2005, 10–12; Lee

---

6   Full text available at http://www.un.org/WCAR/durban.pdf (accessed 7 October 2012).

and Piper 2013; OECD 2013). Irregular migrants appear to be the most vulnerable because, due to their unauthorized situation, they are more amenable to silently accept any kind of abuse.

Recent official moves toward the criminalization of irregular migration, with little to no consideration for the reasons and circumstances of such irregularity, constitute a real challenge to the recognition of inalienable human rights beyond the validity of a visa. The criminalization of unauthorized migration provides a legal basis for the confinement of clandestine migrants, overstayers and other visa violators in real prisons called 'detention centres'. The Global Detention Project, an interdisciplinary research endeavour sponsored by Zennstrom Philanthropies and the Swiss Network for International Studies, counted more than 200 detention centres in Europe between 2009 and 2010.[7] Claims about the inadequateness and improper use of detention centres in the EU have been raised by several civil society groups (Rodier and Saint-Saëns 2007; Flynn 2011).[8]

External–Territorial Borders

In the globalized world, receiving countries and regions tend to 'externalize' their borders, introducing new requirements and influencing the migration legislation, policies and practices in the countries of transit and departure. For many migrants and tourists, the crossing of the international border begins at the receiving country's diplomatic post in their home country with the request for a visa. Being in nature a mere endorsement made on a passport by the proper authorities allowing the bearer to leave for a given country, a visa is not an assurance of admission, since upon arrival, immigration officers can still deny the entry of a visa holder on a legal basis. In recent years, the application screening process has generally become longer and more meticulous. The verification and examination of information and documents and the introduction of technological improvements resulted in a considerable increase of visa fees, which are often charged to applicants (IOM 2012). Applications via mail and e-mail have become a normal custom and individual follow-ups are often not allowed. Personal interviews with candidates, entrusted to consuls or visa officers, have become a normal practice and great discretion is granted to the interviewing officers.

Bilateral agreements between countries of origin and destination of international labour migration may represent an effective measure to assure decent working and living conditions to migrants (GCIM 2005, 69–70). Nonetheless, with the introduction of specific conditions and criteria, they are used by some receiving countries to externalize the screening and selection of immigration candidates. Moreover, bilateral agreements often try to influence the drafting of emigration policies in the countries of origin, giving way to concessions (for instance discriminative salaries, limited labour rights, compulsory HIV tests and so on) that jeopardize the civil liberties of their citizens. Finally, the decision to sign agreements with some countries and not with others or to include in the deal some categories of migrants and others not reveal clear intentions of differentiated exclusion. In recent times, other elements of bilateral relationships have been used to regulate international migration flows. There are cases where the commitment of sending countries to development aid has been bound to their assurance of stricter emigration control and compulsory reacceptance of deported migrants (Pécoud and de Guchteneire 2007, 3–4).

---

7   See www.globaldetentionproject.org (accessed 14 September 2014).

8   See also the position statement of JRS Europe, available at http://www.detention-in-europe.org/ (accessed 8 October 2012).

There is also a 'physical' dimension of the externalization of national/regional borders. The identification of 'transit countries' has catalyzed the attention of many migration policymakers in North America and Europe, who multiplied their efforts to enact mechanisms able to set immigration controls beyond the real national/regional borders. A study of Kimball (2007) presents the cases of Mexico and Morocco. On the basis of empirical evidence, the author shows how the United States and the European Union exert significant pressure on Mexico and Morocco respectively to restrict transit migration. Nevertheless, Kimball (2007, 148–9) argues that these transit countries' restrictive immigration measures went beyond simple compliance with their northern neighbours' objectives. Mexico and Morocco have already embraced a 'transit migration police role' from which they believe to gain in terms of economic and regional integration. The intention of externalizing the national borders – and all the related immigration problems and humanitarian concerns – seems more evident in the case of the European Union, where all member states have engaged in the constitution of an 'immigration security belt' involving several countries in the South and East:

> Furthermore, negotiations between the EU, single member states and the actual and future neighbors are designing the construction of a 'security belt' against undocumented migration. Readmission-agreements, safe third-country regulations, and lists of secure countries of origin are some of the instruments used by the administrations of the EU and its member states to move the problem of undocumented immigration away from the center of the Union. As well, southern neighbors like Morocco, Algeria and Libya as well as eastern neighbors such as Ukraine are being integrated into the safeguard system against undocumented migration – and therefore are being asked to strengthen their border controls and to readmit those migrants who entered the EU via their territories. (Alscher 2005, 1)

In the last decade, some receiving countries decided to externalize their national borders setting up initial immigration checks at the international airports of sending or transit countries. Such 'deployment' of immigration controls poses a series of questions about international spaces, national sovereignty and dynamics of reciprocity. Moreover, the airline carriers have been progressively entrusted more immigration screening responsibilities. According to the Convention on International Civil Aviation,[9] airlines' check-in operators are asked to assess the validity of passports and visa and flight assistants have to distribute and explain to passengers immigration and custom forms. Airline companies are also responsible of the conveyance of improperly documented travellers. In the United States, airline staffs are tasked to retrieve the departure forms of travellers. For in-bound flights to New Zealand, airline carriers are requested to check evidence of passengers having sufficient funds for their stay, as per the requirement of New Zealand's immigration law.

Towards the Ethicization of the Human Mobility–Borders Nexus

As is diffusely shown above, the policies and practices discussions resulting from the complex relationship between human mobility and borders, considered in their national, internalized and external-territorial dimensions, lead us to pose a series of ethical questions that deserve answers. Even in a world dominated by economistic considerations, human mobility cannot be dealt with

---

9   Available at http://www.icao.int/publications/Documents/7300_cons.pdf (accessed 8 October 2012).

in the same way as the import or export of commodities or services and provisions. The fact that 'people' and not 'goods' are at stake compels the undertaking of a deep ethical reflection that would clarify the frequent confusion between goals and means. This chapter would like to contribute to this reflection through the identification of some principles that should guide the ethical judgment addressing human mobility and border questions.

Respect for Equal and Inalienable Human Rights

The Universal Declaration of Human Rights, approved and adopted by the UN General Assembly on 10 December 1948, can be fairly considered as an (almost) world ethic platform. The pertinent ethical principle is the respect for every single right contained in the declaration. All the rights are not 'assigned to' but 'recognized for' every person equally and inalienably. The 1948 declaration was elucidated and codified into two international covenants: the International Covenant on Civil and Political Rights and the International Covenant on Economic, Social and Cultural Rights. Both documents were approved by the United Nations on 16 December 1966 and entered into force on 23 March 1976. There are other UN conventions aiming at underlining the rights of specific categories of people, like the Convention on the Rights of the Child (1989). Among those that are more closely dealing with human mobility, one should mention the Convention Relating to the Status of Refugees (1951) and the International Convention on the Protection of the Rights of All Migrant Workers and Members of Their Families (1990).

Declarations, covenants and conventions state the inalienability and inviolability of the rights that belongs to every single human being according to the principle of non-discrimination. Respect for the rights represents a 'universal' criterion to judge the 'fair' behaviour of individuals, groups, societies and states, that is, a true ethical principle. Beside the notorious universal consensus apparently granted to the human rights contained in the above-mentioned documents, so to recognize for them a kind of sacred value, there are evident limitations that should be considered.

Firstly, the Enlightenment roots of the Declaration of Human Rights as well as the environment that determined their rational definition clearly betray their 'Western' connotation. Within the UN gatherings, the representatives of eastern states repeatedly debated the universality of the human rights listed in the declarations, noting that some of these rights contradict non-Latin and non-Anglo-Saxon cultural and historical traditions.

Secondly, the human rights recognized in the 1948 declaration seem to be inspired only by the Enlightenment ideals of 'freedom' and 'equality'. The third Enlightenment ideal, meaning 'fraternity', appears visibly neglected. This ideal, which as a paradox insinuates more globalizing pretensions in the field of human relationships, may serve to ground complementary ethical principles like the coresponsibility in the world stewardship, universal solidarity and global citizenship.

Thirdly, the ethical principle of respect for human rights is 'limited' to the protection of the holders aiming at preventing – and/or punishing – the violation of their fundamental rights. Such principle sets minimum levels of 'humanity'. However, it does not express at all the duty of individuals, groups, societies and states to promote the welfare and the holistic development of every person, family and collectivity beyond such levels.

Even with these limitations, respect for human rights constitutes an important paradigm towards the formulation of ethical judgements. The real extent of the human rights' universality may be enhanced through the dialogue with non-Latin and non-Anglo-Saxon traditions, looking for common elements.

Promotion of Human Dignity

The concept of 'human dignity', which is an inheritance of the ancient Roman world, historically developed into two different meanings. On the one side, human dignity indicates the special position of the human being within the universe. On the other side, it simply refers to one's status in public life. In both cases the expression signifies an element of differentiation/supremacy within the contest of reference. However, while in the first case the meaning is universalistic and absolute, that is, inherent to all human beings, in the second case it is particularistic and relative, that is, it refers to an individual person as far as he/she acts consistently with his/her status (Becchi 2009).

The ethical principle of the promotion of human dignity clearly refers to the first meaning. Christian philosophy grounds such dignity on the creation of all human beings in the 'image of God' (see Gen. 1:27). Nevertheless, even without reference to the transcendence, human dignity can be conceptualized. An important author of modern *Jus Naturalism*, von Pufendorf (2009), grounded human dignity on the concept of freedom, in which all human beings are able to choose and self-impose norms that are considered universal. With or without transcendental reference, the promotion of human dignity is translated into the advancement of what constitutes human beings as unique, different and inimitable persons, being the latter either their identity in the image of God or their ability to choose.

From a pure ethical perspective, human beings can never be 'means' to achieve a 'goal' other than their own full personal fulfilment. Everything that may obfuscate the divine image or diminish the ability to choose is to be avoided. On the contrary, any action aiming at reaffirming and enhancing the uniqueness, diversity and inimitableness of the human being should be promoted. The proactive dimension of the principle of the promotion of human dignity may somehow compensate the above-mentioned third limitation of the principle of respect for human rights. However, it should be noted that also in this case the conceptual development of human dignity is marked by a prominent Western connotation.

Superiority of the Common Good

The superiority of the 'common good' is one of the major principles guiding contemporary politics. Nonetheless, there is no universal consensus on the definition of the common good. Some schools of thought (for instance that of Machiavelli) argue that the common good is constituted by the conditions needed for the welfare and development of the society or the state understood as an independent entity from the people who compose it. For other schools (for instance that of Hegel) the common good is the welfare and development of the majority of people constituting a given community. In both cases the superiority of the common good over the individual good is evident and justifies the sacrifice of individuals where necessary.

According to the Christian perspective, the 'common good' is the sum of conditions that allow the full fulfilment of the society as well as the people who compose it. It is the ultimate goal of the social actions of individuals and institutions. The common good is the reason for the different forms of social groupings, from family to state (Pontifical Council for Justice and Peace 2004, 164). Since the relational dimension of human beings is essential to their personal fulfilment, the latter depends deeply on the fulfilment of 'others', that is, of people they relate to. The lack of other people's fulfilment ultimately affects one's interpersonal relationships and therefore one's own fulfilment. The principle of the superiority of the common good over the individual good is based on the fact that the good of all is also the good of every person involved. And this principle applies to different social groups, from family to state (165).

The principle of the superiority of the common good is the foundation of 'distributive justice', which regulates the relationship between the individual and society as a whole. Distributive justice deals with the 'distribution' of the common good according to what a person deserves. The right of individual persons, members of a given society, is not determined on the basis of an equitable assessment, but on a series of subjective conditions: attitudes, contributions, merits, needs, responsibilities and functions. Distributive justice does not give everyone the same, but it gives to each what is due, with the aim of achieving a greater common good. Institutions must consider the subjective conditions prior to the distribution of the common good, so that each member of society might be equally fulfilled. The growth of the common good depends on the realization of all individuals. Distributive justice is entrusted to the institutions. The concrete determination of the distributive justice is not easy, because it doesn't respond to the criterion of arithmetic equality. Such exercise requires the deep and fair knowledge of the concerned individuals, clear and wise judgment and selfless foresight.

The first limitation of this principle is the different interpretation of the common good within the same Western reflection. The way in which the state and society interact with individuals varies depending on what is referred to as the goal of the action, that is, the common good. It is therefore necessary to read the superiority of the common good in the light of the first two principles, meaning respect for human rights and the promotion of human dignity.

Universal Destination of Goods and Solidarity

The ethical principle of solidarity is the common heritage of the great religions and many secular ethics. Some religious doctrines grounded this principle on the universal destination of goods, which corresponds to a precise determination of the supreme creator or a cosmic order that guides human history.

Many secular ethics explicitly refer to philanthropy and solidarity as excellent virtues, but these generally don't represent a moral obligation for individuals. It should be added, however, that solidarity is considered as a 'duty of civilization' in many parts of the world. Although historically solidarity has often been limited to the inner groups, the significant increase in concrete demonstrations of solidarity among countries and groups of people over the last several decades gives us reason to be optimistic. It is worth recognizing the important contribution of modern technology to 'globalizing' local concerns and demands of aid, enhancing considerably the spread of solidary actions to and from every place on earth (Sánchez Sorondo 2002).

Deeply rooted in the Jewish tradition, the Christian doctrine states the inviolability of private property but understands private wealth as a gift of God to be used for the good of the entire humankind. The words of Jesus of Nazareth emphasize the contingency of material goods together with the duty to share them with those in need. The selfish use of private wealth deserves divine punishment. The first Christian community in Jerusalem is presented in the Acts of the Apostles 2:44–5 as a case study of distributive justice: 'All the believers were together and had everything in common. They sold property and possessions to give to anyone who had need.'

According to the Islam, the whole creation is for the good of humanity and private wealth is to be considered as a gift, which reveals a predilection by Allah. There is a moral duty of solidarity with the less fortunate, which is grounded on the values of sharing and solidarity (Aala Maududi 2012). Hinduism and Buddhism teach that all goods meet a specific universal design and are distributed according to the role that each person plays in society. Every 'commendatory' person must practice charity and solidarity (Sivananda 1997; Sivaraksa 1987; Jotika of Parng Loung 2012).

Confucianism considers the difference between the rich and the poor as an opportunity for interpersonal cooperation. Charity and solidarity should normally mark the relationship between government leaders and the people they are responsible for and between one person and another, according to the possibilities of each one (Riva 2007, 426–30).

Global Stewardship and Co-responsibility

The ethical principle of co-responsibility is based on the awareness of the duty of all the members of humankind towards the proper stewardship, sound use and harmonious development of the natural and environmental resources. With no prejudice to the principle of subsidiarity, the principle of co-responsibility imposes the obligation to act in aid to whom – individuals, groups, societies and states – are not able, for different reason, to fulfil the obligation alone.

The processes of globalization led the different secular ethics to consider in their discussions the undeniable reality of the global impact of local decisions and actions. Humanitarian and ecological issues are seen as more related to each other, with a growing emphasis on both the individual and the collective responsibility in building the future of the world. The same processes of globalization revealed the increasing interdependence of domestic economies. Due to this interdependence, the international cooperation to develop other countries is increasingly seen in terms of international investment. For the societies and the states that recognized and adopted it, the principle of co-responsibility is a major step forward from the principle of solidarity. In fact it does not depend on a philanthropic sensitivity, which is a sign of civilization, but rather on the acceptance of the duty that derives from understanding the global beneficial impact of every humanitarian and ecological choice.

According to the Christian understanding, which is mediated from the Jewish tradition, the task to govern the universe was entrusted to humankind though an explicit divine disposition at the beginning of the world. The biblical account of the creation of the universe clearly states that the 'Creator', God, chose human beings as stewards of the whole world, which was intended to provide them with the resources necessary to survive and grow (see Gen. 1:26–31 and 2:8–17). Several biblical texts state that God is the true 'owner' of creation; humanity is only the administrator who must be ready to give an account of his administration to the owner at any time. The preaching of Jesus of Nazareth further clarified the way such stewardship should be performed. The Christian interpretation of the 'likeness of God', in which the first human beings were created (see Gen. 1:26), explains every human being's obligation to adopt the same loving care towards creation that the Creator has.

According to the Qur'an, Allah entrusted the administration of the whole creation (*amana*) to humankind. All human beings must be accountable directly to Allah, who, in his mercy, guide and assist them in the management of the *amana* (Denny 1998).

The Hindu and Buddhist cosmology argues that there are forces in the natural world that interact with life itself. The human race has no privileged position in the cosmos and must show great respect for the preservation of all living species. In this sense, humankind's stewardship is to be understood as part of the 'global responsibility' (and global co-responsibility), which is the task of every living being (Chapple and Tucker 2000; Batchelor and Brown 1992).

The Confucian view of the world is basically 'antropocosmic': the sky (a driving force), the earth (the nature) and human beings interact intrinsically. In this global interaction, the relationship between human beings and nature is determined by a strict moral responsibility, which collective dimension is highly stressed (Tucker 1998, 1).

Global Citizenship

The principle of global citizenship has not received much attention in traditional ethical reflections. However, due to the growing relevance of the phenomenon of human mobility in recent decades, nowadays it represents an interesting object for scholars. It is not a new principle, as its roots are found in the most ancient religions and philosophies. Most of the secular ethics insist on the concept of global fraternity, highlighting the fact that one's nationality, which can be given and taken away, cannot be a real source of identity. In particular, the socialist tradition considers national belonging as a denial of the universal brotherhood of human beings (Weiner 1996, 176).

The constitution of the modern European states that formalized the difference between citizens and foreigners is intrinsically bound to the history of Christianity. However, it is worth noting that Christian faith itself promotes a sense of belonging that transcends national borders and even those of the immanent world. Such Christian 'universalism' is based on two main ideas: (1) the true homeland for Christians is not in this world, but in the world to come, that is, the Kingdom of Heaven or Paradise; (2) every human being is called to be a universal citizen in the 'Kingdom of God' inaugurated by Jesus of Nazareth (Tirimanna 1999).

Original Muslim traditions point to an Islamic world, *Ummah Islamia*, that exists beyond the borders of countries. Within the *Ummah Islamia* freedom of movement must be guaranteed to all believers, as citizens of the Islamic world. This freedom of movement has been codified in Article 23 of the Islamic Declaration of Human Rights (1981).[10] The geographical expansion of the *Ummah Islamia* also entails the expansion of this 'global citizenship' (Aldeeb Abu-Sahlieh 1996).

According to the Hindu and Buddhist scheme of evolution/involution and the related theory of transmigration of souls, the whole cosmos is the natural theatre of human existence. Countries are just historical and contingent determinants of one of the possible existences. In accordance with its universalistic spirit, the very concept of nation and homeland does not seem to belong to the Hindu worldview (Jain 1994; Kawada 2001).

The Confucian ethic proposes a holistic and changing cosmology. Individual happiness, which is the ultimate goal of the acts of every human being, cannot be achieved without universal peace. Therefore wars and rivalries should be abolished in order to establish the 'Big Unity' of the world (Confucius Humanitarianism 2012).

The ethical principle of global citizenship is the ethical principle that most strongly undermines the contemporary immigration policies, which two main pillars are the national sovereignty and security of nationals, as well as, on the background, ideas around 'national identity' constructing internalized borders.

Conclusion

This reflection on national, internalized and external-territorial borders only reaffirms the paradox of a globalizing world which gets more compartmentalized, a world where goods and

---

10   See http://www.alhewar.com/ISLAMDECL.html (accessed 6 October 2014).

capital seem to enjoy greater freedom than human beings, who are supposed to be the owner of the former. From the ethical perspective borders are contested by the principles that guide ethical judgment. Far from offering a solution to the paradox, this chapter tried to identify some principles composing an initial ethical paradigm, which should serve to assess the complex contemporary scenario of migration and refugee policies at the global, regional and national levels. Such ethicization of the human mobility–borders nexus envisions a more consistent and humanized politics. In the name of the principle of democracy, all stakeholders, that is, governments, academia, trade unions, NGOs and migrants, should be involved in this exercise. Some limitations of the proposed ethical platform have been mentioned. Some others may be found in the selection of religious and philosophic traditions. Further reflection is deeply needed, as well as the engagement of concerned thinkers of diverse backgrounds.

References

Abdallah, J. 2012. "Human Rights and State Security: The Conflicting Features of International Migration." *E-International Relations*, 19 September. Accessed 6 October 2012. http://www.e-ir.info/2012/09/19/human-rights-and-state-security-the-conflicting-features-of-international-migration/.

Aala Maududi, S.A. 2012. "The Economic Principles of Islam." *Islam101*, 9 October. http://www.islam101.com/economy/economicsPrinciples.htm.

Aldeeb Abu-Sahlieh, S.A. 1996. "The Islamic Concept of Migration." *International Migration Review* 30 (1): 37–57.

Alscher, S. 2005. "Knocking at the Doors of 'Fortress Europe': Migration and Border Control in Southern Spain and Eastern Poland." CCIS Working Paper 126. San Diego: Center for Comparative Immigration Studies.

Amnesty International. 2009. "Detention on Christmas Island." Accessed 9 October. http://www.amnesty.org.au/refugees/comments/20442/ (site discontinued).

_____. 2012. *Choice and Prejudice: Discrimination against Muslims in Europe*. London: Amnesty International.

Asis, M.M.B. 2005. *Preparing to Work Abroad: Filipino Migrants' Experiences prior to Deployment*. Quezon City: Scalabrini Migration Center.

Baggio, F. 2007. "Migrants on Sale in East and Southeast Asia: An Urgent Call for the Ethicization of Migration Policies." In *Mondialisation, migration et droits de l'homme: Un nouveau paradigme pour la recherche et la citoyenneté/Globalization, Migration and Human Rights: A New Paradigm for Research and Citizenship*, edited by M.C. Caloz-Tschopp and P. Dasen, 715–64. Brussels: Bruylant.

_____. 2010. "Fronteras nacionales, internalizadas y externalizadas." In *Migraciones y fronteras: Nuevos contornos para la movilidad internacional,* edited by M.E. Anguiano Téllez and A.M. López Sala, 49–73. Barcelona: Barcelona Centre for International Affairs.

Batchelor, M., and K. Brown, eds. 1992. *Buddhism and Ecology*. London: Cassell.

Battistella, G. 2007. "Migration without Borders: A Long Way to Go in the Asian Region." In *Migration without Borders: Essays on the Free Movement of People*, edited by A. Pécoud and P. de Guchteneire, 199–220. Oxford: Berghahn Books.

_____. 2008. "Irregular Migration." In *World Migration Report 2008: Managing Labour Mobility in the Evolving Global Economy*, edited by the International Organization for Migration, 201–33. Geneva: IOM.

Becchi, P. 2009. "Il dibattito sulla dignità umana: tra etica e diritto." *Persona e danno*, 27 April. Accessed 9 October 2012. http://www.personaedanno.it/index.php?option=com_content&view=article&id=29797.

Boca, C. 2012. "I barconi della speranza: Diritti e doveri." In *Mediterraneo: Crocevia di popoli*, edited by F. Baggio and A. Skoda Pashkja, 99–112. Roma: Urbaniana University Press.

Chapple, C.K., and M.E. Tucker, eds. 2000. *Hinduism and Ecology: The Intersection of Earth, Sky and Water*. Cambridge, MA: Harvard University Press.

CIR (Consiglio Italiano per i Rifugiati). 2013. "Rafforzare la protezione per i rifugiati e rispettare obblighi internazionali." *CIR*, 15 October. Accessed 6 October 2014. http://www.cir-onlus.org/index.php?option=com_content&view=article&id=896:cir-rafforzare-la-protezione-per-i-rifugiati-e-rispettare-obblighi-internazionali&catid=13:news&Itemid=143&lang=it.

Confucius Humanitiarism. 2012. "What is Confucianism?" Accessed: 9 October. http://terpconnect.umd.edu/~tkang/welcome_files/what.htm.

Cornelius, W.A., T. Tsuda, P.L. Martin, and J.F. Hollifield, eds. 2004. *Controlling Immigration: A Global Perspective*. Stanford: Stanford University Press.

Denny, F.M. 1998. "Islam and Ecology: A Bestowed Trust Inviting Balanced Stewardship." *Earth Ethics* 10 (1): 10–11.

Directorate-General for Research. 1997. "European Union Anti-Discrimination Policy: From Equal Opportunities between Women and Men to Combating Racism." Working Paper 12. Luxembourg: European Parliament. Accessed 7 October 2012. http://www.europarl.europa.eu/workingpapers/libe/102/default_en.htm.

ENAR (European Network Against Racism). 2007. "Religious Discrimination and Legal Protection in the European Union." Fact sheet 34. Brussels: ENAR.

Fireside, H. 2002. "The Demographic Roots of European Xenophobia." *Journal of Human Rights* 1 (4): 469–79.

Flynn, M. 2011. "Immigration Detention and Proportionality." Global Detention Project Working Paper 4. Geneva: Global Detention Project.

Galossi, E., and M. Mora. 2012. "Employment Discrimination against Migrant Workers in the Italian Labour Market." *Migrationspolitisches Portal*. Accessed 7 October. http://www.migration-boell.de/web/migration/46_1815.asp.

GCIM (Global Commission on International Migration). 2005. *Migration in an Interconnected World: New Directions for Action; Report of the Global Commission on International Migration*. Geneva: GCIM.

GFMD (Global Forum on Migration and Development). 2008. "Civil Society Days Report to the Government Meeting, Manila 2008." Accessed 18 October 2012. http://www.mekongmigration.org/mmn/?p=230.

Grant, S. 2005. "International Migration and Human Rights." Paper prepared for the Policy Analysis and Research Programme of the Global Commission on International Migration. Geneva: GCIM.

ILO (International Labour Organization). 2007. *Equality at Work: Tackling the Challenges*. Geneva: International Labour Office.

IOM (International Organization for Migration). 2008. *World Migration Report 2008: Managing Labour Mobility in the Evolving Global Economy*. Geneva: IOM.

———. 2011. *World Migration Report 2011: Communicating Effectively about Migration*. Geneva: IOM.

IOM (International Organization for Migration). 2012. "Visa System." Accessed 9 October. http://www.iom.int/jahia/Jahia/about-migration/managing-migration/passport-visa-systems/visa-systems/cache/offonce;jsessionid=54BE028764B215F5676BBD444D452D13.worker01 (site discontinued).

Jain, G. 1994. *The Hindu Phenomenon.* New Delhi: UBS Publishers.

Jotika of Parng Loung. 2012. "The Function of Wealth in Buddhism." Accessed 9 October 2012. http://www.khmerbuddhism.com/profiles/blogs/the-function-of-wealth-in (site discontinued).

Kawada, Y. 2001. "The Importance of the Buddhist Concept of Karma for World Peace." In *Buddhism and Nonviolent Global Problem-Solving: Ulan Bator Explorations,* edited by G.D. Paige and S. Gilliatt, 103–14. Honolulu: Center for Global Nonviolence.

Kimball, A. 2007. "The Transit State: A Comparative Analysis of Mexican and Moroccan Immigration Policies." Working Paper 150. San Diego: Center for Comparative Immigration Studies.

Lee, S., and N. Piper. 2013. *Understanding Multiple Discrimination against Labour Migrants in Asia: An Intersectional Analysis.* Berlin: Friedrich Ebert Stiftung.

OECD (Organisation for Economic Co-Operation and Development). 2013. *International Migration Outlook 2013.* Paris: OECD Publishing.

Papademetriou, D.G. 2005. "The Global Struggle with Illegal Migration: No End in Sight." *Migration Information Source,* 1 September. Accessed 7 October 2012. http://www.migrationpolicy.org/article/global-struggle-illegal-migration-no-end-sight/.

Pécoud, A., and P. de Guchteneire. 2007. "Introduction: The Migration without Borders Scenario." In *Migration without Borders: Essays on the Free Movement of People,* edited by A. Pécoud and P. de Guchteneire, 2–30. Oxford: Berghahn Books.

Pontifical Council for Justice and Peace. 2004. *Compendium of the Social Doctrine of the Church.* Vatican City: Libreria Editrice Vaticana.

Pufendorf, S. 2009. *Two Books of the Elements of Universal Jurisprudence.* Indianapolis: Liberty Fund.

Riva, E. 2007. *Manuale di filosofia: Dalle origini a oggi.* Torino: Ernesto Riva.

Rodier, C. and I. Saint-Saëns. 2007. "Contrôler et filtrer: Les camps au service des politiques migratoires de l'Europe." In *Mondialisation, migration et droits de l'homme: Le droit international en question/Globalization, Migration and Human Rights: International Law under Review,* edited by M.C. Caloz-Tschopp and P. Chetail, 620–37. Brussels: Bruylant.

Sánchez Sorondo, M. 2002. *Globalizzazione e solidarietà.* Vatican City: Pontificia Academia Scientiarum.

Shields, J., K. Rahi, and A. Scholtz. 2006. "Voices from the Margins: Visible-Minority Immigrant and Refugee Youth Experiences with Employment Exclusion in Toronto." CERIS Working Paper 47. Toronto: Joint Centre of Excellence for Research on Immigration and Settlement.

Sivananda, S.S. 1997. *All About Hinduism.* 6th ed. Uttar Pradesh: The Divine Life Society.

Sivaraksa, S. 1987. *Religion and Development.* Bangkok: Thai Inter-Religious Commission for Development.

Tibe-Bonifacio, G.L.A. 2005. "Filipino Women in Australia: Practising Citizenship at Work." *Asian and Pacific Migration Journal* 14 (3): 293–326.

Tirimanna, V. 1999. "La chiesa e il superamento delle frontiere." *Concilium* 35(2), 119–32.

Tucker, M.E. 1998. "Confucianism and Ecology: Potential and Limits." *Earth Ethics* 10 (1): 1–9.

Vogel, R.D. 2006. "Harder Times: Undocumented Workers and the U.S. Informal Economy." *Monthly Review* 58 (3): 6–18.

Weiner, M. 1996. "Ethics, National Sovereignty and the Control of Immigration." *International Migration Review* 30 (1): 171–97.

Webber, F. 2009. "Crusade against the Undocumented." *Institute of Race Relations*, 5 February. Accessed 14 April 2009. http://www.irr.org.uk/2009/february/ha000011.html.

Wrench, J., A. Rea, A., and N. Ouali, eds. 1999. *Migrants, Ethnic Minorities and the Labour Market: Integration and Exclusion in Europe.* London: Macmillan.

Zegers de Beijl, R. 2000. *Documenting Discrimination against Migrant Workers in the Labour Market: A Comparative Study of Four European Countries.* Geneva: International Labour Office.

PART III
PLACES OF TRANSFER
AND TRAJECTORIES

Chapter 13

Doing Borderwork in Workplaces: Circular Migration from Poland to Denmark and the Netherlands

Marie Sandberg and Roos Pijpers

The 'Code Orange' alert issued by the Dutch minister of Social Affairs, Lodewijk Asscher, in the fall of 2013, urging EU member states to join forces in a battle against the unwanted consequences of the migration of 'low-skilled' workers from new EU member states (*De Volkskrant*, 17 August 2013) illustrates his frustration with the lack of control over a complex and rapidly changing migration system. In particular, the role of market actors in sustaining migration flows from Poland and other EU accession countries as well as the construction of immigrant communities is proving to be extremely difficult to understand, let alone manage (Garapich 2008). Migration scholars are confronted with the challenge of naming and labelling developments, a situation which results in combined notions of temporary and transnational migration such as 'lasting temporariness' (Grzymała-Kazłowska 2005) and other terms which remain open for further discussion.

The threshold approach coined by Van der Velde and Van Naerssen (2011) offers several opportunities to rethink and broaden our understanding of what motivates Polish and other Eastern European workers to migrate. Importantly, the approach is constituted by 'thresholds' accompanying each of these stages: the mental border threshold, the locational threshold and the trajectory threshold. A key feature of the threshold approach is the combination of a bounded rationality perspective with a more structuralist understanding of migration. People are 'driven' to consider cross-border mobility because of relative differences and changes in societal structures on both sides of the border. These may be related to changes in government regulation, job opportunities and social structures, and they challenge a posture of indifference towards mobility. Van der Velde and Van Naerssen (2011, 219) argue that 'in future elaborations of this approach the diversity of international mobility (commuting, labour and pension migration, international tourism) … has to be addressed as well.' This chapter aims to discuss two specific forms of international mobility: temporary and circular migration. It does this by offering 'hands-on' insights into how circular migrants overcome different thresholds through the enactment of multiple borders in different yet related ways. Therefore, this chapter is focused on the kinds of activities and practices that can be regarded as examples of doing 'borderwork' (Rumford 2006). Borderwork designates a process in which the border is continuously enacted, sustained and transcended in and through practices such as those of the labour migrants. In particular, borderwork in workplaces will be considered as sites where the daily experiences of migrant mobility can be observed. The idea is to look at how migratory practices are rendered possible and desirable through not only the independent choices of individuals but rather through specific 'assemblages' of a heterogeneous set of objects, devices and settings.

This chapter brings together two workplace environments, in Denmark and the Netherlands, focusing on Polish labour migrants working in these environments. By examining the similarities and differences between these two cases, the aim is furthermore to address comparative approaches to studies on borders and circular migration practices. Often, however, comparison is made on the basis of assumptions and components of a *tertium comparationis* identified prior to the comparison in question (Sørensen 2008). The chapter will therefore discuss issues of 'comparativism' versus 'comparison as achievement', where the latter aims at establishing the comparative sites as outcomes rather than a starting point.

Intra-European Labour Migration: Regulations and Flows

The transitional restrictions on free movement of labour applying to Polish workers lasted in the Netherlands from 2004 until 2011, and in Denmark from 2004 until 2009.[1] Restrictions for workers from Romania and Bulgaria were lifted in 2014; a number of countries including Germany, the UK and the Netherlands imposed these restrictions for the maximum amount of time. This meant that workers from Romania and Bulgaria could only formally access these countries through a valid work permit, which is the most common way for states to regulate the labour market entry of non-natives, or through bilateral (seasonal) labour agreements.

In retrospect, the transitional restrictions imposed before the 2004 EU enlargement, when eight Central and Eastern European countries joined, can be said to have functioned as a repel factor, especially in the case of Germany (Van der Velde and Van Naerssen 2011). Because the route to Germany, one of the first countries to impose the restrictions and one of the last to drop them, was closed off for a relatively long period of time, Ireland, the UK and Sweden became more popular as destinations. In this sense, Denmark and the Netherlands have shared a status as alternatives to Germany insofar as labour migration from the 2004 accession states is concerned.

It has been notoriously difficult to assess the volume and impact of migration from these countries to Germany and surrounding states such as the Netherlands and Denmark, which has taken place despite the existence of transitional restrictions. The majority of investigations have focused on labour migration from Poland, by far the largest country to enter the EU in 2004. However, publications documenting the Polish emigration context around the time of the 2004 EU enlargement round suggest that this impact should be evaluated in light of the fact that labour migration between Poland and Western Europe occurred before that time as well (Kępińska 2005; Garapich 2008). In the words of Kępińska (2005, 28): 'Perhaps inertia, networks, established employment relationships, and entrenched demand patterns serve to sustain flows independently of the underlying institutional environment.' Illegal employment aside, the establishment of contacts between individual agricultural employers and so-called 'German Poles' played an important role in the labour migration patterns to Germany. This term refers to people living in parts of Poland that formerly belonged to the Prussian Empire who are eligible for a German passport if they comply with a number of criteria, notably proof of German ancestry. Since the creation of the European single market in 1993, citizens of any EU-15 member state have been entitled to free movement of labour and may seek paid employment in any other member state without a work permit. Hence, a German passport implied access to EU-15 labour markets. Also, long-term bilateral agreements between Poland and EU-15 countries have resulted in the onset of specific migration flows, for instance in the migration of Polish nurses (Pool and De Lange 2004). The overwhelming attention to emigration from Poland and the possible socio-economic impact of this emigration has

---

1 Denmark imposed the transitional restriction, the so-called 'Østaftalen' (Eastern Agreement) in order to secure that the labour migrants coming from Eastern Europe worked in accordance with national pay and working conditions to avoid what has later been commonly designated as social dumping (Hansen and Hansen 2009). In the Netherlands, similar arguments led to the restrictions.

somewhat overshadowed the fact that Poland still receives a significant flow of labour migrants from other Eastern European countries like Ukraine and Belarus (Kindler 2012). The mobility across European borders can thus more accurately be imagined as a chain of migration flows rather than a one-directional East-West movement.

Assemblages and Borderwork

In scrutinizing the various practices of the Polish labour migrants this chapter applies the concepts of 'assemblages' and 'borderwork'. The concept of assemblage relates to the works of Deleuze and Guattari ([1987] 2004) and is conceived as a material-discursive heterogeneity made up of various entities, devices and materials that are configured in such a way that it becomes practiceable (or temporarily) stable (Irwin and Michael 2003; see also Law 1991).[2] The concept designates a provisional mode of ordering that is not necessarily part of any larger structure or order. However, when exploring assemblage one looks for the particular composition of a specific assemblage rather than emphasizing aspects of 'messiness' (Law 2004). Assemblage is thus used here as a heuristic tool that can help depict the role of material-discursive entities in the process of rendering migratory practices feasible, be it the particular regulations and directives within the relevant policy areas, the technology used when borders are crossed, or the role of family members who stay behind but play an important part in making the migratory ventures possible. In other words, assemblage is applied with the aim to expand the range of entities and actors involved in circular migration due to the fact that it takes more than one wo/man's choice to make a migratory practice possible (Sandberg 2012b). Hence, the assemblage approach is intended to contribute to a broadening of the usual explanatory schemes of migratory decision-making processes as argued in the threshold approach.

With the concept of 'borderwork' British sociologist Rumford (2006, 2008) suggests a way to conceptualize a general dispersal of the world's borders. From being concrete manifestations of formal arrangements between states placed at the very outskirts of the territory, borders have become spatially displaced and dispersed within society. The ongoing processes of making and remaking international borders is therefore no longer an endeavour exclusively reserved for states. Borderwork takes place, is negotiated and 'worked upon' also within the realm of everyday life practices. Borderwork is thus in principle carried out everywhere by everyone, such as when hotelkeepers and internet café owners document accommodation dates and data traffic (Rumford 2008).

In border studies the concept of borderwork reflects a wider turn to practice within contemporary social theory (Andersen and Sandberg 2012). Increasingly, it is considered necessary to analyse the many different ways borders are socio-spatially constructed, distributed and enacted in everyday life in order to understand the various de- and re-bordering processes of contemporary Europe (see for instance Van Houtum 2005; Van Houtum and Van Naerssen 2002; Linde-Laursen 2010; Löfgren 2008; Sandberg 2009a, 2009b, 2012a).

The Polish circular migrants from the two cases in this chapter can be said to perform borderwork as they conduct their work-lives across national borders. It will be shown how the Polish labour migrants use the loopholes of national regulations and convert them into windows of opportunity for their circular migratory practices. Hence, this analysis shows little evidence of the fact that society has become increasingly securitized ('securitization of everyday life'), which is obviously

---

2   In this chapter we combine the concept of assemblage with related ways of thinking about heterogeneous connections, such as within the field of science and technology studies and actor network theory.

an important point that can be drawn from the works of Rumford. However, he also encourages the investigation of cases of borderwork performed by citizens, which cannot immediately be linked to the state and the security discourse it supports. In this chapter emphasis is placed on the influences of mobility on the daily lives of these migrants and the pursuit to maintain a level of well-being through the use of intra-European borders. These cases hopefully contribute to a further understanding of how 'borders are not necessarily the enemy of mobility' (Rumford 2008, 9).

Borderwork in Workspaces

This study focuses on borderwork as it is performed in workspaces, including physical worksites (assembly hall, construction site, office, home and so forth), work practices and workplace governance: the whole of 'work organization, firm-level and worksite industrial relations/human resource (IR/HR) practices, local labour market regulation and employment law' (Rutherford and Holmes 2013, 117). Previous studies of workspaces have proven very productive in understanding practices and transformations originating from within and outside these places. In a classic ethnographical study, Burawoy (1979) conceptualizes the workplace as an internal state in which worker rights and duties are the outcomes of bargaining processes between the management and representatives of workers and unions. Under the influence of managerial tactics and peer pressure, workers often end up benefiting the firm by outperforming production norms and targets. Others, including geographers Peck (1996) and Samers (1998), have argued that workplace governance is embedded within different varieties of capitalism and that the functioning of local labour markets also depends in part on underlying mechanisms of urbanization and localization. As a result, different workplaces are embedded within spatially different, uneven environments.

The performance of borderwork in workspaces is not a new phenomenon. In the Danish context there is a long tradition of strong occupational boundaries and disciplinary demarcations (Nielsen 2004). This is reflected in collective wage agreements, among others, which are often exclusively designed for a specific sector or occupation. The presence of migrant workers at the workplace is also not something new, although the specific ways in which the present-day East-West labour migration system changes and challenges practices in workspaces indeed might be a novelty (see Nielsen and Sandberg 2014). The labour migration of the 1960s was based on labour recruitment agreements between Western European countries (such as Germany) and south and Eastern European countries such as Italy, Greece, Turkey and the former Yugoslavia. These 'guest worker' agreements secured among other things work permits and the legal status of the migrants' residential status in the destination country (Körber 2012). The present-day East-West labour migration differs from the agreements of the 1960s because the former have been subject to the aforementioned transitional restrictions that were introduced in most of the 'old' EU member states regulating the numbers of labour migrants and the issuing of work permits for the 'new' EU member states. One obvious consequence is that until 2011 labour inspection in restricting countries has been extra vigilant about illegal employment of citizens from new member states. After 2011, this concern was limited to the employment of nationals from Romania and Bulgaria. Other concequences will be discussed in the remainder of this chapter.

Comparing Denmark and the Netherlands

The following cases are based on two research projects, one focused on a Danish construction site located outside of Copenhagen, Denmark, and the other on farm work in a Dutch border region.

The projects were conducted independently of each other. Similarities and intrinsic differences between the two projects inspired a further investigation into comparative approaches to studies on borders and circular migration practices. However, as argued by Sørensen (2008), comparative research designs often fail to explain the ground(s) on which the comparison is made as well as the formats used for comparing, which she characterizes as a *tertium comparationis* – a kind of ready-made and undisputed comparativism. The following section will therefore work with 'comparison as achievement', which means that comparison is stressed as an active doing. As Niewöhner and Scheffer (2010, 2) have argued, a comparison is something which is produced or *built* into research objects in order for them to be compared, hence the comparison stands 'as the result of the ethnographic inquiry, not [as] its natural starting point'. Sørensen (2008) further proposes the term multi-sited comparison in order to untangle the understanding of the *tertium comparationis* as one single site for investigation.[3] Adopting the multi-sited comparison approach is not an easy way out. As Sørensen (2008, 327) points out:

> Experiences of similarities and differences in fieldwork cannot count as comparative analysis in themselves; these experiences constitute only sampling of the sites to compare. Furthermore, it is important to note that multi-sited comparison cannot be based on comparisons between already completed sets of data. Instead, it must involve a process of further sampling, leading to further investigation, to incorporating more data, including the ethnographer's responses to other related sites, and thereby thickening the descriptions overall.

Two cases are brought together here that were not initially designed to be compared; however, we will demonstrate our attempt to explore the comparative 'working together' of the two cases in systematic ways. Accordingly, rather than presenting the two cases in any symmetrical way, we have selected the most interesting answers to the research questions, with each case thus providing a story of borderwork seen through particular workplace transformations.

The first research project scrutinizes how migratory practices are made possible and desirable among a group of Polish circular migrants working within the building and construction industry in Denmark. The practices of these labour migrants can be characterized as circular because the migrants work in Denmark on a temporary basis, while their families are in Poland.[4] The ethnographic material was collected during autumn 2011 and spring 2012 and consists of participant observations from the construction site, in-depth interviews with members of the Polish team, their Danish colleagues, managers on site and representatives of Danish labour unions and employers' associations.

The setting of the second project is the Lower Rhine (*Nederrijn/Niederrhein*) border region between the Netherlands and Germany, a predominantly rural area that has been a destination for Polish labour migrants since the opening of the internal market in 1989. Here, the focus has been on migratory practices developed by employers, recruiters and specialized consultancies as the most important representatives of a migration industry which has developed around labour migration from new EU member states in a 'dialectical' relation with regulatory authorities.

---

3   Sørensen's term 'multi-sited comparison' is inspired by Marcus' (1998) 'multi-sited ethnography', which designates ethnographic fieldwork as encompassing not one delimited area of investigation, but rather of connections between practices, objects and people.

4   The first case presented forms part of a continuing, larger research project on Polish labour migrants in the area of Copenhagen, Denmark, conducted by Nielsen and Sandberg of the Ethnology Section of the Saxo Institute University of Copenhagen (see Nielsen and Sandberg 2014).

Empirical material was collected between January and December 2005 and consists of in-depth interviews with employers, labour recruiters, labour inspectors, policymakers and legal experts, as well as personal communication with migrants.

In Situ

We begin with the circular migrants living and working on a temporary basis in Denmark. They return regularly to Poland, where several of them have families. The case addresses the following questions: How do these migrants overcome the locational and mental thresholds and trajectories as described in the threshold approach? In which particular ways are these migrants doing 'borderwork'? The following ethnographic examples illustrate how Danish labour market regulations are managed by the migrants, the employer and the labour union. This managing, it is argued, contributes to a slight but steady transformation of the workplace. Additionally, our intention is to demonstrate that in applying the concepts of borderwork and assemblage to the practices of the Polish circular migrants, the scope of analysis is widened. The focus is not only the mobile individuals but also the specific and heterogeneous assemblages of objects, settings and devices that support the migrants' choices and dispositions and in that sense act as allies of the migratory process and of the managing of the different thresholds.

The Construction Site

The construction site is located in the outskirts of a middle-sized town about a one-hour drive from Copenhagen, Denmark. It is run by a Danish entrepreneur, and the company has around 200 employees of which 150 are Polish. On this particular project a team of eight Polish workers is building a nursing home along with a cluster of residential facilities for handicapped people. On site there is a range of other subcontractors and subconstruction firms working together with employees from various staffing agencies. Some teams are of mixed nationality, others are Danish or Polish, and there is also a small group of Serbs employed through a staffing agency.

The eight workers on the Polish team are all men between 25 and 50 years of age. At the time of inquiry all eight were members of Danish labour unions and had unemployment insurance. However, this degree of unionization and social security is certainly not the case in general among labour migrants in Denmark, and it might be due to the fact that this particular construction site is affiliated with a public institution – the local municipality (cf. Hansen and Hansen 2009). As compared to other construction sites, the members of the Polish team describe the conditions at the construction site as luxurious. According to the workers, examples of these favourable conditions include safety boots and working clothes provided by the company, a mobile workers' hut with facilities for lunches and breaks, and a locker for each employee which is heated in order to dry soaked working clothes during the day.

The 'social container' is the Polish team's name for the mobile workmen's hut. This place is an important meeting point during lunch and breaks; this is where prices in Poland and Denmark are compared and the next journey back to Poland is planned. The Polish workers carefully calculate when they have to renew their special tax licences so that they can keep using their Polish cars in Denmark. Some of the workers live in close proximity in Poland and usually travel together in one car when returning to Poland every fourth week. The rapport among these Polish working migrants is evident during breaks in the 'social container'. The space is filled with jokes and mutual friendly teasing, and everyone has a nickname appropriate to their place and position in the group.

The Flexible Migrant Worker

According to the building manager on-site, who is a Danish employee of the construction company, it is a great advantage for the company that the workers from the Polish team are willing to do any type of work at the construction site. Apparently, this is unlike the Danish workers who adhere to professional levels (cf. Nielsen 2004). If the floor needs to be wiped, he explains, the Polish team members do it without hesitation. Marcin,[5] for example, was educated at a technical school in a medium-sized city in northern Poland. However, on the construction site he is doing all-around tasks, such as running the wheel loader, supplying new building materials and covering the buildings with plastic to protect against rainy weather. Together with the rest of the Polish team, Marcin performs the role of migrant 'flexi-worker': he is hard-working, flexible and mobile (Pijpers 2010). Further, the brand of flexibility is also used by the Polish workers themselves as a strategic self-representation, together with the fact that they are migrants from Poland – so being flexible and being Polish are intrinsically linked. However, as we shall see, this particular brand is only possible to maintain because of other kinds of borderwork undertaken by the Polish circular migrants.

Karol as Mediator

Karol is the foreman of the Polish team and lives in a dormitory next to the construction site where he shares a room with one of his colleagues. The room is approximately 12 square metres, has two single beds, a toilet with bath, a fridge and access to a shared kitchen. Karol shares the monthly rent of 450 euro with his room-mate; importantly, free internet access is included in the price. Every evening Karol accesses Skype in order to chat with his wife and two sons, 12 and 18 years old. In Denmark, Karol earns approximately twice as much as he could make in Poland. Karol works 46 hours, Monday to Saturday, for three weeks; then he returns to his family in Poland for one week. As long as he is married and can prove that he supports a household in Poland, he can get a tax refund from the Danish tax authorities.[6]

As a foreman Karol earns slightly more than the rest of the Polish team, and included in his salary is a free mobile phone. He spends most of his working day ensuring that the other workers are on task and supplying sufficient building materials. Also, he functions as a contact person between the Polish team and the Danish manager on-site. This mediating function further encompasses assisting the Polish workers in filling out and renewing forms and papers and facilitating any necessary trips to the hospital. Perhaps Karol's most important function is keeping track of the working hours of the Polish team. According to Danish law a minimum of 11 hours of rest a day is obligatory.[7] However, in order to be able to return to Poland every third week, the Polish team has made a special agreement with the employer and the Danish labour union that they can at times work up to 25 per cent more than the average for three weeks in order to take the fourth week off. Importantly, the workers on the Polish team are not allowed to exceed 46 weekly hours during the three weeks. Therefore, Karol calculates weekly time accounts on each worker to make sure their hours stay within the limit.

---

5   For purposes of anonymity original names have been replaced with pseudonyms.

6   Since January 2013 the maximum amount of the tax refund connected to double households is, however, reduced in Denmark from DKK 50,000 to DKK 25,000 a year.

7   See http://arbejdstilsynet.dk/da/regler/bekendtgorelser/h/sam-hvileperiode-og-fridogn-324.aspx, Chapter 2, Article 3 (accessed 26 October 2014).

The flexible organization and structuring of the workforce is thus not only supported by the circular migrants but also by the labour unions and employers. Hence, the special managing of working hours is another example of how workplaces undergo a slight but steady transformation impacted by labour markets within Europe.

Borderwork Is Not a One-Man Show

When encountering Karol and his colleagues at the Danish construction site, questions unavoidably arise: Why do they choose this way of life? What does it take to separate from the well-known routines at home in order to live in a small, shared room in a dormitory without the company of family? It became increasingly clear when learning about Karol's life and working conditions that more than one wo/man's choice is entailed in making a migratory practice possible. Rather, specific assemblages constituted by material-discursive entities, settings and devices take part in rendering migratory movements such as Karol's feasible. In the case of this Polish team, important co-constitutors of the circular migratory practices are for example the cars transporting the Polish workers (often in groups and packed with cheap groceries) to and from Poland; social networks including those built during breaks in the 'social container'; and employers who not only hire workforce but also, as in this case, help out by providing temporary accomodation. Equally important are the family members who stay behind and can to a large extent be regarded as participants in the migrant practices (see Nielsen and Sandberg 2014). Hence, in order to grasp the rationales of various migratory movements it is necessary to go beyond an explanation that focuses solely on acts of choice, as also suggested in the threshold approach (see Van der Velde and Van Naerssen 2011).

Recruiters as Borderwork Professionals

We now turn to the case of the Lower Rhine border region, where the focus is on institutional and private actors as part of broader research on the facilitation and settlement of migration in the area. One of the key research questions is how and with what result the temporary staffing industry has professionalized in the course of the last decade or so. Notably, this process of professionalization has included the introduction of quality marks: recruitment agencies with a quality mark conform to labour laws and agree to properly arrange housing and transport for their migrant workers. These and other efforts have certainly helped to not only improve the circumstances of workers but also to improve the image of the industry as a whole. However, controversies remain. For example, when the industry proposed the introduction of a collective wage agreement applying to temporary workers from outside the Netherlands (other EU member states), labour unions heavily criticized the idea. They argued that the agreement specified lower gross salary costs for migrant workers, which would give employers hiring migrant workers cost advantages over hiring domestic workers (Pijpers 2008). In addition, in order to stay ahead of competitors, recruiting agencies try to organize work in such a way that time intervals between jobs are minimized to ensure that neither time nor money is wasted. For that reason, workers stay where they are housed and wait to be transported to the next worksite.[8]

---

8   This is a matter of managerial tactics: according to their contracts, migrants have to be available for work a certain number of hours or on certain days, but since work assignments tend to come on short notice, they are often called to work without warning. Effectively, therefore, migrants do not have the freedom to leave their housing.

The Cross-Border Region

Other more complicated practices exist. Polish workers who are mobile across borders partially leave the Polish social welfare system and also partially enter the system of their destination country/countries. Within national systems, taxes and social security are inextricably linked: in general, increases in social security payments are connected to decreases in taxes. Within an EU context, however, synchronization between national systems is lacking (Essers and Willems 2005). As a consequence, it is not always clear which system is to receive payments for taxes and social security and/or is responsible for worker benefits, and within what time frame after the arrival of the migrant workers. This becomes even more complicated in pending cases of temporary or circulatory mobility.

In the Lower Rhine area, this ongoing lack of harmonization between national systems is used by employers and recruiters of labour migrants to save and even gain money. To understand why this is the case, it is important to know that Poland has bilateral taxation agreements with both the Netherlands and Germany. Through this agreement, Polish migrant workers are not liable to pay Dutch or German income taxes if working only a maximum continuous period of 183 days. Anticipating this restraint, a number of large employers in the region have established production sites in Germany in order to move groups of people around between the Dutch and German sites once the 183-day limit is nearing in either country. This allows them to work year-round without having to pay the higher Dutch or German taxes, saving the employer or recruiter labour costs. The consequence is that most workers stay for several years if not longer, alternating short returns to Poland with much longer periods of time in the region.

Employers and recruiters alike may request the help of consultancy firms specializing in accounting and tax advice. Building on detailed legal knowledge acquired through years of consulting, these firms are well equipped to navigate the tax, social security and labour regulations that concern Polish employment. A young accountant working for a Lower Rhine-based office of a large multinational consultancy firm explained that for him the challenge is always to find an answer to the following question: 'How can I, with Dutch legislation at hand and all the possibilities and facilities that it offers, how can I optimally make use of it in order to … create a maximum net revenue of those Poles … against the lowest possible labour cost' (interview 2005). In this way, an employer or recruiter is handed a tailor-made solution to a specific question regarding the hiring or employment of Polish workers.

The Courtroom

As part of the free movement of services in the EU, Poles who are self-employed are entitled to offer services in the Netherlands to Dutch clients without a work permit, no matter whether they are based in the Netherlands or in Poland. However, self-employed migrants run a relatively high risk of being exploited (cf. Pool 2011). For example, Dutch farmers may 'sell' the 'right to harvest' to subcontractors from Poland, who become formally entitled to carry out this work as a service. In practice, however, the farmers act as conventional employers, dictating what the workers do. Under these circumstances, self-employed migrants may have to accept precarious working conditions including low pay, long working hours and irregular shifts.

Since the main task of labour inspection is to find and prove cases of illegal employment, disagreements about the legality of employment accounting practices are widespread. When labour inspectors encounter a case in which the work permit requirement may be unlawfully avoided, they take it to court – or they issue a fine which is then challenged by the employer or recruiter in question. Thus, there is a grey area between legal and illegal employment of Polish workers and other workers from new member states that is negotiated and judged in the courtroom. In this space, labour inspection as a key implementing organization of migration policy confronts creative entrepreneurs represented by consultants who are in turn represented, in particularly complicated cases, by specialized lawyers. When labour inspection loses, a new standard is set for other employers and recruiters.[9] When it wins, the employment situation is discontinued and the construction managers return to the drawing board. However, as a labour inspector asserts:

> Well, they live through every lawsuit, so when, at some point, they stand before a judge and a judge does not accept certain things, then they often learn how to circumvent that, or arrange it differently. In that respect, it is increasingly difficult [for labour inspectors to succeed in court]. (interview 2005)

These examples tentatively support the idea that labour migration 'emerges in relation to legal categories and is not simply dictated by them' (Black 2003, 34). Meanwhile, in outsourcing the organization of work and employment to recruitment agencies, and/or the payrolling to specialized firms, employers lose sight of the process of hiring and employing Polish migrant workers. According to one interviewed employer, a consequence of the growing importance of sophisticated legal and accounting knowledge is that 'the one who monitors the legal matters of the business and does the payments, those constructions with Poles, he also gets a large share. And he binds himself to the employers' (interview 2005).

Bending and Stretching Rules – Establishing a *Tertium Comparationis*

The specific kinds of borderwork performed at the two workplaces constitute a *tertium comparationis*. A negotiation of prevailing labour market regulations is observed in both cases. In the Danish case, it has been shown how the Polish circular migrants perform borderwork through the bending and stretching of existing regulations and rules, which includes the special managing of working hours in order to 'save up' hours for the weeks off in Poland; the skilled use of knowledge on local tax regulations; the use of social networks to maintain a base in Poland and compare prices; and last but not least the role of Karol as a mediator, both between the Polish team and the Danish employers as well as between the labour unions and the Danish authorities. Together with a continuous self-branding as 'flexible workers', this rule bending contributes to keeping the Polish workers attractive in the job markets in question.

This case illustrates how circular migration is made possible through specific assemblages of heterogeneous actors. Considering the perspective of how migration is facilitated, the Dutch case has subsequently shown that employers, recruiters and associated actors also engage in the yielding of rules. In the Netherlands, the employment of Polish workers has become a highly specialized affair, with the interaction between the migration industry and regulating authorities resulting in a cumulation of knowledge and practices. Actors who know how to manage the web of rules and for whom 'trial and error' is part and parcel of daily business have the potential to make substantial profits.

---

9   Initially, these cases are taken to regional courts. Only at later stages, if either party chooses to appeal, are they taken to national courts.

An interesting difference between the two cases is that, while bending and stretching, in the end the Polish workers featured in the Danish case still play by the rules, while the migration industry actors in the Netherlands have the capacity, to some extent at least, to make the rules. Reflecting on these differences, the issue of power inequality arises. Polish workers may resort to labour unions for help and advice, but, both in Denmark and the Netherlands, unions have strong incentives to advocate playing by the rules when migrant workers are viewed as competitors to domestic workers. As the Danish case shows, the unions have certainly helped to make possible the circulatory aspect of this particular migration flow, enabling the workers to return home for a short visit every month. Also, a report on exploitation in the Netherlands shows that individual migrant workers who are looking for help to get out of a precarious employment situation are often satisfied to have their own problem redressed, without filing a lawsuit (FairWork 2012). Even if standing up for one's rights in court or joining forces would ultimately benefit a group of migrant workers *as a whole*, according to the report this hardly ever occurs.

Further, key to understanding the elongating of rules as a *tertium comparationis* is another aspect of the assemblage: the transitional restrictions on free movement of labour, which were the outcome of rhetorical debates on the real and imagined 'threats' posed by immigration from new EU member states. As became evident from the Dutch case, some of the practices developed in order to sustain and continue a migration flow which had already emerged before the enlargement of the European Union in 2004. In this respect, the restrictions have functioned as strategic sites of opportunity, pushing migrants and employers alike into finding 'creative' solutions.

Conclusion

Despite the obvious differences between the two cases presented here, we argue that they can be paralleled and juxtaposed to establish a common ground for comparison. Our proposal designates this endeavour as 'comparison as achievement', due to the fact that it necessitates, first, an explanation of the significant differences – such as farm work as opposed to the construction industry, a rural setting versus a (sub)urban site, and so on. Second, the comparison requires a systematic rethinking of the two cases in question. There was no opportunity here to incorporate more data as proposed in the earlier mentioned quote by Sørensen; however, this achievement confirms one of the core thoughts behind the idea of multi-sited research, namely that the fieldwork does not stop when the empirical gathering of material has formally ended. Rather it continues and reaches deeply into the crafting of analytical findings based on the material (cf. Coleman and Collins 2006).

How do the findings speak to Van der Velde and Van Naerssen's threshold approach (2011)? Circulatory migration seems to fit the approach quite well, although compared to temporary or (semi-)permanent migrants, circular migrants go through the decision-making phases more often and presumably more quickly as well. When Karol and his team return to the construction site at the end of the fourth week, they do not have to make the same material and emotional investments they made when first going to Denmark. While they surpassed the mental border threshold only once, they cross the locational and trajectory thresholds repeatedly – but without much effort. At some point, they may decide to go to different countries or even alternate countries, depending on changes in location or other factors that are significant to them.

Drawing on the findings of the two case studies, the idea that the migration decision-making process, especially for circular migration, involves a rhythm, a back and forth which can partly be traced back to migrants' own preferences but also, and crucially, to employment terms and conditions, sector-specific agreements, transitional restrictions, and so on, can be added to the threshold approach. This, in turn, points to a mode of ordering which finds expression in the continued dominance of national welfare systems, national labour market regulation and national employment law. It also manifests itself in the existence of institutional 'mismatches' between the various national systems in the EU and in loopholes which render certain migratory practices possible. Finally, some institutional barriers are higher or stronger than others, depending, for example, on (seemingly) small changes in regulation or on larger material-discursive shifts. Therefore, the rhythm of circular migration is not determined by notions of closure and/or loopholes only; it depends on contingently opening and closing windows of opportunity as well. This contingency, it is argued, can be well understood when investigating along assemblage lines in addition to using the threshold approach.

Acknowledgements

The authors wish to thank the Nijmegen Centre for Border Research, Radboud University, the Netherlands, for making a stay as guest researcher possible for Marie Sandberg and for providing a frame for collaboration between both authors and the crafting of this article. A paper version of this chapter was presented at the EastBordNet Conference 'Relocating Borders: A Comparative Approach', 11–13 January 2013, in Berlin.

References

Andersen, D.J., and M. Sandberg. 2012. "Introduction." In *The Border Multiple: The Practicing of Borders between Public Policy and Everyday Life in a Re-Scaling Europe*, edited by D.J. Andersen, M. Klatt, and M. Sandberg, 1–19. Farnham: Ashgate.

Black, R. 2003. "Breaking the Convention: Researching the 'Illegal' Migration of Refugees to Europe." *Antipode* 35 (1): 34–54.

Burawoy, M. 1979. *Manufacturing Consent: Changes in the Labor Process under Monopoly Capitalism.* Chicago: Chicago University Press.

Coleman, S., and P. Collins, eds. 2006. *Locating the Field: Space, Place and Context in Anthropology.* Oxford: Berg Publishers.

Pool, C.M., and T. de Lange. 2004. "Vreemde handen aan het bed: De werving van Poolse verpleegkundigen in Nederland." *Migrantenstudies* 20 (3): 130–44.

Deleuze, G., and F. Guattari. (1987) 2004. *A Thousand Plateaus: Capitalism and Schizophrenia.* London: Continuum.

Essers, G. and S. Willems. 2005. *Vrij verkeer van diensten: De gedetacheerde werknemer.* Brussels: Federatie Nederlandse Vakbeweging.

FairWork 2012. *Verborgen slavernij in Nederland.* Amsterdam: Stichting Fairwork. Accessed 18 October 2012. http://www.fairwork.nu.

Favell, A. 2008. "The New Face of East-West Migration in Europe." *Journal of Ethnic Migration Studies* 34 (5): 701–16.

———. 2009. "Immigration, Migration, and Free Movement in the Making of Europe." In *European Identity*, edited by J.T. Checkel and P.J. Katzenstein, 167–89. Cambridge, MA: Cambridge University Press.

Garapich, M.P. 2008. "The Migration Industry and Civil Society: Polish Immigrants in the United Kingdom Before and After EU Enlargement." *Journal of Ethnic and Migration Studies* 34 (5): 735–52.

Grzymała-Kazłowska, A. 2005. "From Ethnic Cooperation to In-Group Competition: Undocumented Polish Workers in Brussels." *Journal of Ethnic and Migration Studies* 31 (4): 675–97.

Hansen, J.A., and N.W. Hansen. 2009. *Polonia i København: Et studie af polske arbejdsmigranters løn-, arbejds- og levevilkår i Storkøbenhavn.* Copenhagen: LO-Dokumentation.

Irwin, A., and M. Michael. 2003. *Science, Social Theory and Public Knowledge.* Maidenhead: Open University Press.

Kępińska, E. 2005. "Recent Trends in International Migration: The 2005 SOPEMI Report for Poland." CMR Working Papers 2/60. Warsaw: Warsaw University.

Kindler, M. 2012. *A Risky Business? Ukrainian Migrant Women in Warsaw's Domestic Work Sector.* Amsterdam: Amsterdam University Press.

Körber, K. 2012. "So Far and yet so Near: Present-Day Transnational Families." *Ethnologia Europaea: Journal for European Ethnology* 42 (2): 12–25.

Law, J., ed. 1991. *A Sociology of Monsters: Essays on Power, Technology and Domination.* London: Routledge.

_____. 2004. *After Method: Mess in Social Science Research.* London: Routledge.

Linde-Laursen, A. 2010. *Bordering: Identity Processes between the National and Personal.* Farnham: Ashgate.

Löfgren, O. 2008. "Regionauts: The Transformation of Cross-Border Regions in Scandinavia." *European Urban and Regional Studies*, 15(3): 195–209.

Marcus, G.E. 1998. "Ethnography in/of the World System: The Emergence of Multi-Sited Ethnography." In *Ethnography through Thick and Thin: Multi-Sited Ethnography*, edited by G.E. Marcus, 79–104. Princeton: Princeton University Press.

Nielsen, N.J. 2004. *Mellem storpolitik og værkstedsgulv: Den danske arbejder – før, under og efter Den kolde krig.* Copenhagen: Museum Tusculanums Forlag.

Nielsen, N.J., and M. Sandberg. 2014. "Between Social Dumping and Social Protection: The Challenge of Creating 'Orderly Working Conditions' among Polish Circular Migrants in the Copenhagen Area, Denmark." *Ethnologia Europaea: Journal for European Ethnology* 44 (1): 23–37.

Niewöhner, J., and T. Scheffer, eds. 2010. *Thick Comparison: Reviving the Ethnographic Aspiration.* Leiden: Brill Publishing.

Peck, J. 1996. *Workplace: The Social Regulation of Labor Markets.* New York: Guilford Press.

Pijpers, R.A.H. 2008. "Circumventing Restrictions on Free Movement of Labour: Evidence from a Dutch-German Border Region." In *Migration and Mobility in an Enlarged Europe: A Gender Perspective*, edited by S. Metz-Göckel, M. Morkvasic, and A. Senganata Münst, 225–46. Opladen: Barbara Budrich Publishers.

_____. 2010. "International Employment Agencies and Migrant Flexiwork in an Enlarged European Union." *Journal of Ethnic and Migration Studies* 36 (7): 1079–97.

Pijpers, R., and M. van der Velde. 2007. "Mobility across Borders: Contextualising Local Strategies to Circumvent Visa and Work Permit Requirements." *International Journal of Urban and Regional Research* 31 (4): 819–35.

Pool, C. 2011. *Migratie van Polen naar Nederland in een tijd van versoepeling van migratieregels.* The Hague: Boom Juridische Uitgevers.

Rumford, C. 2006. "Theorizing Borders." *European Journal of Social Theory* 9 (2): 155–69.

_____ . 2008. "Introduction: Citizens and Borderwork in Europe." *Space and Polity* 12 (1): 1–12.

Rutherford, T. and J. Holmes. 2013. "(Small) Differences That (Still) Matter? Cross-Border Regions and Work Place Governance in the Southern Ontario and US Great Lakes Automotive Industry." *Regional Studies* 47 (1): 116–27.

Samers, M. 1998. "'Structured Coherence': Immigration, Racism and Production in the Paris Car Industry." *European Planning Studies* 6 (1): 49–72.

Sandberg, M. 2009a. "Grænsens nærvær og fravær: Europæiseringsprocesser i en tvillingeby på den polsk-tyske grænse." PhD diss. Copenhagen: University of Copenhagen.

_____ . 2009b. "Performing the Border: Cartographic Enactments of the German–Polish Border among German and Polish High-School Pupils." *Anthropological Journal of European Cultures* 18 (1): 107–28.

_____ . 2012a. "Border Orderings: The Co-Existence of Border Focusing and Letting Border Issues Take the Back Seat at the German–Polish Border." In *The Border Multiple: The Practicing of Borders between Public Policy and Everyday Life in a Re-Scaling Europe,* edited by D.J. Andersen, M. Klatt, and M. Sandberg, 119–140. Farnham: Ashgate.

_____ . 2012b. "Karol's Kingdom: Commentary for Imagined Families in Mobile Worlds." *Ethnologia Europaea: Journal for European Ethnology* 42 (2): 87–93.

Scheffer, T. 2008. "Creating Comparability Differently: Disassembling Ethnographic Comparison in Law-in-Action." *Comparative Sociology* 7 (3): 286–310.

Sørensen, E. 2008. "Multi-Sited Comparison of 'Doing Regulation'." *Comparative Sociology* 7 (3): 311–37.

Van der Velde, M., and T. van Naerssen. 2011. "People, Borders, Trajectories: An Approach to Cross-Border Mobility and Immobility in and to the European Union." *Area* 43 (2): 218–24.

Van Houtum, H. 2005. "The Geopolitics of Borders and Boundaries." *Geopolitics* 10 (4): 672–9.

Van Houtum, H., O. Kramsch, and W. Zierhofer, eds. 2005. *B/ordering Space*. Aldershot: Ashgate.

Van Houtum, H., and T. van Naerssen. 2002. "Bordering, Ordering and Othering." *Tijdschrift voor economische en sociale geografie* 93 (2): 125–36.

Chapter 14

Between the New World and the Old World: Changing Contexts of Exit and Reception in the Bolivia–Spain Migration Corridor

Gery Nijenhuis

Migration is a common phenomenon among the Latin American population. Although precise figures are hard to find, some authors estimate that there are 30 million Latin American and Caribbean migrants (Durand 2009, 35; Durand and Massey 2010), representing approximately 15 per cent of all migrants worldwide. The preferred destination of most Latin American 'would-be' migrants was traditionally the US, which at the start of the new millennium was home to approximately 78 per cent of all Latin American migrants (Durand and Massey 2010, 20). However, the number of Latin Americans in Europe has increased significantly since the 1990s.[1]

Two phases in Latin American migration to Europe are generally distinguished (Pellegrino 2004; Durand 2009). The first is commonly referred to as the phase of political exile: from the late 1960s to the 1980s mainly higher-educated Latin Americans escaped from the bureaucratic authoritarian regimes in Chile, Brazil and Uruguay and found safe havens in Europe. The second phase started around the late 1980s and consisted largely of economic (labour) migrants. These were initially mainly higher-educated professionals from Brazil, Argentina and Chile, who came as regular migrants and already had job contracts. Then in the mid-1990s, there was a huge increase in the number of irregular migrants arriving in Europe in search of more income and a better future. Many Colombians, Peruvians, Ecuadorians and later on also Bolivians found their way to Spain, since access to Spain (and to a lesser extent also to Portugal and Italy) was fairly easy as Latin Americans did not require a visa.

The reason for the increasing popularity of southern Europe, and of Spain in particular, among Latin American migrants was a combination of factors, such as the economic downturn in what had previously been popular destinations (in the case of Argentina), economic growth in Europe, more rigid border controls in the wake of 9/11 and the implementation of the North American Free Trade Agreement (see also Pellegrino 2004). The contexts of exit and reception have thus changed, leading to new migration patterns and trajectories. This implicitly implies that the concept of borders has another meaning to these migrants. Certain borders may have become more restricted, such as the Mexican-US border, while others have become more permeable, due to new visa arrangements and demands in labour markets.

This contribution explores the changing contexts of exit and reception for one specific migration flow, namely that of Bolivian migrants to Spain, and the consequent shifts in the meaning of

---

1   Precise information on the number of Latin American migrants in Europe is lacking (Peixoto 2012; Pellegrino 2004), not only due to differences in sources and definitions at country level, but also – and perhaps more importantly – because of the large number of irregular migrants, mainly in southern European countries, who are not registered and thus not counted.

'borders' as a social construct. As such, it adds to our empirical knowledge regarding the threshold approach (Van der Velde and Van Naerssen 2011). The focus is on the locational threshold, although the indifference and trajectory thresholds naturally also come into play. This locational threshold is specifically meant to explain why migrants choose certain geographical destinations. In this contribution, however, a time dimension is added to the threshold to explain how changes at both ends of the corridor modify the working of the threshold. The factors in the country of origin and the country of destination that explain the (im)mobility of Bolivians are discussed, and the keep and repel factors are identified. The questions addressed concern the perceptions Bolivian migrants have of the borders they cross, or have to cross: To what extent can changes in the contexts of exit and reception be linked to these perceptions? And what is the impact of economic crisis and more rigid migration policies?

First, a brief interpretation of the concepts of borders and (im)mobility is presented. This is followed by a description of the background of Bolivian migration to Spain. The empirical section pays attention to the main flows and the contexts of exit and reception. Finally, the main findings and their link to the threshold approach that is central to this volume are discussed. This contribution is based on interviews with Bolivian migrants in Barcelona and Madrid in the period 2007–10, and in two sending areas: the village of Rincon (Santa Cruz department) and the city of Santa Cruz, Bolivia.

Borders, (Im)mobility and Contexts of Exit and Reception

In this chapter, (im)mobility, the concepts of borders, and contexts of exit and reception form the main analytical framework. The academic debate on borders and that on international migration were traditionally strictly separated, as borders were often understood as rather passive and fixed boundaries that demarcate the territory of the state and thus hinder transnational activities (Gielis 2009). As such, state borders steer the frequency and intensity of transnational activities through government policies that can result in high, but also low, barriers to transnational activities. In recent border studies, however, borders are also considered social constructs: they enable and determine the shape of transnationalism, namely the multiple connections and exchanges of people and institutions across the borders of states (Vertovec 2009). People do not passively experience borders as rigid lines that divide states; on the contrary, they are actively involved in processes of ordering, in combining 'here' and 'there'. Thus, borders 'should not been seen as "lines of division" but rather as "lived spaces"' (Ernste et al. 2009, 582). Borders are experienced by migrants, not only in the sense that they obstruct or enable transnationalism, but also because they are an integral part of the transnational experience itself (Gielis 2009). Therefore, borders create migrants, but migrants also create borders: borders connect the 'here' and the 'there', and have an active role in shaping this transnational relationship. In this sense, mobility and transnationality are intrinsically linked to processes of debordering and rebordering.

The discussion on borders takes place against the background of the broader debate on the 'age of migration' – an age that is dominated by globalization, increasing migration flows and widespread transnationalism. However, apart from the seemingly 'borderless world' there are some countervoices that critically argue that there are several signs of increased immobilities. *Mobilities* – a journal dedicated to the study of movement in the broadest sense of the word – was launched in 2006. Its publication was justified by the observation that 'it seems that a new paradigm is being formed within the social sciences, the "new mobilities" paradigm' (Hannam et al. 2006, 2). In the new journal's introductory article, the initiators describe how 'mobility has become an evocative keyword of the twenty-first century.' They consider mobility a trigger, a driver that is central to many social, economic and political processes. They put the concept of immobility parallel to the concept of mobility by quoting Skeggs (2004): 'Mobility is a resource to which not everyone has an equal relationship' (3). As such, they point towards the need to study these immobilities: potential flows of people, goods, capital and information that are excluded. Turner (2010, 247) observes in this respect an 'emerging mobility divide', referring to those who cannot get a visa or work permit because their educational background does not match the labour market, or their religious or ethnic profile makes them 'unwanted' for security reasons.

This also points towards the importance of the 'contexts of exit and reception', a concept introduced by Portes and Rumbaut (2006) in their studies on migrant incorporation in the US. Bringing the different contexts of exit and reception to the fore is useful to give a face to various contextual factors. The context of exit encompasses the main motivations and aspirations of the migrant, the economic conditions in the area of origin and the attitude of the sending country towards migrants. Factors to be considered for the context of reception are the societal response, the labour market and the opportunities for integration in the receiving country (Portes and Rumbaut 2006; Menjívar 2001). Defined in this way, contexts of exit and reception explicitly form part of locational thresholds and as such offer an interesting framework for analysis.

Methodology

This contribution is based on fieldwork conducted in the period 2007–10 in Barcelona and Madrid, and in two sending areas: the village of Rincon (Santa Cruz department) and the city of Santa Cruz, Bolivia. It draws on a range of research methodologies, including interviews with 80 Bolivian migrants in Barcelona and Madrid, interviews with representatives of NGOs working with Bolivian migrants in both cities and interviews with sending households in Rincon and Santa Cruz.

Over this period of four years, every year two to three weeks were spent in Spain, interviewing Bolivian migrants – both 'new' cases and follow-up interviews. Topics addressed during these interviews were the socio-economic characteristics of the migrants, their migration history and trajectories, labour and housing mobility in Spain, and their links with their region of origin and their relatives.

The respondents in Spain were selected via a multiple entry points approach. The project started with four contacts (two men and two women) who originate from different parts of Bolivia. These networks were gradually expanded by means of the snowball method. This led to interviews with a broad variety of Bolivians in Spain: recent arrivals as well as those who had been living in Spain for a longer period and were therefore more established. The selection covered both genders and all age groups. Besides Bolivian migrants, also representatives of NGOs that work in the field of migration, integration and social welfare were interviewed, as were representatives of Bolivian migrant organizations in Spain. In Bolivia, a selection of the sending households were interviewed about their relationship with the migrant in Spain, the inflow of remittances and their contacts with the migrant.

Bolivian International Migration

Until the 1980s, Bolivian migration patterns were dominated by internal, mainly rural-urban migration flows with a permanent character. Starting in the 1980s, interdepartmental migration flows increased; these also included rural-rural mobility flows. Well-known examples are circular migration to the coca-producing region of Cochabamba and migration to the sugar cane plantations in the eastern lowlands. International migration started relatively late, that is, in the 1980s and 1990s. The most important destinations within the region were Argentina and Brazil. In both countries, Bolivians mostly worked in the garment and construction sectors; others worked in agriculture. Regarding South-North migration, the United States was the most important destination. After Argentina's economic crisis in 2001 and the 9/11 events in the United States, Bolivians opted for other destinations (Dulón 2008; Nijenhuis 2010).

Reliable and precise figures on Bolivian emigration, which is mainly driven by the desire to improve living conditions (Farah and Sánchez 2001; Roncken and Forsberg 2007), are difficult to obtain. Figures provided by the Bolivian Department for Statistics show that in 2001, 0.1 per cent of the Bolivian population lived abroad (INE 2001). However, since irregular migration and unofficial border crossings are not taken into account, many authors working on Bolivian emigration consider this a gross underestimation (Farah and Sánchez 2001). A review of immigration figures that are available in the main destination countries (see Roncken and Forsberg 2007) revealed that over 2.5 million Bolivians are involved in international migration, representing more than 25 per cent of the country's entire population. According to these estimates (IBCE 2008), some 1 to 1.5 million Bolivians live in Argentina, almost 1 million live in the United States and more than 500,000 live in Brazil (Roncken and Forsberg 2007).

Spain is a relatively new destination for Bolivians. The number of Bolivians in Spain increased from 6,000 in 2001 to almost 250,000 in 2008, before decreasing to 178,000 in 2012 (see Figure 14.1). These figures are based on the information available in the *Padrón Municipal* (municipal population registers), which provide fairly reliable figures, including those on irregular migrants, as most migrants, regular and irregular, register with the municipality. By registering, they gain

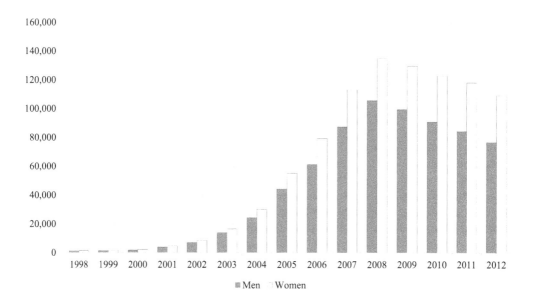

Figure 14.1   Development of the number of Bolivians in Spain according to gender, 1998–2012

Source: INE 2013.

access to free social services, such as education and health, although the access to free health care is heavily debated. The municipal registers are not open to the police or linked to immigration officials, which is why many irregular migrants do not hesitate to register.

The large majority (70 per cent) of Bolivian migrants in Spain are irregular: most overstayed their allowed visit deadlines and do not hold residence permits (see also ACOBE 2011). Women are overrepresented in the migration flow to Spain, which corresponds with the global trend towards a feminization of migration flows. In 2012, the female to male ratio was one to four.

A Quick Scan of the Research Population

The large majority of the respondents arrived after 2000 as economic migrants. Only a few have been in Spain for a longer period; most of these respondents had studied in Spain and then started a career there. Migrants who arrived after 2007 are scarce, as a result of the introduction of the visa requirement for Bolivians in April 2007. Most of the respondents came on their own, but already had relatives in Spain. Almost a quarter had acquired job contracts before they came to Spain, and a further quarter had good prospects of getting a job. Overall, the migrants have a relatively high educational background: almost all completed secondary education, and a considerable number completed higher and/or professional education (see Table 14.1). Most migrants had no previous migration history, but some (mainly males) had worked in Argentina, Chile or Brazil. With respect

Table 14.1   Main characteristics of respondents (*N* = 80)

Arrived in Spain	
– before 2000	2%
– 2000–03	44%
– 2004–07	50%
– Since 2008	4%
Migration status	
– Regular	12%
– Irregular	88%
Gender	
– Male	44%
– Female	56%
Age group	
– Under 20	4%
– 20–35	64%
– 36–50	28%
– 51–65	2%
– Over 65	0%
Education	
– Secondary, uncompleted	16%
– Secondary, completed	48%
– Higher education	36%

Source: the author.

to their spatial trajectories, all but five had arrived at Madrid's international airport and then moved to Barcelona (the five exceptions had arrived via Italy, the Netherlands or Germany) (INE 2013).[2]

Bolivian Migration to Spain: Changing Contexts of Exit and Reception

Four phases of exit and reception contexts can be distinguished based on the development of the number of Bolivian migrants in Spain, Spanish migrant policies and practice, and economic development (combined with labour market development) in Spain and Bolivia. These phases are pre-2000, 2000–March 2007, April 2007–08 and post-2008. The following sections discuss the phases and the implications of these changes for the migration experience and the perception of borders among Bolivians in Spain.

Pre-2000: 'Consolidated' Professionals

Relatively few Bolivians moved to Spain prior to 2000, particularly when compared to other Latin Americans, such as Colombians and Ecuadorians. The Bolivians who did migrate to Spain in this period are generally higher educated, who went to Spain to study, then got jobs or married and consequently stayed on. Overall, this group is relatively well integrated and fairly 'consolidated': all respondents in this group had permanent jobs (and many had held their jobs for a long time) and owned houses or apartments. The majority are married, half of them to Bolivians and the other half to Spaniards. They consider themselves permanent migrants and do not plan on returning to Bolivia. They arrived in Spain legally and now hold residence permits. Crossing borders is just a routine for most of them: they visit Bolivia every two or three years, and other European countries occasionally. Their children were born and raised in Spain, and the identity of the children is Spanish. Of course, many of them still adhere to their Bolivian cultural practices, and organize cultural events – such as carnival – to celebrate Bolivian culture, but their future is in Spain: 'What do you mean, would I like to return to Bolivia permanently? Oh no, we belong here, in Spain. I have my family here, my kids were born here, my job is here. Although Bolivia is my country of origin, the country where I was born, my life is here, in Spain' (female, in Spain since 1995).

El Milagro Económico Español: 2000–2007

In 2000, there were 3,723 registered Bolivians in Spain; by 2007, there were 200,749 registered Bolivians – an increase of more than 5,000 per cent. Several factors explain this dramatic increase. A first important driver was the economic crisis in Argentina, which until 2000 had been one of the most popular destinations for Bolivian economic migrants. The economic crisis of 1999–2001 implied that many Bolivians had to return to Bolivia, where they started looking for other destinations.

A second, and probably more important reason was the *milagro económico español* (Spanish economic miracle), as it is commonly referred to. From 1996, Spain experienced unbridled economic growth that was accompanied by an increasing demand for low and unskilled labour in

---

2   According to data from the Spanish Institute for Statistics, Barcelona is home to 23 per cent of all Bolivians registered in Spain; another 23 per cent live in Madrid, 9 per cent live in Valencia and 7 per cent live in Murcia (data from 2012, available at http://www.ine.es/jaxi/menu.do?type=pcaxis&path=/t20/e245/&file=inebase, accessed 3 October 2014).

construction, the hotel and restaurant sector and agriculture. This was supported by the presence of a relatively large informal economy, which contributed to favourable labour perspectives for migrants; irregular migrants could quite easily find a job in agriculture or domestic service.

Third, ageing played a role, directly by the release of labour positions and indirectly by the increasing demand for care for the elderly. The fact that from the end of the 1990s onwards Spanish women increasingly entered the labour market – which until then had been a rather unusual phenomenon in traditional Spanish society – substantially increased the demand for carers, not only for their children but also for the elderly, as this was traditionally considered a task for daughters or daughters-in-law. As a consequence, the Latin American nanny or domestic servant became a common sight on the streets of many large Spanish cities.

The large majority of Bolivian migrants live in the major cities – Madrid, Barcelona and Valencia – where the demand for construction workers and private carers is the highest. Other important destination areas were Murcia and Andalusia, where the agricultural sector offered employment to men (picking oranges, tomatoes and grapes). In this period, moving to Spain was a relatively easy exercise, as Bolivians did not need visas to enter Spain as tourists: they only needed to buy a ticket and board the plane. That it was a relatively easy exercise was also depicted by our respondents in both Spain and Bolivia:

> As all young men headed for Spain, and reported very positively about the numerous job opportunities, I thought: 'Why not give it a try?' So I booked a ticket, and took off for Spain. But you know, it wasn't that good, in fact. Living over there was quite stressful, you have to work very hard, so I returned home after six months. It was a nice experience, and it's good that I now understand what they are talking about, but life is so much better here, it is much more '*tranquila*'. (Return migrant, male, 39 years, Rincon)

Although the literature often portrays the decision to move as a well-thought-out one, intensively deliberated process, and some people indeed took their time, many respondents said that it had not taken a long time to decide to move. It had often been a rather impulsive act: people had decided within a very short time frame: 'Of course, we talked about moving to Spain, everybody, but when my sister called and said that she might have a job, I was in Spain within two weeks' (female, in Spain since 2006). As such, the border was mainly a mental, psychological issue, rather than a physical barrier.

The main motive for Bolivians to migrate was economic: they wanted to make a lot of money in a relatively short period. Almost all migrants who arrived in this period, and particularly those who arrived after 2005, had a clear idea of their objective: they wanted to save a considerable amount of money within a couple of years, then take the money home with them and build a house, start a business, and so on. Although it was not that difficult at the time to find a job, some migrants did find it difficult adjusting to the different way of living and working in Spain, thus pointing towards a more mental, psychological border:

> My first weeks here were not the best weeks of my life, as I felt uncertain and not at ease. I was disappointed to see that there were hardly any green areas in the neighbourhood, that you had to spend hours in the metro to reach the neighbourhood where you work, and that you really have to work very hard. There is hardly any time for relaxation. (Female, in Spain since 2005)

For some of the Bolivian women who became domestic servants, cleaning the house of a family and taking care of the children was a disappointing experience, particularly since some

Bolivian women realized that they had left their children behind in order to take care of another woman's children so that she could work. On the other hand, some of the women – even those with children – had the time of their lives in Spain: they excitedly talked about their work, the families they worked for and the things they did after work; they could travel and experience many new things, all without being under the control of their parents. Migration thus also led to the accelerated emancipation of Bolivian women, and crossing the border held the promise of freedom.

At the time, entering Spain as a tourist was not difficult. However, after overstaying the 90 days they were allowed, most women became irregular migrants. Some Bolivian women experienced this new status directly, for example when they had to cross a formal border. One woman told about the proposed holiday trip of the family she worked for:

> The first year, the family went to the mountains, in Catalonia, by car, so I could quite easily join them. The second year, however, they booked a villa in the Canary Islands, and it was only when they wanted to book tickets that they realized that I couldn't join them and take care of their children and the meals, as I couldn't catch a flight without a valid ID. (Female, in Spain since 2002)

In early 2005, the Spanish government announced a new regularization procedure – the sixth in just nineteen years – to cope with the enormous increase in the number of irregular migrants and to prevent their marginalization and exploitation. The eligibility criteria for regularization were a track record in employment and housing, and a certificate of good conduct.[3] Of the 700,000 requests for regularization that were received, 577,923 were approved, and these people were granted one-year residence permits (see also Sabater and Domingo 2012). The 47,000 Bolivian applications represented 6.7 per cent of all applications, a relatively small share compared to the total number of irregular Bolivians in Spain. This can be explained by the relatively recent inflow of Bolivians, due to which only a small proportion met the regularization requirements. Several respondents said that they wanted to apply for regularization, but could not collect all the documents that have to be submitted. This applied in particular to the employer's declaration, as some employers refused to provide one.

Interestingly, the consolidated Bolivians, who arrived before 2000, seem to distinguish themselves from Bolivians who arrived in the 2000s. They consider themselves different, which might also relate to class and ethnic differences:

> Our position is a bit double. On the one hand, we're really concerned about these people: they come here without anything, just a bag with a few clothes, and they're being exploited, they share a small apartment with 20 people. Really, these are not human conditions. On the other hand, we feel a bit, well, how to put it correctly, ashamed, as the media portrays them as Bolivians who are involved in criminal activities, poorly educated. We do not fit the image created by the newspapers: we are well integrated, we are well educated, we pay our taxes. We're not that kind of people. (Male respondent, in Spain since 1993)

To a certain extent, this attitude might also exemplify certain class and ethnic differences, as some of the Bolivians who arrived before the new millennium belong to the middle and upper class.

---

3   This regularization was strongly opposed by other EU countries, who accused Spain of not taking into account the overall EU interests (see also Nijenhuis 2011).

Restricted Access and Economic Downturn: 2007–2008

Under pressure from the EU (the European Parliament mandated visas for Bolivians in December 2006), Spain issued a visa requirement for Bolivian citizens from 1 April 2007. This accelerated Bolivian immigration in the first three months of 2007, not in the least spurred by airlines and the tourist industry, which had spotted a commercial opportunity. There were no less than 15 direct flights a day from Bolivia to Spain in this period; the planes were full when they left and almost empty when they returned to Bolivia. Flight tickets, each costing US $1,000, were sold out months in advance and Bolivians had to queue for a passport at the offices of the migration service.

In the months leading up to the introduction of the visa requirement, border controls started to become more intense and strict, and Bolivians who emigrated in this period were nervous about entering Spain, particularly through an airport: 'They [airport immigration officers] asked me about my plans, where would I stay, which cities did I plan to visit, in which hotels would I stay? How much money could I spend? And why did I not carry a travel guide? I was so glad when I finally got the stamp!' (male, in Spain since 2007).

The changing attitude at the Spanish borders inspired some NGOs in Bolivia to offer a quick training in 'how to behave like a tourist'. They advised Bolivians to always take a tourist guide, a camera and sufficient dollars, and to wear trousers whose legs could by unzipped above the knee.

The last week of March 2007 was extremely busy at Santa Cruz airport, with hundreds of Bolivians trying to get on one of the last flights to Madrid. At the same time, the main carrier from Bolivia to Spain (Lloyd Aereo Boliviano) had severe financial problems, and on the eve of 1 April had to suspend all flights. This had dramatic consequences for the 1,500 Bolivians who had been 'lucky' enough to buy tickets: some of them had spent over US $10,000 on tickets, but had to stay in Bolivia and shoulder the burden of crippling debt (see also *El Mundo*, 2 April 2007).

Although it had become more difficult to enter Spain, the economic prosperity in the country continued, and for those Bolivians who were there, it was not difficult to get a job, especially if they were women. Many respondents experienced social mobility and improved their labour and housing positions considerably. Some of the respondents had managed to regularize their status, and to rent their own apartment or even buy one. In this period, Bolivian families that had been in Spain for quite a long time (some of them for as long as seven years) were interviewed, and they revealed a new phenomenon: parents who would like to return to Bolivia, and offspring who do not even want to consider it. As one woman explained:

> We – my husband and I – have thought about it, and we really think it's a good idea to return to Bolivia. We can sell our apartment here, we have quite some savings, so we could start a hotel in Bolivia, in Cochabamba, where we are from. It's been a good experience, and we've been able to profit from it, but we'd like to spend the last years of our working life in the country we were born in, where our family is, and where we want to die. But our children, 14 and 16, they don't want to return, they don't understand our position at all: why opt for a poor country, while everything you want is here in Spain; a job, a nice house, a car, nice clothes? It makes me sad; I thought they would love to go there as well, but this will divide the family. (Female, in Spain since 2001)

Deep Economic Recession: Post-2008

This relatively optimistic period came to an end as the financial crisis of 2008 evolved into a deep economic crisis, and unemployment rates increased dramatically. In the last quarter of 2010, almost 20 per cent of the population was unemployed; among migrants, the rate was as high as 30 per cent (Nijenhuis 2011). The situation was even worse for migrants under the age of 25: some 45 per cent of them did not have a job. At first, Bolivians did not seem to worry a lot: most of them still had jobs, and if they lost them, there would be something else. As a male respondent, in Spain since 2006, said: 'Why would I return to Bolivia? ... The crisis? There's always a crisis in Bolivia! This one in Spain will pass, so I will just wait for that!'

However, it was not long before Bolivians experienced the crisis both directly, since they received lower wages, and indirectly, as there was more competition and labour conditions worsened considerably. Prior to 2007, a Bolivian domestic servant earned approximately 800 euros a month. Since 2007, most of the women had to accept a 25 per cent decrease, to 600 or even 500 euros a month. This reduction is directly linked to the increase in unemployment, also among employers, and the increasing cost of living in Spain. Sometimes, however, employers seem to use the crisis as an excuse to pay lower wages. In addition, a wage reduction is not always announced, but just happens to be the case on payday. Due to their irregular status, most women are unable to oppose such practices:

> I knew that he [the employer] did not have a very stable position, but it still came as a shock, also because I'd worked the same hours that month. On payday, he only paid me 400 euros, half of my salary, and said, 'I am sorry, but I can't afford to pay you more this month, perhaps next month.' Well, I was shocked: how did he expect me to pay my rent? I was entitled to more money, I worked for it! But he simply replied, 'If you're not content, then you should look for another position.' And of course, he knows very well that there are no other jobs. I tried to negotiate, working fewer hours for that money, but he does not give any room, so, in fact, I am trapped. (Female, in Spain since 2004)

For men, the situation appeared to be worse, as there was almost no demand for labour in construction or agriculture. Forced unemployment often implied asking other relatives in Spain, or friends, for support. Couples of which the woman still had a job were the least vulnerable, whereas single men really faced difficult times. The flexibility of the social networks among Bolivians was tested; in good times people did not hesitate to support each other, but many relationships were not crisisproof. As many Bolivians lost their jobs or suddenly found their job to be less secure, they had to rely on their networks to lend them money to pay the rent. Initially, this was not a structural phenomenon, and people financially supported relatives and friends. However, they increasingly experienced that the money was not repaid – because there was still no job, or only a part-time one. As most of those lending money realized that they too could end up jobless and without a source of income or access to social security, they soon started refusing to give any financial support, even to relatives. This, of course, was not always received well:

> My sister approached me last week for the second time: could I lend her 500 euros again, like last month? Her husband lost his job, and she now has only 30 hours a week. I felt awful, but I said that I couldn't, I just don't have the money. She already borrowed some, and didn't return it. She is annoyed, why don't I want to help her out, she is family after all ... I will stick to my position. Next month, I could lose my job, and who would help me out? (Female, in Spain since 2006)

Most irregular Bolivians who had lost their jobs considered returning to Bolivia, but since getting back into Spain again would be very difficult, if not impossible, they opted, at least initially, to stay put. However, after a couple of months, they decided to return after all, sometimes after a period of seven years in Spain; they were not indifferent to returning and they had passed the first threshold:

I'll leave, perhaps next week or the week after. I've tried all kinds of things, but I can't find a job; there are hardly any. I now live with some friends, but there's hardly any room, and you can't continue to live on the budget of somebody else. A few weeks ago I fell ill, and needed some medical assistance, but I don't even have the money to buy the necessary pills. I already have some debts to people, so the first thing I need to do is return the money to them. (Male, in Spain since 2001)

Not only irregular Bolivians were affected, also Bolivians with residence permits had trouble maintaining their newly acquired status of settled, successful migrants. There are numerous examples of Bolivians who had lived in Spain since the early 2000s, had jobs, had been regularized, and had got better jobs and even bought houses. They too, however, had now lost their jobs. According to some, this had been a selective process, as employers first fired the foreign workers.

In the end, returning to Bolivia was for many Bolivian migrants the logical and in fact only option: the locational threshold was the border of return. It is estimated that 17 per cent of the 250,000 Bolivians living in Spain in 2008 returned to Bolivia in the post-2008 period (INE 2013); this percentage is smaller than previously thought, which might also be related to the limited economic opportunities in Bolivia (IOM 2011). Although the Spanish government provided a voluntary return scheme in 2010, only very few Bolivian migrants made use of it. According to interviews with local NGOs in Spain, some Bolivian households in Spain decided that the woman would remain in Spain, as she still had a job, while the man and the children returned to Bolivia, which is also reflected in the demographic data (see Figure 14.1). Instead of returning to Bolivia, some Bolivians tried other trajectories and moved to Italy – which has a relatively large Bolivian community – or to northern Europe. Others returned to Bolivia, and then migrated to Brazil, which is an upcoming 'market' for Bolivian labour migrants.

Final Comments

This contribution has explored the perceptions of borders among Bolivian migrants in Spain, and the way these perceptions are linked to the prevalent context of exit and reception. For this purpose four phases of exit and reception for Bolivian migrants in Spain have been distinguished: the period before 2001, that from 2001 to 2007, that from 2007 to 2008, and that after 2008. In only a relatively short period, the context of exit and reception for Bolivian migrants in Spain changed considerably. Following the introduction of the visa requirement for Bolivians in April 2007 in line with EU policies, the border between Spain and Bolivia as a physical demarcation line changed from a relatively low barrier with hardly any restrictions on entry, into a high barrier. The economic crisis in Spain that started in 2007, due to which unemployment figures increased dramatically, is perceived as a similar high barrier. For 'would-be' migrants who are still in Bolivia, Spain is no longer a realistic option, and for those Bolivians in Spain who lost their jobs, it is the main motive to cross the border again, to return – a difficult decision, since opportunities for work in Bolivia are also scarce.

Apart from these more tangible meanings of the border between Spain and Bolivia, some other interpretations of borders were observed, involving different groups. An example of this are the processes of 'bordering' and 'othering' within the Bolivian community in Spain, illustrated in the border between 'consolidated Bolivians' and 'newcomers'. The former consider themselves distinct from the second group, a process in which also hierarchy and class play a role (see also Van Houtum and Van Naerssen 2002). This also has spatial implications: the consolidated Bolivians live in different neighbourhoods, attend different schools and participate in different Bolivian sociocultural organizations. Another example of the border as a more social construct is the border between those in Spain and those who remain in Bolivia: the migrants in Spain are considered more progressive, modern and advanced.

What does this case tell us about the role of thresholds, and in what way can the threshold approach, and in particular the locational threshold, be applied to explain and analyse the Bolivia–Spain migration corridor? The locational threshold came quite clearly to the fore, as certain push and pull factors can be observed: a lack of alternative income sources in Bolivia and the initial huge demand for labour in Spain appear to be critical push and pull factors, respectively, in the decision to move. Another factor of importance are migration policies, such as the high accessibility of Spain before the crisis – which was a clear pull factor. Moreover, the presence of social networks in Spain, through which Bolivians could get access to jobs and housing, was a pull factor.

In addition to these push and pull factors, certain keep and repel factors were also encountered. Interestingly, these factors can be found on both sides of the border, acting simultaneously as a keep and repel factor. From the perspective of Bolivians in Bolivia, the fact that many of them could no longer enter Spain because of visa requirements is a clear repel factor. However, it was conceived as a keep factor from the perspective of Bolivians without permits in Spain: since returning to Bolivia would imply that they would not be able to re-enter Spain, they decided to stay put. Finally, the observed border as a social construct – between those in Spain and those who remain in Bolivia, as well as the perceived differences between the consolidated Bolivians and the newcomers – adds an interesting new dimension to the threshold approach that is central to this volume, namely that of social class.

In conclusion, the case of Bolivian migrants in Spain has shown that the contexts of exit and reception – interpreted here as locational thresholds – are dynamic: they depend on changing migration policy frameworks, economic processes and labour market developments. These changing contexts in turn have an impact on the barrier function of borders, which can enable mobility and transnational activities, but can also result in immobility. As such, this draws attention to the fact that migration is a highly dynamic process.

References

ACOBE (Asociación de Cooperación Bolivia España). 2011. *"La experiencia del retorno ... ": Estudio del caso boliviano*. Madrid: ACOBE.

Dulón, R. 2008. *Migración transnacional de Bolivianos y Bolivianas a la Argentina y su impacto en comunidades de origen: Informe final convocotaria chorlavi*. Sucre: Fundación Pasos.

Durand, J. 2009. "Processes of Migration in Latin America and the Caribbean (1950–2008)." Human Development Reports Research Paper 2009/24. Washington, DC: United Nations Development Programme.

Durand, J., and D.S. Massey. 2010. "New World Orders: Continuities and Changes in Latin American Migration." *The Annals of the American Academy of Political and Social Science* 630 (1): 20–52.

Ernste, H., H. van Houtum, and A. Zoomers. 2009. "Trans-World: Debating the Place and Borders of Places in the Age of Transnationalism." *Tijdschrift voor economische en sociale geografie* 100 (5): 577–86.

Farah, I.H., and C.G. Sánchez. 2001. "Bolivia: An Assessment of the International Labour Migration Situation; The Case of Female Labour Migrants." GENPROM Working Paper 1. Geneva: International Labour Office.

Gielis, R. 2009. "Borders Make the Difference: Migrant Transnationalism as a Border Experience." *Tijdschrift voor economische en sociale geografie* 100 (5): 598–609.

Hannam, K., M. Sheller, and J. Urry. 2006. "Editorial: Mobilities, Immobilities and Moorings." *Mobilities* 1 (1): 1–22.

IBCE (Instituto Boliviano de Comercio Exterior). 2008. "Bolivia: Migración, remesas y desempleo." *Comercio Exterior* 16 (159).

INE (Instituto Nacional de Estadística). 2001. "Resultados censo nacional de población y vivienda 2001." Accessed 13 September 2012. http://datos.ine.gob.bo/binbol/RpWebEngine.exe/Portal?&BASE=CPV2001COM (site discontinued).

_____. 2013. "Padrón: Población por municipios." Accessed 30 May. http://www.ine.es/jaxi/menu.do?type=pcaxis&path=/t20/e245/&file=inebase.

IOM (International Organization for Migration). 2011. *Perfil migratorio de Bolivia.* Buenos Aires: IOM.

Menjívar, C. 2001. "Latino Immigrants and Their Perceptions of Religious Institutions: Cubans, Salvadorans and Guatemalans in Phoenix, Arizona." *Migraciones Internacionales* 1 (1): 65–88.

Nijenhuis, G. 2010. "Embedding International Migration: The Response of Bolivian Local Governments and NGOs to International Migration." *Environment and Urbanization* 22 (1): 67–79.

_____. 2011. "De Europeanisering van het Spaanse migratiebeleid." *Geografie* 20 (6): 14–16.

Padilla, B., and J. Peixoto. 2007. "Latin American Immigration to Southern Europe." *Migration Information Source*, 28 June. Accessed 20 September 2012. http://www.migrationpolicy.org/article/latin-american-immigration-southern-europe.

Peixoto, J. 2012. "Back to the South: Social and Political Aspects of Latin American Migration to Southern Europe." *International Migration* 50 (6): 58–82.

Pellegrino, A. 2004. "Migration from Latin America to Europe: Trends and Policy Challenges." IOM Migration Research Series. Geneva: International Organization for Migration.

Portes, A., and R.G. Rumbaut. 2006. *Immigrant America: A Portrait.* 3rd rev. ed. Los Angeles: University of California Press.

Roncken, T., and A. Forsberg. 2007. *Los efectos y consecuencias socioeconómicos, culturales y políticos de la migración internacional en los lugares de origen de los emigrantes Bolivianos.* La Paz: Programa de Investigacion Estratégica en Bolivia.

Sabater, A., and A. Domingo. 2012. "A New Immigration Regularization Policy: The Settlement Program in Spain." *International Migration Review* 46 (1): 191–220.

Skeggs, B. 2004. *Class, Self, Culture.* London: Routledge.

Turner, B.S. 2010. "Enclosures, Enclaves, and Entrapment." *Sociological Inquiry* 80 (2): 241–60.

Van der Velde, M., and T. van Naerssen. 2011. "People, Borders, Trajectories: An Approach to Cross-Border Mobility and Immobility in and to the European Union." *Area* 43 (2): 218–24.

Van Houtum, H., and T. van Naerssen. 2002. "Bordering, Ordering and Othering." *Tijdschrift voor economische en sociale geografie* 93 (2): 125–36.

Vertovec, S. 2009. *Transnationalism.* London: Routledge.

Chapter 15

Boats, Borders and Ballot Boxes:
Asylum Seekers on Australia's Northern Shore

Graeme Hugo and Caven Jonathan Napitupulu

Australia is one of the world's countries most influenced by migration. At the 2011 population census, 26 per cent of the population were foreign-born and 19 per cent were Australia-born with at least one parent born overseas. Moreover, in June 2011 there were about 900,000 persons who were overseas visitors or temporary residents (DIAC 2012b). There is a strong tradition of migration with a multiplicity of channels for moving to Australia as a settler, temporary resident or visitor with there being 318 separate visa categories (DIAC 2011a). As only New Zealanders[1] are not required to obtain a visa before entering Australia, considerable control is exercised over who can enter the country. Moreover, Australia's island geography and relative isolation facilitate control of its borders, assisted by increasingly sophisticated electronic surveillance. Hence clandestine, undetected entry into Australia is extremely limited. Undocumented immigrants to Australia are of three main types – via boat arriving on Australia's northern shores, arrival without a visa at Australian airports and not meeting the requirements of visas by overstaying and/or working while on a non-working visa. The focus of this chapter, and the dominant contemporary public discourse within Australia, is on the first of these.

The geographical approach adopted in this book argues that borders and trajectories act as important thresholds in the decision to migrate and in the migration process. This chapter demonstrates the importance of this argument to understand asylum seeker migration to Australia. Migration is, implicitly at least, usually conceptualized as a direct movement to a destination but the actual experience of many movers, especially those in irregular situations, is that the trajectory of their movement is more complex. The journey is disjunctive and can take a significant amount of time as migrants are delayed in transit locations.

The chapter begins by putting the boat arrival flows in the context of all Australian international migration. It is shown that the number of asylum seekers in Australia, although increasing in recent years, is small both in relation to all Australian migration and as a proportion of the number of global asylum seekers. Nevertheless, the 'boat people' issue dominates the national discourse on migration, has become a focus of political debate and was a central issue in recent federal elections. The trajectories taken by asylum seekers migrating to Australia are then examined with some discussion of the role that Indonesia plays as a transit country. This leads to a discussion of how the migration process operates among asylum seekers moving to Australia. It is argued that perceptions of the Australian border are not only significant in the decision-making processes of asylum seekers but also in shaping Australian public opinion and attitudes toward migration and asylum seekers.

---

1 Under the 1973 Trans-Tasman Travel Arrangements New Zealanders have the right to live and work in Australia.

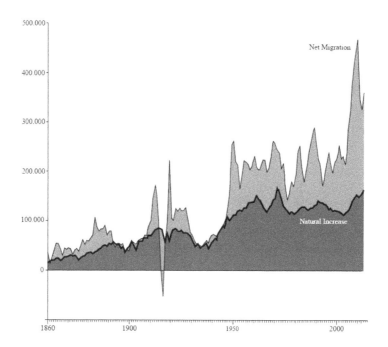

Figure 15.1　Natural increase and net migration, 1860–2012

Sources: ABS, *Australian Demographic Statistics*, various issues; Borrie 1994.

Asylum Seekers in the Context of Australian International Migration

Figure 15.1 shows that net migration has been a major contributor to Australia's population growth over the last 150 years but it has been especially significant since World War II. As a result half of Australia's contemporary population is a first- or second-generation immigrant. Prior to World War II Australia's immigration was predominantly from the British Isles so that the population was overwhelmingly Anglo-Celtic. Subsequently, however, there has been a widening of the origin countries from which Australia has drawn its immigrants. Hence, Table 15.1 shows that Australia is now characterized by significant cultural diversity although English ancestry still is dominant. It is especially important to note in the table that there are 67 birthplace groups which have more than 10,000 persons in Australia and 133 with more than 1,000 persons. This indicates that Australia has significant communities from a majority of the world's countries which serve as the base for establishing international connections, and in many cases, the creation of active corridors of substantial population movement.

The contemporary Australian migration system is a highly planned one comprising comprehensive permanent and temporary migration components (Hugo 2012). The former comprises skill, family and refugee-humanitarian elements, the levels of which are fixed at the beginning of each year by the federal government.[2] Immigrants include persons who are selected

---

2　In 2011–12 the number planned for was 185,000 and the outcome was within two of this number (see http://www.immi.gov.au/about/speeches-pres/transcript-opening-statment-4.htm, accessed 6 November 2014).

Table 15.1   Indicators of Australian diversity, 2011

Indicator	Per cent
Born overseas	26.1
Born overseas in non-English speaking (NES) country	16.6
Australia-born with an overseas-born parent	18.8
Speaks language other than English at home	19.2
Ancestry (multiresponse) in a NES country (2006)	26.0
Non-Christian religion	22.3
Indigenous population	2.6
No. of birthplace groups with 10,000+ persons	67
No. of birthplace groups with 1,000+ persons	133
No. of indigenous persons	548,369

Source: ABS 2011 Census, unpublished data.

for settlement while outside Australia (59.7 per cent of the 2010–11 total) and persons who apply for, and obtain permission, to settle when they are already in Australia as a visitor or with another temporary visa. There are separate planning quotas for each visa category and the criteria for selection in the family and skill categories are very restricted with the latter being on the basis of a so-called points test assessment. The refugee-humanitarian category has been capped at around 13,000 per annum for recent years. Figure 15.2 shows how this has been comprised of 'offshore' and 'onshore' settlers. The former have made up the majority until recent years when the increase in Irregular Maritime Arrivals (IMAs) who successfully applied for refugee status meant that a larger share of the 13,000 places went to 'onshore' settlers. Since the planning level quotas are fixed[3] the number of 'onshore' settler places are at the expense of 'offshore' settlers.

Australia has long had an emphasis on attracting permanent settlers to the country and a strongly expressed opposition to attracting temporary and contract workers. During the labour shortage years of the 1950s and 1960s Australia's migration solution to the problem contrasted sharply with that of European states like Germany and France when it opted to concentrate on attracting permanent migrants to meet worker shortages rather than contract workers. However, since the mid-1990s attitudes have changed and it has been recognized that in the context of globalized labour markets it is essential to have mechanisms to allow non-permanent entry of workers in certain skilled groups. Consequently Australia has introduced a suite of temporary residence visa categories.

---

3   In 2011–12 this was fixed at 13,750; in 2012–13 this was increased to 20,000. This was part of a series of measures aimed at reducing the flow of asylum seekers to Australia (Chris Bowen.net 2012). However, when the new government was instilled in 2013 it made the 2013–14 quota 14,000.

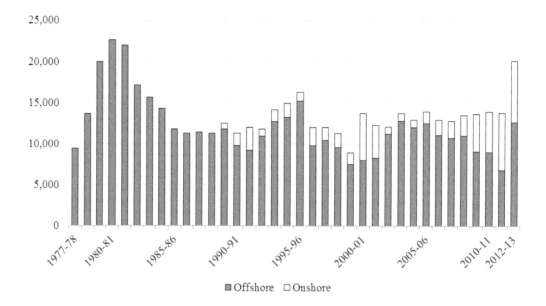

**Figure 15.2   Humanitarian Programme visa grants,
            1977–78 to 2012–13**

Sources: DIAC 2012b, 2012c, 2013.

The Temporary Business visa (known as subclass 457) is similar to the H1B visa in the United States in that it is initiated by employers but it is not capped. The number of 457s in Australia peaked in mid-2009 at 155,673 and at the end of 2011 was 128,602 (DIAC 2012d). The programme is even more focused on skill than the permanent migration programme, being confined to the managerial, professional, paraprofessional and trades occupation categories, and research has shown it has been generally quite successful (Khoo et al. 2007). However, the 457 programme has come under intense scrutiny in recent times with some employers being accused of misusing the scheme to displace Australian workers, especially in some regional areas.

The largest, and most rapidly increasing inflow of temporary migrants with the right to work in Australia has been of foreign students. Mid-2012 there were 307,050 foreign students resident in Australia with 77.4 per cent being from Asia. The Working Holiday Maker (WHM) programme has also reached record levels of 154,148 arrivals in 2007–08, doubling in the last ten years and increasing by 15 per cent over the previous year. The WHM programme is a reciprocal one which allows young people (aged 18–30 years) from 19 countries to have working holidays in Australia for periods of up to one year. WHMs fill some important niches in the labour market such as in harvesting, tourist activity, restaurants and so on, which in many other countries are sectors involving irregular migrants.

One important aspect of temporary migration is that through temporary residents applying for resident status it contributes to the settlement migration programme. As Figure 15.3 shows, temporary migration is now a significant contributor to net overseas migration gain.

The key issue from the perspective of this chapter is that there are a multitude of legal channels for obtaining entry to permanent settlement in Australia. Moreover, one of the principles of Australia's immigration programme is the absence of discrimination on the basis of nationality,

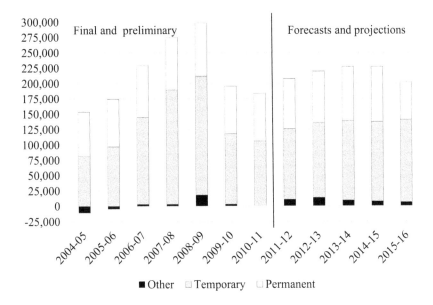

**Figure 15.3   Components of net overseas migration,
2004–11 actual and 2011–16 projected data**

Source: DIAC 2012e.

religion, ethnicity or race.[4] However, it is important to note that the ability of many groups to access these channels is limited because of several reasons. For one thing, both permanent and temporary migration programmes strongly focus on the selection of highly trained and skilled groups. Secondly, there are significant English language requirements in most categories. Thirdly, family migration has become increasingly restricted and now is predominantly partner migration. And fourthly, access to offshore selection as refugee-humanitarian settlers is limited because of the small numbers of places made available by the Australian government compared with the global number of refugees as well as the focus on particular source countries.

Moreover, Australia's isolated and island nature means that these constraints on who can enter Australia are strongly and effectively enforced. Hence Australia's borders, including its northern border, represent major barriers to immigration.

Recent Asylum Seeker Migration to Australia

In light of the barriers presented by Australia's borders, what are the avenues which have been used by people seeking to move to Australia outside of the regular channels? Firstly, there is the *clandestine entry*: while there is little information on this it is clearly very limited due to the isolated and island nature of the country. Secondly, one can *overstay* one's visa. In recent decades

---

4   A small exception is the Seasonal Agricultural Worker programme which is targeted at migrants from a selected number of Pacific Islands (Hugo 2011a).

Table 15.2   Estimated overstayer population, 30 June 2011: top 10 source countries

Citizenship	Overstayers, 30 June 2010	Overstayers, 30 June 2011	Change (%)	Proportion of overstayers in 2011 (%)
China (excludes SARs and Taiwan)	7,490	8,070	7.7	13.8
United States	*5,010*	*5,080*	*1.4*	*8.7*
Malaysia	3,890	4,260	9.5	7.3
United Kingdom	*3,470*	*3,610*	*4.0*	*6.2*
India	2,200	3,290	49.5	5.6
South Korea	*2,570*	*2,730*	*6.2*	*4.7*
Indonesia	2,460	2,580	4.9	4.4
Philippines	2,430	2,400	-1.2	4.1
Thailand	1,660	1,790	7.8	3.1
Vietnam	1,550	1,670	7.7	2.9
Other	21,130	22,970	8.7	39.3
Total*	53,860	58,400	8.5	100.0

Notes: OECD countries are italicized.
* Due to known errors in overstayer data, the total is not the same as the sum of the data presented.
Source: DIAC 2012a, 58.

the number of overstayers has averaged around 60,000. Table 15.2 shows that in 2011 the largest numbers of overstayers were from China and the largest increase was from India. Some 72 per cent of overstayers are former visitors and 17 per cent are students.

A third group of irregulars are people who enter Australia under a visa category which does not permit them to work but *they nonetheless do*. The government has substantial compliance detection activities and in 2010–11 enforced 10,175 compliance-related departures compared with 8,825 a year earlier (DIAC 2012a, 59). A fourth avenue comprises *unauthorized air arrivals*. The numbers in this category are limited by the sophisticated electronic system in place which checks the visa of all persons on flights departing for Australia when intending passengers check in at the origin. Figure 15.4 shows the numbers in recent years have fluctuated between 937 in 2002–03 and 2,058 in 2004–05.

The avenue of entry which has attracted most attention is *unauthorized boat arrivals* on Australia's northern border. These arrivals are not clandestine, however. These migrants seek to be detected and claim refugee status. Figure 15.4 shows that these numbers have fluctuated over the last few decades but have peaked around the year 2000 and have reached record levels

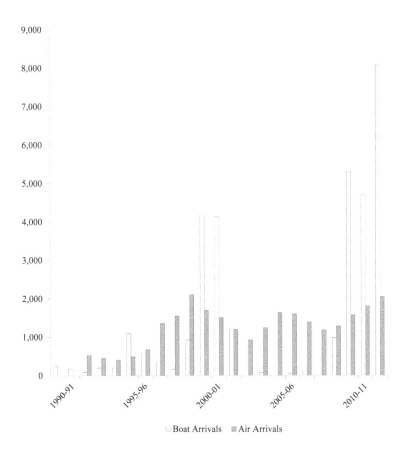

Figure 15.4   Unauthorized arrivals in Australia, 1989–90 to 2011–12

Sources: DIMIA 2002, 2004, 2005; DIAC, *Annual Report*, various issues; Phillips and Spinks 2012.

in recent years. The largest influx was reached in 2011–12 with the arrival of 111 illegal entry vessels and 8,371 asylum seekers (DIAC 2012c, 168). The initial arrivals were from Indo China, mainly Vietnam, and began in the late 1970s and continued through to the early 1990s. In the mid-1990s the numbers were small and the migrants came mainly from China but in the late 1990s the flows from the Middle East, especially Iraq, Iran and Afghanistan, began peaking. They dried up with the introduction of the 'Pacific Solution'[5] (Jupp 2002, 193–196) but began again in 2008–09 and reached record levels in 2011–12. In recent years Sri Lanka has become an important origin.

Boat arrivals of asylum seekers have dominated the public discourse on migration issues in recent years in Australia and this is especially the case with media coverage. It is argued

---

5   The Pacific Solution was introduced by the conservative Howard government in 2001. It involves the interception of asylum seekers before they enter Australia's 'migration zone'. They are then sent to detention centres in selected Pacific countries where they are interned while a decision is made on their claim for refugee status.

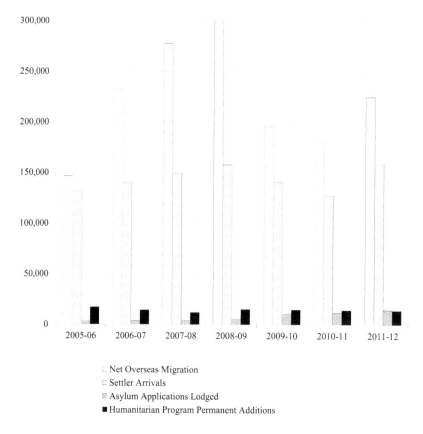

Figure 15.5 Migration and asylum applications in Australia, 2005–06 to 2011–12
Sources: ABS 2012; DIAC 2012c, 2012f.

elsewhere (Hugo 2011a) that with respect to asylum seekers, the Australian media has not only reflected swings in public opinion on migration but has played an important role in shaping common attitudes as well as those of influential policymakers. Media has shaped contemporary public opinion on migration in negative ways. A study of the United States argues that while 'individual articles and broadcasts about immigration may have been entirely accurate, the cumulative effect of US media coverage has distorted the underlying realities of immigration' (Suro 2009, 186). This distortion has been achieved through three major tendencies within the way the media covers immigration and these are common to many countries of immigration: episodic coverage (surges of coverage when there is a particular migration-related event, usually of a negative nature), a focus on illegality and a lack of context.

In Australia these arguments have some resonance in the domination of media coverage of migration issues by the asylum seeker issue (Cartner 2009). Asylum claims in 2008 numbered 5,020. This was only 1.3 per cent of all asylum seekers worldwide (UNHCR 2011, 6) and is only a tiny fraction of the net overseas migration gain in 2008–09 of 315,686 (ABS 2011, 11). Despite this, media coverage in Australia on immigration in 2008 was overwhelmingly focused on the arrival of asylum seekers in boats on Australia's northern shores. Figure 15.5 shows how tiny the number of asylum seeker applications in Australia has been in recent years

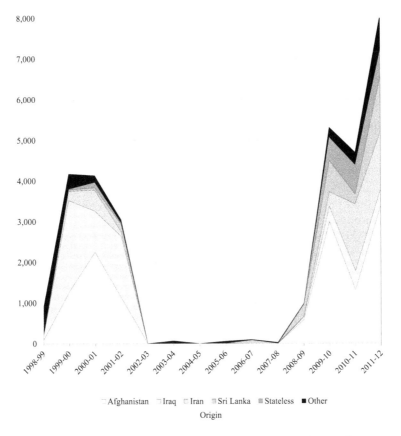

Figure 15.6   Irregular maritime arrivals in Australia, 1998–99 to 2011–12

Source: Australian Government 2012, 89.

compared with the number of permanent settler arrivals and the contribution of net overseas migration to population increase.

The comparative numbers in Figure 15.5 are important since much of the Australian public believes that the migration intake is dominated by refugee-humanitarian settlers in general and asylum seekers arriving by boat on Australia's northern shore in particular. These relativities are also important when the proportion of global asylum seekers that Australia receives is examined. In 2009–10 and 2010–11 Australia received 3.5 per cent and 2.7 per cent of all global asylum applications (UNHCR 2013). Yet the public perception is quite different, largely because of political and media discussions and distortions.

Who are these asylum seekers? Figure 15.6 shows that in recent years Afghanistan has been the major source of asylum seekers with Iran, Iraq and Sri Lanka also being significant. In 2011–12 Sri Lankan Tamils have made up an increasingly large proportion of asylum seeker arrivals with more than 1,500 arriving in 2012 before August – seven times the number in the previous year (*Australian*, 15 August 2012). There is a predominance of males among asylum seekers although the proportion of females increased from 5.7 per cent in 2008–09 to 16.8 per cent in 2010–11 (DIAC 2011c, 27). The largest group are young adults aged 18–40 (68.3 per cent). Children aged 17 or less comprised 21 per cent (DIAC 2011b, 28).

Trajectories of Migration

As Van der Velde and Van Naerssen (2011, 220) point out, one of the shortcomings of much thinking about international migration is that it assumes migrants move directly from their place of origin to a destination. A significant exception to this has been studies of forced migration – both empirical and theoretical. The classical works of Kunz (1973, 1981) are among the most insightful theoretical studies. Figure 15.7 is a diagrammatic depiction of Kunz's theory of forced migration. Central to this is the fact that most forced migrants are not able to move directly to a place of permanent resettlement due to the largely unplanned and unanticipated nature of the move and the sudden circumstances which precipitate the move. Accordingly they often move to a temporary haven which Kunz describes as a 'midway to nowhere' situation to emphasize the precariousness, uncertainty and temporariness of their stay there. Figure 15.7 indicates that from their place of temporary asylum they either return to their homeland when conditions improve or move on to a place of permanent resettlement. This movement often occurs in waves. The diagram shows that once resettled, they often attempt to bring their family to join them either from the homeland or from a place of temporary refuge. In forced migration the intermediate transit locations often assume major significance.

In recent years there has been a growing literature on transit locations which are of importance, not only to forced migration but also to other situations where the movers are uncertain of their final destination or where there are difficulties or barriers in entering the intended country of destination (Içduygu 1996, 2000; Papadopoulou 2004; Mavris 2002; Ivakhniouk 2004; Sushko 2003; Hamood 2006). As Van der Velde and Van Naerssen (2011, 221) have described: 'It is becoming acknowledged that it is increasingly difficult to distinguish between countries of destination, origin and transit. These are often unclear notions for the migrants themselves, particularly when they have to travel large distances.' In many modern migrations, especially those of asylum seekers, migration trajectories can extend over long periods of time and a number of intermediate locations before a final (intended or unplanned) destination is reached and this is certainly the case for asylum seekers arriving by boat on Australia's northern border.

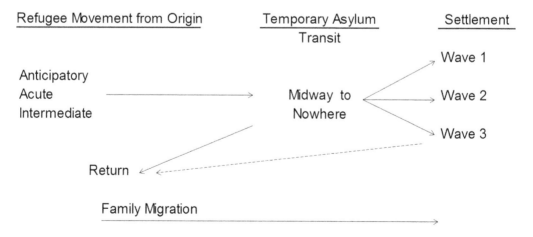

Figure 15.7   Expansion of the models of Kunz

Source: developed by the authors from Kunz 1973, 1981.

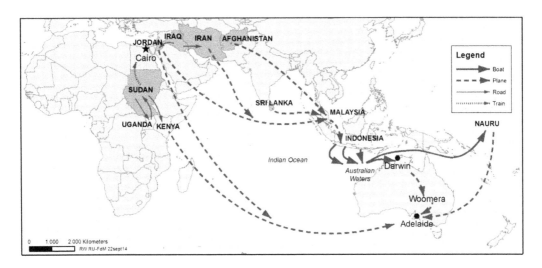

Figure 15.8 Map of routes taken by refugees to Australia
Source: Adapted from Hinsliff 2006.

Studies by the Department of Immigration and Multicultural and Indigenous Affairs (DIMIA 1999) indicate that asylum seekers in Australia had spent varying periods of time as transit migrants. The main transit countries identified in those interviews were as follows: Firstly, in the flows from the Middle East and Afghanistan the three main staging points have been Malaysia, Thailand and Indonesia. Maley et al. (2002, 13) point out that 'Malaysia became a key transit stop partly because it allows visa-free access for people of the Islamic faith. About 80 per cent of all Asylum Seekers from Iraq and Afghanistan pass through Malaysia en route to Australia.' From Malaysia the asylum seekers fly directly to Australia or travel to Indonesia from where they either fly to Australia or move to small fishing villages and ports in southern Java or Nusa Tenggara from where they board boats to take them to the northern coast of Australia. Secondly, Seoul in South Korea has also become a significant staging point for active people smuggling into Australia, bringing in people from Iran and Algeria (DIMIA 1999, 22). Thirdly, Chinese undocumented migrants leave from Fujian ports in southern China. They usually enter Australia with genuine travel documents and enter under visitor visas but overstay and work in contravention of visa conditions. A few enter illegally by boats travelling directly to Australia but this flow has not been significant in recent years.

The trajectories followed by the major groups of contemporary persons seeking asylum in Australia are shown in Figure 15.8. The most important paths of asylum seekers from Sudan, Iraq, Iran, Afghanistan and Sri Lanka are shown. In all cases they generally fly to Malaysia as tourists since Malaysia offers visas upon arrival to nationals of more than 60 countries in order to facilitate tourism (Missbach and Sinanu, 2011, 73). From Malaysia they travel to Indonesia which is the taking-off point for the final leg – a boat trip to Australia. These corridors of movement have become well established and a complex industry of interconnected agents has developed along the route to facilitate migration. Many asylum seekers from the largest origin, Afghanistan, initially move to camps in Pakistan from where they negotiate with a people smuggler. Some asylum seekers directly travel to Indonesia, especially those like Iranians who can obtain a 30-day tourist visa on arrival. Some asylum seekers initially move to Thailand which has long been a hub

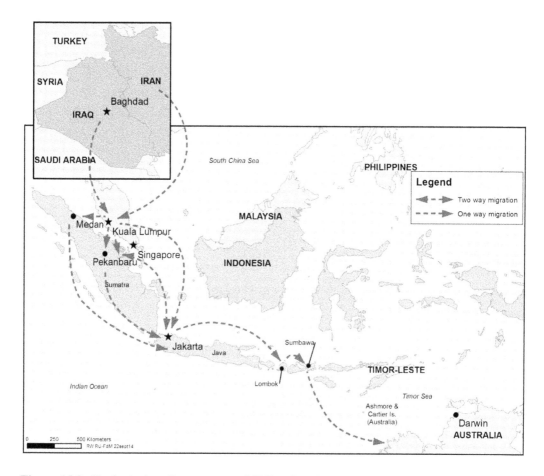

Figure 15.9   Trajectories of movement of 40 Iraqi asylum seekers, 2008

Source: unpublished map supplied by Directorate General of Immigration, Jakarta, Indonesia.

for trafficking in the Asian region (Skrobanek et al. 1997). Many of the asylum seekers initially arriving in Malaysia, clandestinely travel by boat to Indonesia.

There is substantial boat traffic between Indonesia and Malaysia and an established migration industry linking them with more than two million Indonesians working as international labour migrants in Malaysia, many of them undocumented (Hugo 2011b; Jones 2000). In many cases the people smugglers, whom the asylum seekers negotiate with in their home country, only get them as far as Malaysia or Indonesia and it is then up to the asylum seekers to negotiate a passage to Australia with agents based in Malaysia or, especially, Indonesia.

Figure 15.9 depicts the routes taken by 40 Iraqi asylum seekers in travelling to Australia in 2008. Before trying to get to Australia, some flew directly from Kuala Lumpur to Jakarta while others travelled by sea either directly to Jakarta or through Sumatra.

Neither Malaysia nor Indonesia are signatories to the 1951 Convention relating to the Status of Refugees or the 1967 Protocol and neither country has a legislative framework or system for the protection of refugees (Missbach and Sinanu 2011, 68). However, the United Nations High Commissioner for Refugees (UNHCR) have substantial offices in both countries and many asylum seekers register with the UNHCR office in either or both countries. Table 15.3 shows the numbers officially registered in recent years. The UNHCR and the International Organization for Migration (IOM) are very active in supporting and assisting asylum seekers in both countries. Indeed, Missbach and Sinanu (2011, 74) found that many people smugglers encourage asylum seekers to register with the UNHCR on arrival in Indonesia.

Table 15.3   Indonesia, Malaysia and Thailand: asylum seekers, 2001–11

	COUNTRY OF ASYLUM		
	Indonesia	Malaysia	Thailand
1996	9	8	198
1997	35	8	681
1998	61	570	828
1999	18	30	582
2000	373	25	361
2001	806	252	343
2002	237	1,571	1,050
2003	68	9,205	2,657
2004	59	10,322	1,044
2005	58	10,838	32,163
2006	265	9,186	18,424
2007	211	6,851	13,484
2008	353	9,323	12,578
2009	1,769	10,267	10,255
2010	2,071	11,339	10,250
2011	3,233	10,937	13,357

Source: UNHCR, *Statistical Yearbook*, various issues.

The borders of Malaysia and Indonesia present little problem to asylum seekers as they seek to make their way to Australia. Certainly, they are at risk of detection and detention as well as experiencing exploitation at the hands of unscrupulous agents, police, officials and other groups in Malaysia and Indonesia but they generally are able to reach the take-off points for the boat journey to Australia in the Indonesian provinces of West Java and Nusa Tenggara Timor. However, the Australian border presents a more significant and daunting barrier.

One dimension of this is the boat voyage from Indonesia to the northern border of Australia. The boat journey is a hazardous one and the boats tend to be in poor condition and overcrowded.

Several shipwrecks have been reported which claimed the lives of many asylum seekers (DIAC 2012c). There are two major routes taken by the boats (see Figure 15.8). One from the south coast of Java (usually in the province of West Java) to the Australian Christmas Island located some 400 km south of Java. The second from the province of East Nusa Tenggara, some 140 km to Ashmore Reef, also Australian territory. As mentioned, these journeys are not intended to be clandestine forays into Australia but the asylum seekers aim to be intercepted by Australian authorities on the sea or when they land on Christmas Island or Ashmore Reef, where their refugee status will be assessed or from where they will be transferred to Australia to await assessment.

One of the features of trajectories as delineated by Van der Velde and Van Naerssen (2011, 221) is that often, when migrants travel over long distances, the place of destination can be 'unclear and insecure' so that the 'journey is disjunctive and it can take months or even years before it is accomplished, if at all'. This has some resonance with the trajectories of asylum seekers moving to Australia. However, it does seem certain that most asylum seekers do set out with the definite intention of settling in Australia. There is little uncertainty at the outset of the journey as to what the eventual destination is. This certainty has its basis in two planks. The first is the fact that many asylum seekers have family and friends in Australia who have not only been sending information back home but also often provide the funds to pay for the journey and offer assistance once they arrive in Australia. Table 15.4 shows the size of the communities of some of the key asylum seeker groups in Australia at the last three censuses. Clearly they have grown rapidly over the last decade. These communities not only act as important anchors that shape where the asylum seekers intend to go, but also actively support and encourage the movement. Secondly, the establishment of well-trodden corridors of migration between the four countries and Australia by the migration industry certainly has an impact in establishing the spatial pattern of migration. There is no evidence of asylum seekers settling permanently in Malaysia and Indonesia.

Table 15.4   Growth of communities of major asylum seeker groups in Australia

	2001	2006	2011	Average Annual Growth Rates	
				2001–06	2006–11
Birthplace					
Iran	18,789	22,549	34,455	3.72	8.85
Iraq	24,832	32,520	48,170	5.54	8.17
Sri Lanka	53,461	62,256	86,413	3.09	6.78
Afghanistan	11,296	16,751	28,599	8.20	11.29
Ancestry					
Tamils	7,706	8,897	19,426	2.92	16.90
Afghan	12,410	19,414	29,474	9.36	8.71
Iraqi	11,190	16,763	28,001	8.42	10.81
Iranian	18,798	23,575	36,168	4.63	8.94

Source: ABS 2001, 2006 and 2011 Population Censuses, unpublished data.

The Migration Process

Most of the substantial theoretical literature on migration relates, at least implicitly, to movements which are to some extent voluntary while that concerned with forced migration is limited. It is important to acknowledge that, while it is possible to recognize some movements as being totally forced or voluntary, many movements contain elements of both (Hugo 1996). While for most asylum seekers coming to Australia the major reasons impelling their migration were insecurity and fear of violence, war and conflict in the origin country, there were clearly also elements of choice involved in the decision-making process. Whether or not people move is clearly influenced by the social network of the potential movers and the extent to which they have family and friends in Australia. This is important both from the perspective of supply of information regarding the destination and assurance of support at the destination, but also the financing of the move itself. The nexus between the community in Australia and asylum seekers is important. Asylum seekers are in constant contact with their Australian contacts before and during the migration process, often using mobile phones. Agents are also a very important element in the asylum migration process. Very often the agents are of the same nationality or ethnicity as the asylum seekers themselves. Almost all asylum seekers rely to some extent on people smugglers at least at some stage of the process. As Missbach and Sinanu (2011, 73) have pointed out: 'Over the last ten years, efficient smuggling routes, routines and networks have developed stretching from the home countries in the Middle East and Central/South Asia over diverse transit countries including Malaysia and Indonesia.' In most cases it seems that people smugglers do not arrange the complete journey from the origin country to Australia but one or more parts of the movement. They can arrange introductions to agents to facilitate the next leg while in other cases asylum seekers were under the impression that they had paid for the entire trip to Australia only to be abandoned in Malaysia or Indonesia. The overall picture which emerges is not of an integrated international structure of tightly linked elements between origin and international destination but one described by Missbach and Sinanu (2011, 66) as 'loose, temporary, acephalous networks' and by Içduygu and Toktas (2002, 46) as 'a loosely cast network consisting of hundreds of independent smaller units which cooperate along the way'. Despite the public perception of the dominance of 'Mr. Big' figures in people smuggling, the industry is dominated by smaller, independent agents, travel providers, bureaucrats, police, lawyers, accommodation providers and so on. Certainly there are major criminals and large-scale operators involved. It is clear too that much of the organization goes across countries. Recent prosecutions of people smuggler organizers in Australia testify to this. In addition, countrymen from the key origin countries are involved as well as nationals in transit countries. Missbach and Sinanu (2011) argue that international organizations like the UNHCR and IOM are drawn into the structure as well. Indeed, they argue that people smugglers have integrated these agencies into their operations since agents encourage intending asylum seekers to register with either the UNHCR or IOM on arrival in Indonesia. This often provides the asylum seekers with access to support while they are in Indonesia.

Government officials and police in transit countries also play a role in the networks. Missbach and Sinanu (2011, 74) point out that there is a contrast between Malaysia and Indonesia in this respect: 'Unlike in Malaysia, where asylum seekers face massive repression by the local police and immigration authorities even if they hold UNHCR documents, the Indonesian authorities normally accept these documents.' Clearly, throughout the corridors of movement corrupt officials are a key element in the people smuggling process.

The personal stories of asylum seekers are replete with heartbreaking incidents of exploitation, denial of human rights, loss, abuse and even exposure to personal injury and death. The dangers that many asylum seekers face cannot be underestimated and the personal tragedies that so frequently occur not only must be an important part of the narrative of asylum seekers, they can and do influence policy. It needs to be said also, however, that these corridors of movement do contain networks of support and communication which facilitate and support the migration. The availability of mobile phones, email and so on allow asylum seekers to maintain contact with friends and family at home and in the destination at all stages of the mobility process. Moreover, there clearly has been the development of a range of support mechanisms in transit countries like Indonesia (Missbach and Sinanu 2011). These include communities (such as in the former hill station area of Puncak in West Java) who are developing a function as dormitories for asylum seekers awaiting resolution of their refugee status and resettlement or awaiting the departure of a boat for irregular entry to Australia. These dormitories have also become the places where people smugglers have the opportunity to recruit new clients.

Policy in Australia

The most substantial barrier or threshold faced by asylum seekers arriving on Australia's northern shores is Australian government policy. The arrival of 'boat people' on Australia's northern shores resonates with long-held fears of a northern invasion from the densely populated nations of Asia. These fears have a long history in the European settlement of Australia and lay behind the fact that the legislation of the infamous White Australia Policy was one of the first actions of the newly established Australian parliament formed from the separate colonies on the continent in 1901 (Price 1974). This was partly a response to the significant immigration of Chinese in the second half of the nineteenth century (Choi 1975). In the period between World War I and World War II these fears about the 'northern hordes' were fanned in a strong national debate, especially in the 1920s between boosters of an 'Australia Unlimited' push, which saw Australia as having enormous potential and a need for rapid population growth in order to realize that potential (Hugo 2011c; Borrie 1994; Powell 1984), and a group led by the geographer Taylor, who pointed to the significant environmental constraints on agricultural and other economic development. This debate was prosecuted not only in the political and academic arenas but in the community more widely. An important aspect of the 'Australia Unlimited' view is that if Australians do not massively expand its population and economic development there will be non-White groups from the north who will invade and do it for them (Powell 1984, 89). These fears were further fanned in World War II when Japanese southern expansion included bombing of Darwin in northern Australia and a submarine attack in Sydney Harbour.

Accordingly there is a deep-seated, long-established element in the psyche of some European-origin Australians about an invasion from the north and this is to some degree played upon by some Australian media and populist politicians. It was an element in the early post-war push to expand immigration from Europe but, as Viviani (1984) has demonstrated, it has been especially associated with the waves of boat people arrivals beginning with those from Vietnam in the later 1970s – although, as was demonstrated earlier, the flows since the later 1990s have been dominated by groups other than East and Southeast Asia (Middle East, Afghanistan, Sri Lanka). This historical paranoia of a feared 'northern invasion' has been a strong element in shaping public opinion and policy.

Accordingly, the reduction of the flow of asylum seekers arriving on Australia's northern shores has been a major political issue in Australia for more than a decade. Governments have experimented with a number of policies which have sought to deter intending asylum seekers to make the hazardous boat journey from Indonesia to Australia. These tactics have been of two major types: First, they involve a range of controversial policies of 'dealing with' asylum seekers once they arrive in Australia. These include a policy of mandatory detention while awaiting the outcome of applications for refugee status, reducing the possibilities to bring family members to Australia and reducing access to government support systems in Australia. Second, the tactics include efforts to divert intending asylum seekers before they arrive on Australian soil and are able to claim asylum. This largely involves the 'Pacific Solution' referred to earlier (see Footnote 5).

These efforts have been the subject of intense debate within Australia which involved a mix of the longstanding paranoia of northern invasion, concern about the human rights of the asylum seekers, the opportunism of politicians, the post-9/11 dominance of security concerns and, as is the case in many migration controversies bigotry, racism and self-interest. The debate has become highly politicized with misrepresentation, exaggeration and emotion playing important roles.

The controversial policy of mandatory detention began in the 1990s but the strictest measures were introduced by the conservative Howard government (1996–2007) in October 2001 which 'harnessed anti-immigrant sentiment to electoral advantage' (Foulkes 2012, 2) and included a number of initiatives in order to deter asylum seekers coming to Australia (Mares 2001; McMaster 2001; Jupp 2002):

- Mandatory detention while awaiting UNHCR assessment
- Excising Australia's external territories (including Christmas Island and Ashmore Reef) from the Australian migration zone
- Offshore processing of all asylum seekers in Pacific islands
- Introducing Temporary Refugee Protection Visas which were granted to 'onshore' refugees. These visas had more limited rights and ran for only three years.

These hard measures evoked strong criticism at home and abroad but it is apparent from Figure 15.4 that, indeed, the boat arrivals largely dried up during that period.

The incoming Labour government in 2007 abolished the last two of these measures. However, increasing numbers of asylum seeker arrivals and increasing community pressure saw a progressive hardening of government attitudes. Eventually this led to a return of offshore processing (Koleth 2012). Initially in 2010 the government announced that it would process asylum seekers in East Timor but this was rejected by the Prime Minister of Timor Leste. Their efforts to establish offshore processing in Malaysia (which is not a signatory to the UN Refugee Convention) was defeated by the High Court of Australia and the lack of support from other political parties.

One of the strategies employed by Australia to deter boat arrivals has been increasing investment in building regional capacity, especially in Indonesia and Malaysia (Chris Bowen. net 2012). 'Under pressure from Australia, Jakarta has, however, gradually increased its punitive surveillance of refugees and criminalised people smuggling,' as Missbach and Sinanu (2011, 82) have pointed out.

Eventually the government set up an expert panel to provide advice and recommendations on policy options (Australian Government 2012). It sought to be a 'circuit breaker' to the political stalemate regarding how Australia should deal with asylum seekers. The panel produced 22 recommendations which can be divided into two categories. The first set of changes sought to encourage the use of regular established migration pathways. A major part of this was an increase in the annual refugee-humanitarian quota from 13,500 to 20,000. The changes also involved enhanced efforts of collaboration and capacity building in both the major source countries and transit countries (Malaysia and Indonesia). The second set of measures sought to discourage the use of irregular migration routes to Australia. This involved enhanced offshore processing in Nauru and Papua New Guinea and seeking to establish such operations in Malaysia, as well as reducing access of asylum seekers with refugee status to the family reunion part of the migration programme. It also raised the contentious issue of 'turning back the boats' which occurred on a few occasions under the Howard government. It suggested that while at present such should not be practised, it did not rule it out in the future if conditions would exist to do such in a lawful and safe manner.

The recommendations were accepted by the Gillard Labour government but irregular maritime arrivals continued. The incoming conservative government in 2013 not only maintained the policies of offshore processing but also instituted a policy of turning boats back to Indonesia. Hence at the time of writing there had not been any asylum seeker boats able to reach Australia's shores since the instalment of this new government.

It is apparent then that asylum seeker policy remains an important and divisive issue within Australia. At present the two major political parties both have approaches that involve elements which have drawn criticism, both internal and external, indicating mandatory detention of asylum seekers and offshore processing. Hence, the Australian border presents a significant challenge to asylum seekers seeking to settle in Australia and changes in government policies have had significant effects on asylum seeker flows.

Conclusion

Australia is a migrant state with a generally positive attitude to migration. However, that attitude has been based historically, in part, on an understanding that the government has been able to control who comes to Australia and when. The arrival of asylum seekers challenges this and has begun a questioning in the community of the government's ability to manage migration in the national interest. In response to these concerns both main political parties have taken a hard line with boat people arriving on Australia's northern shores. This has provoked the criticism from within and outside Australia that the state is avoiding its responsibilities under the United Nations Refugee Convention. In fact, the bulk of asylum seekers who have been assessed have eventually been granted refugee status. However, the Australian community continues to distinguish between migrants (including refugees) who came to Australia as part of a planned immigration programme and have widespread community support and those who take the initiative themselves to move as asylum seekers. The current situation is a delicately poised one. There is a real risk not only that media and populist politician-led opposition will lead to increasingly strict efforts to deter asylum seekers but also that it will lead to a wider dissolution of support for other forms of migration. A poll taken in 2012 found that 51 per cent of Australians wanted to 'close' Australia's borders compared with 41 per cent in 2005 (*Herald Sun*, 22 May 2012).

This case study of asylum seeker migration to Australia has demonstrated the utility of a geographical approach examining the elements of people, borders and trajectories in understanding the dynamics of international migration. Intending asylum seekers to Australia perceive the border as open and the Australian government has tried a range of approaches in order to change these perceptions. This disjuncture in perceptions of the border has produced distinctive trajectories and patterns of migration. The intending asylum seekers follow a well-developed corridor of migration along which a number of institutions, individuals and businesses have developed that facilitate their movement. Networks with previous generations of migrants settled in Australia help fund, arrange and facilitate the migration. The last part of the journey presents most challenges for the migrants. The dangerous boat journey from Indonesia to Australia has claimed hundreds of lives and successive Australian governments have introduced a range of border control policies which seek to deter migrants. Most asylum seekers decide to undertake the journey with the intention of settling in Australia and in this sense the locational threshold is passed at an early stage and is stable through the journey. Ultimately, the most substantial barrier and threshold faced by asylum seekers arriving on Australia's northern shores is Australian government policy.

References

ABS (Australian Bureau of Statistics). 2011. *Australian Demographic Statistics, September Quarter 2010 (cat. No. 3101.0)*. Canberra: ABS.

_____. 2012. *Australian Demographic Statistics, June Quarter 2012 (cat. no. 3101.0)*. Canberra: ABS.

Australian Government. 2012. *Report of the Expert Panel on Asylum Seekers*. Canberra: Australian Government.

Borrie, W.D. 1994. *The European Peopling of Australasia: A Demographic History, 1788–1988*. Canberra: Australian National University.

Cartner, J.M. 2009. "Representing the Refugee: Rhetoric, Discourse and the Public Agenda." MSc. Thesis. Fremantle: University of Notre Dame Australia.

Castles, S., and M.J. Miller. 2009. *The Age of Migration: International Population Movements in the Modern World*. 4th ed. Basingstoke: Palgrave Macmillan.

Choi, C.Y. 1975. *Chinese Migration and Settlement in Australia*. Sydney: Sydney University Press.

Chris Bowen.net. 2012. "Refugee Program Increased to 20,000 Places." *Chris Bowen Media Centre*, 23 August. Accessed 19 September 2014. http://www.chrisbowen.net/media-centre/media-releases.do?newsId=6237.

DIAC (Department of Immigration and Citizenship). *Annual Report*, various issues. Canberra: DIAC.

DIAC (Department of Immigration and Citizenship). 2011a. *List of Visa Classes and Subclasses*. Canberra: DIAC.

_____. 2011b. *Asylum Trends – Australia: 2010–11 Annual Publication*. Canberra: DIAC.

DIAC (Department of Immigration and Citizenship). 2012a. *Trends in Migration: Australia 2010–11; Annual Submission to the OECD's Continuous Reporting System on Migration (SOPEMI)*. Canberra: DIAC.

_____. 2012b. *Population Flows: Immigration Aspects 2010–11 Edition*. Canberra: DIAC.

_____. 2012c. *Annual Report 2011–12*. Canberra: DIAC.

_____. 2012d. *Immigration Update: July to December 2011*. Canberra: DIAC.

_____. 2012e. *The Outlook for Net Overseas Migration: June 2012*. Canberra: DIAC.

_____. 2012f. *Asylum Trends – Australia: 2011–12 Annual Publication*. Canberra: DIAC.

_____. 2013. *Annual Report 2012–13*. Canberra: DIAC.

DIMIA (Department of Immigration and Multicultural and Indigenous Affairs). 1999. *Protecting the Border: Immigration Compliance*. Canberra: DIMIA.

_____. 2002. *Unauthorised Arrivals by Air and Sea, Fact Sheet 74*. Canberra: DIMIA.

_____. 2004. *Unauthorised Arrivals by Air and Sea, Fact Sheet 74*. Canberra: DIMIA.

_____. 2005. *Managing the Border: Immigration Compliance 2004–05 Edition*. Canberra: DIMIA.

Foulkes, C.D. 2012. "Australia's Boat People: Asylum Challenges and Two Decades of Policy Experimentation." *Migration Information Source*, 11 July. Accessed 18 September 2014. http://migrationpolicy.org/article/australias-boat-people-asylum-challenges-and-two-decades-policy-experimentation.

Hamood, S. 2006. *African Transit Migration through Libya to Europe: The Human Cost*. Cairo: American University in Cairo.

Hinsliff, J. 2006. "Integration or Exclusion? The Resettlement Experiences of Refugees in Australia." Submitted for degree of PhD, Discipline of Geography and Environmental Studies. Adelaide: University of Adelaide.

Hugo, G. 1996. "Environmental Concerns and International Migration." *International Migration Review* 30 (1): 105–31.

_____ . 2011a. *Economic, Social and Civic Contributions of First and Second Generation Humanitarian Entrants.* Canberra: DIAC.

_____ . 2011b. "Migration and Development in Malaysia: An Emigration Perspective." *Asian Population Studies* 7 (3): 219–41.

_____ . 2011c. "Geography and Population in Australia: A Historical Perspective." *Geographical Research* 49 (3): 242–60.

_____ . 2012. "National Migration Planning Processes in Australia." IOM MRTC Working Paper Series 2012–01. Seoul: IOM Migration Research and Training Centre.

Içduygu, A. 1996. "Transit Migrants and Turkey." *Review of Social, Economic and Administrative Studies* 10 (1–2): 127–42.

_____ . 2000. "The Politics of International Migratory Regimes: Transit Migration Flows in Turkey." *International Social Science Journal* 52 (165): 357–67.

Içduygu, A., and S. Toktas. 2002. "How Do Smuggling and Trafficking Operate via Irregular Border Crossings in the Middle East? Evidence from Fieldwork in Turkey." *International Migration* 40 (6): 25–54.

Ivakhniouk, I. 2004. "Analysis of Economic, Social, Demographic and Political Basis of Transit Migration in Russia: Moscow Case." Paper presented at the Regional Conference on Migration, 30 September–1 October. Istanbul: Council of Europe.

Jones, S. 2000. *Making Money Off Migrants: The Indonesian Exodus to Malaysia.* Hong Kong: Asia 2000.

Jupp, J. 2002. *From White Australia to Woomera: The Story of Australian Immigration.* Cambridge, MA: Cambridge University Press.

Khoo, S.-E., P. McDonald, C. Voigt-Graf, and G. Hugo. 2007. "A Global Labor Market: Factors Motivating the Sponsorship and Temporary Migration of Skilled Workers to Australia." *International Migration Review* 41 (2): 479–509.

Koleth, E. 2012. "Asylum Seekers: An Update." Briefing Paper 1/2012. New South Wales: NSW Parliamentary Library Research Service.

Kunz, E.F. 1973. "The Refugee in Flight: Kinetic Models and Forms of Displacement." *International Migration Review* 7 (2): 125–46.

_____ . 1981. "Exile and Resettlement: Refugee Theory." *International Migration Review* 15 (1–2): 42–51.

Maley, W., A. Dupont, J.-P. Fonteyne, G. Fry, J. Jupp, and T. Do. 2002. "Refugees and the Myth of the Borderless World." Working/Technical Paper Keynotes 02. Canberra: Australia National University.

Mares, P. 2001. *Borderline: Australia's Treatment of Refugees and Asylum Seekers in the Wake of the Tampa.* Sydney: University of New South Wales Press.

Markus, A. 2011. "Immigration and Public Opinion." Paper prepared for the Productivity Commission Roundtable "A 'Sustainable' Population? – Key Policy Issues," 21–22 March. Canberra: Productivity Commission.

Mavris, L. 2002. "Human Smugglers and Social Networks: Transit Migration through the States of Former Yugoslavia." New Issues in Refugee Research Working Paper 72. Geneva: United Nations High Commissioner for Refugees.

McMaster, D. 2001. *Asylum Seekers: Australia's Response to Refugees*. Victoria: Melbourne University Press.

Missbach, A. 2012. "Asylum Seekers in Indonesia: Don't Come, Don't Stay, Don't Go." *The Indonesian Quarterly* 40 (3): 290–307.

Missbach, A., and F. Sinanu. 2011. "'The Scum of the Earth?' Foreign People Smugglers and Their Local Counterparts in Indonesia." *Journal of Current Southeast Asian Affairs* 30 (4): 57–87.

Papadopoulou, A. 2004. "Smuggling into Europe: Transit Migrants into Greece." *Journal of Refugee Studies* 17 (2): 167–84.

Phillips, J., and H. Spinks. 2012. *Research Paper: Boat Arrivals in Australia since 1976*. Canberra: Parliament of Australia.

Powell, J.M. 1984. "Home Truths and Larrikin Prophets – 'Australia Unlimited': The Interwar Years." In *Populate and Perish: The Stresses of Population Growth in Australia*, edited by R. Birrell, D. Hill, and J. Nevill, 80–99. Sydney: Fontana.

Price, C.A. 1974. *The Great White Walls Are Built: Restrictive Immigration to North America and Australasia, 1836–1888*. Canberra: Australian National University Press.

Skrobanek, S., N. Boonpakdi, and C. Janthakeero. 1997. *The Traffic in Women: Human Realities of the International Sex Trade*. London: Zed Books.

Suro, R. 2009. "America's Views of Immigration: The Evidence from Public Opinion Surveys." In *Migration, Public Opinion and Politics: The Transatlantic Council on Migration*, edited by Bertelsmann Stiftung and Migration Policy Institute, 52–76. Gütersloh: Verlag Bertelsmann Stiftung.

Sushko, O. 2003. *Human Trafficking and Transit Migration as Soft Security Threats in EU-Ukraine Relations*. Ukraine: Centre for Peace, Conversion and Foreign Policy of Ukraine.

UNHCR (United Nations High Commissioner for Refugees). *UNHCR Statistical Yearbook*, various issues.

_____ . 2011. *Asylum Levels and Trends in Industrialized Countries 2010*. Geneva: UNHCR.

_____ . 2013. *Asylum Trends 2012: Levels and Trends in Industrialized Countries*. Geneva: UNHCR.

Van der Velde, M., and T. van Naerssen. 2011. "People, Borders, Trajectories: An Approach to Cross-Border Mobility and Immobility in and to the European Union." *Area* 43 (2): 218–24.

Viviani, N. 1984. *The Long Journey: Vietnamese Migration and Settlement in Australia*. Melbourne: Melbourne University Press.

Chapter 16

African Passages through Istanbul

Joris Schapendonk

Although part of the Turkish city Istanbul belongs, geologically speaking, to the European continent, it is located outside the boundaries of the European Union (EU). Because of its proximity to the EU, Istanbul is an important crossroad for (irregular) migrants who hope to reach the EU. Since the early 1990s, the city has received many migrants from the former Soviet Union, the Middle East and East Asia (Içduygu and Toktas 2002; Içduygu and Yükseker 2008; Akinbingöl 2003; Daniş 2006; Van Liempt 2007). Nowadays, more and more sub-Saharan Africans are moving into this region. While the exact number remains highly uncertain,[1] Istanbul is host to a considerable number of groups of migrants from the Horn of Africa, as well as an increasing number of West and Central Africans, including Nigerians, Sierra Leones, Ghanaians, Congolese and Senegalese. Certainly, not all of these sub-Saharan African migrants are 'in transit' in Istanbul trying to make their way into the EU. On the contrary, many Africans hope for a 'settled life' in Istanbul. Others are studying at the universities whilst still others are commuting as transnational traders between Turkey and their countries of origin. Nevertheless, it is safe to assume that sub-Saharan Africans are increasingly using Istanbul as a gateway to the EU (Brewer and Yükseker 2006; Ozdil 2008; Schapendonk 2011).

This contribution aims to understand the role and function of Istanbul in the (irregular) migration trajectories of sub-Saharan Africans who are heading for Europe. Having passed the phase of indifference and locational thresholds (Van der Velde and Van Naerssen 2011), these migrants face diverse challenges that hinder their journeys en route. The research findings are based on in-depth interviews with sub-Saharan African migrants in and around Istanbul. The insights are used to discuss the thin line between migrants' transit statuses and settlement in the city. Moreover, using a longitudinal research approach (Schapendonk 2011), this chapter elaborates on the different ways migration trajectories evolve through the passage of time. It focuses on migrants' living situations in Istanbul and their changing aspirations in the course of their movements. The empirical insights finally result in a reflection on 'the trajectory', mentioned by Van der Velde and Van Naerssen (2011) as a relevant concept in analysing migration and its decision-making processes. First, however, the contemporary dynamics of sub-Saharan African migration are discussed. As African migration towards the North is one of the most stigmatized forms of human mobility of our times (De Haas 2008), some contextualization is much needed to put the volume and direction of such mobility in a proper perspective.

---

1 However, CNN claims that there are 'tens of thousands' of Africans living in Istanbul (Watson and Bailey-Hoover 2011). There are also estimates of 'thousands of Africans' living in the Kumkapi district alone (*National World*, 5 September 2011). Another indication is the appearance of sub-Saharan Africans in figures on undocumented migration. Between 1995 and 2005, there were 35,000 sub-Saharan Africans apprehended for undocumented migration in Turkey (Brewer and Yükseker 2006).

African Passages to/through Turkey: Fragmented Journeys and Transit Solutions

In studies about African migration, it is important to realize that human mobility is very central in the understanding of social life in this part of the world. The diverse mobilities of, among others, migrants, pastoralists, traders, musicians and Koran students challenge the notion that sedentary lifestyles are the norm (see for instance De Bruijn et al. 2001; Hahn and Klute 2007). When it concerns international mobility, African migration is often associated with an 'exodus' to 'the North' suggesting that almost all African migrants aim to leave the African continent (De Haas 2008). This exodus-like image is misleading as the bulk of African international migrants remains in the African continent and Africa's international migration has in fact a strong regional character. In the West African context for example, it is estimated that almost 85 per cent of the international migrants move across borders within the West African region, mainly in the direction of booming economies such as Ivory Coast, Gabon and Nigeria (Gnisci 2008; see also Spaan and Van Moppes 2006; Adepoju 2008). Only a minority leaves for Europe or the United States. In the southern African region, undoubtedly, it is the modern economy of South Africa that is attracting many migrants from neighbouring countries (Crush and McDonald 2000; Nyamnjoh 2006). A similar regional pattern is found in Central Africa where migrations are catalysed by several violent conflicts in the region. It is, however, commonly known that the majority of these refugees remain within the region (Kraler 2005).

Although regional migration in Africa is the dominant pattern, it is important to note that international migratory patterns within Africa are at the same time diversifying (see for instance Adepoju 2000, 2008; Zoomers et al. 2009; Schapendonk 2012). Regional migrations have expanded to more distant destinations. Post-apartheid South Africa, for example, is an important destination for people from all over Africa (Nyamnjoh 2006). Gabon has received an increasing number of migrants from different countries, including Nigeria and Mali (Adepoju 2000). Moreover, Gadaffi's Libya has been an important destination for migrants originating from West and Central Africa (Brachet 2005; Hamood 2006; Bredeloup 2012a). It remains to be seen, however, if this is still the case after the recent regime change.

This diversification of destinations also holds true in the context of African extracontinental movements. Although colonial and linguistic linkages still provide some axial routes to Europe, the popularity of Europe as a destination seems to decrease, partly because of its harsh border policies (Bakewell and Jonsson 2011). African migrants are increasingly heading for non-Western destinations around the world. Some states and cities in the Middle East, like Lebanon and Dubai, are important destinations for migrants and traders from eastern Africa, including Sudan, Eritrea, Ethiopia and Somalia (Spaan and Van Moppes 2006; King et al. 2010) but, more surprisingly, also for migrants from West Africa. The latter includes among others Ghanaians (Peil 1995), Nigerians (Falola and Okpeh 2008) and Cameroonians (Pelican and Tatah 2009). Furthermore, with the emerging economic and institutional linkages between Africa and China, migration in the direction of Asia has also increased. Although official numbers are lacking, there are some good indicators supporting this. It is noted for instance, that the southern Chinese metropolis of Guangzhou has seen the number of African migrants increase by a third every year since 2003 (Ellis 2011).[2] Today Guangzhou hosts about 100,000 sub-Saharan Africans,[3] many of whom are, however,

---

2   For more information on African mobility to China, see the articles in the Dutch newspaper *Volkskrant* of 25 January 2010 and 31 July 2010.

3   It is estimated that the Guangzhou region houses 30,000 Nigerians of whom more than 50 per cent lack proper documentation (Adepoju and Van der Wiel 2010).

undocumented. The fact that the number of student visa for Africans to China has increased by 40 per cent (in the period 2005–06) also supports the argument that China is an upcoming destination for Africans (Politzer 2008). Next to migration to China, there is also some evidence that Africans are moving to other less traditional destinations such as Israel (Furst-Nichols and Jacobsen 2011), Argentina (*Global Post*, 8 June 2009) and Mexico (Wabgou 2012).

In this context of diversification of destinations, Turkey is not only a 'transit country' for sub-Saharan Africans on their way to Europe, it is also the 'end' destination in its own right. It is an emerging economy attracting all kinds of migrants searching for job opportunities in the formal and informal sectors (Ozdil 2008; De Clerck 2013), as well as for trade possibilities (Steen 2012; Schapendonk 2013). Particularly in the context of African migration, it is important to note that the Turkish government has strengthened political and economic relationships with several African countries in the last ten years (Baird 2011; Özkan and Akgün 2010). This has obviously eased population movements between Africa and Turkey. At the same time, statistics on irregular migration in the Turkish-Greek borderlands suggest that irregular border crossings through Turkey are substantial in terms of numbers. Greece has been confronted with several tens of thousands of migrants who have arrived there by boat (for instance 45,000 in 2006 [Triandafyllidou and Maroukis 2008]; 37,000 in 2012[4]). These migrants have most likely transited through Turkey (Triandafyllidou and Maroukis 2008). Moreover, while there have been 6,600 irregular border crossings recorded in the Evros region (the Greek-Turkish land border) for the year 2009, this number went up to 31,000 in 2010 (*Volkskrant*, 19 October 2010). In this sense, the Evros region is dubbed the 'last door' to the EU (Martino 2010). In response, the Greek authorities built a 12.5-kilometre long barbed wire fence along this section of the borderland. Although the Turkey-Greece route still appears to be important in terms of the number of entrants, it is vital to realize that sub-Saharan Africans account for only roughly 10 per cent of the total number of irregular migrants (Papadopoulos et al. 2014).

All in all, two seemingly contrasting realities are going hand in hand with each other: Turkey as an upcoming destination and as a transit country for Africans. This makes the investigation of migrants' trajectories in Istanbul an interesting but also a challenging task.

Istanbul as a Migration Hub: Migration Industries and 'Transit' Policies

Istanbul, which has officially more than 13 million inhabitants, offers an environment of anonymity where migrants intending to move to Europe can easily hide. The city also hosts a large migration industry that facilitates migrants' onward passages to the EU (see for instance Van Liempt 2007; Içduygu and Toktas 2002). This includes illicit and rather costly services such as the creation and distribution of false travel papers and various smuggling practices, but also conventional services such as the numerous money transfer agencies, cybercafes and travel agencies. The combination of services is vitally important for migrants arranging their onward journeys from Istanbul.

Whilst policymakers and the media tend to stress the involvement of large-scale criminal networks of smugglers and traffickers in irregular migration from Africa to Europe, academic research has indicated that many of these so-called transnational networks are in fact rather fragmented and uncoordinated chains of different actors (Brachet 2005; Hamood 2006). Içduygu and Toktas (2002), who have investigated the migration industry in Turkey, also underline the

---

4   See the map with main migratory routes into the EU, provided by Frontex: http://frontex.europa.eu/ trends-and-routes/migratory-routes-map (accessed 13 February 2014).

importance of locally operating actors in the smuggling practices, which counter the notion of coordinated smuggling networks with mafia-like hierarchical structures.

Thus, Istanbul is *the* place – a migration hub – where migrants meet actors specialized in irregular border crossings to the EU (Suter 2012). The most important routes include the northern route through the Edirne region to Bulgaria and Greece respectively and the sea route via the Izmir region from where various Greek islands can be reached.

Migrants' trajectories to and through Istanbul are not only influenced and shaped by the migration industry, it is also determined by national and local policies. As Turkey has been lobbying hard to join the EU, it aims to harmonize its immigration regulations with that of the EU. The Turkish government has developed the National Action Plan that provides guidelines for this harmonization process. In this light, the so-called transit migration has turned into a highly important policy problem that needs to be 'combated'. The European Commission has strongly encouraged Turkey to strengthen its border controls and to align visa policies with that of the EU (Biehl 2009; Suter 2012). This suggests that migrants' passages from Turkey into the EU may have become more difficult in recent years and that their intermediate stopovers in Turkey are prolonged as a result.

The European Commission also demands Turkey to develop sufficient accommodation facilities for asylum seekers. Until now, Turkey is one of the few countries that apply geographical limitation to the Geneva Convention. This implies that the country only grants asylum to those from Europe. Non-European asylum seekers in Turkey are only granted with 'temporary asylum' until a 'durable solution' is found (see Biehl [2009] for a detailed discussion on Turkey's migration and asylum policy).

An important aspect of the asylum policy in Turkey is its 'satellite city' system. Migrants who apply for asylum in Turkey are sent to one of the thirty satellite cities in the country and they can only leave this place occasionally (Frambach 2011). In general, these cities are relatively remote places. In fact, it is telling that Istanbul, where migrant communities are concentrated and where many informal jobs can be found, is not one of the satellite cities. In the satellite cities, most asylum seekers do not have the permission to work and they live socially isolated lives (Biehl 2009). Hence, for economic and social reasons, many migrants prefer to stay in Istanbul as undocumented or irregular migrants than to go through the strict asylum procedures (see also Suter 2012). The combination of granting temporary asylum and the satellite cities regulations has rather encouraged and enhanced migrants' tendency to use Istanbul as a transit point. It is not an appropriate solution to tackling migrants' precarious situations.

The strengthened policies in Turkey regarding migration can be seen as border-control practices that keep migrants from moving onwards to Europe. It does not, however, deter people from moving towards Europe in the first place. It has only increased the psychological and financial costs of moving, as well as the time spent 'in transit'. As we will see below, this has resulted in some migrants experiencing a daily sense of involuntary immobility (Carling 2002), while others find new opportunities in places they had only intended to pass through in the first place. These policy thresholds, in combination with precarious socio-economic conditions and high living costs, can impact on migrants' aspiration and mobility processes profoundly (Schapendonk 2011; Suter 2012; De Clerck 2013).

Methodology

The empirical insights of this study are mainly based on interviews with sub-Saharan Africans and longitudinal engagement with some of the respondents. They were part of a broader research project focusing on African trajectories to Europe (Schapendonk 2011). Between April and May

2008, 27 sub-Saharan African migrants from various countries were interviewed. The majority of them came from West Africa: Nigeria (12 respondents), Senegal (three), Ivory Coast (two), and Cameroon (one). Among those from Central African countries were Congolese (two), Comorian (two), Sudanese (one), Burundian (one) and Rwandan (one). Additionally, two Somalian migrants were interviewed. The age of the respondents ranged from 22 to 45 years old, and only five of them were female.[5] Some had just arrived in the city, others had been living there for some years – the longest being nine years.

Besides these interviews, six respondents were studied over a period of two years. This longitudinal aspect of the research implies that the courses of migrants' trajectories can be traced by means of telephone calls and internet conversations. Through these translocal engagements, it was possible to grasp (un)expected twists and turns of individual migration processes (Schapendonk 2011). Two of the six longitudinal respondents have also been revisited by the researcher in Heraklion (Greece) and Marseille (France), respectively. In these two places, a week-long ethnographic study was carried out. As a follow-up of these research activities, three master students interviewed a total of 60 African migrants in and around Istanbul (Frambach 2011; Bazuin 2011; Steen 2012). Their findings are used as additional information in this chapter.

The next sections of this chapter discuss the empirical findings and are organized as follows: First, the diversity of routes to Istanbul is discussed. Second, the different situations of migrants living in Istanbul are outlined. It is claimed that the so-called transit situations of migrants are not clear-cut (see also Schapendonk 2012). This section also discusses in detail how migrants' aspirations and migration projects change over time. The third section elaborates on the passages from Istanbul to the EU.

The Diverse Routes to Istanbul

African migrants reach Istanbul in many different ways. At least four different groups can be distinguished. First, there are migrants who move directly to Istanbul without any stopover in any country (see also De Clerck 2013; Suter 2012). While many of these migrants move to the Turkish city with the aim of reaching Europe ultimately, there are also those who intend to stay for a specific period of time in Istanbul without any plan of moving on to the EU (De Clerck 2013). The latter concern, for example, African students who are on scholarship and studying at the universities in the city. The researchers have also met several young African football players who were on probation as players in Turkish football clubs. During their probation, they do their best to obtain a contract with a club. When they fail to do so, however, their situation can become very difficult. As football agents do not usually provide a return flight ticket, many migrants become stuck in Turkey and end up in a vulnerable socio-economic position. Consequently, moving to Europe as an undocumented migrant becomes an alternative. Finally, as noted previously, another group of Africans going directly to Istanbul are the African traders despite the fact that some of them do have highly complex mobility patterns. These 'floating' traders usually stay for short periods of time, such as a certain number of days or weeks. The movements of people involved in this business can be of a long- or short-term character, thus rendering the distinction between 'traders' and 'migrants' rather thin (Steen 2012).

---

5   There is a gender bias in this research. For the male researcher, female migrants appeared to be more difficult to approach. Let alone, to follow-up with them through time and space (for more information, see Schapendonk 2011).

Second, there are migrants who are deliberately undertaking long and fragmented journeys to reach the city of Istanbul (eventually with the aim of moving onwards to the EU). These migrants travel a long way through, for instance, Lebanon and Syria, before reaching the Turkish territory. One example is John, a 25-year-old Nigerian who explained how he moved to Damascus, the capital of Syria, before the war started. He held a three-month valid visa to Syria but he only stayed there for three days as he already knew someone who could transport him to the border with Turkey. However, he was caught by the border guards and was put in a detention centre for several weeks. Altogether, it took him more than a month to travel from Nigeria to Turkey via Syria.

Third, there is a group of migrants who have left their countries of origin without the intention of making Istanbul their destination or even their transit point. More often than not, these migrants already have long migration histories and have travelled to various places before Istanbul came into their frame of reference. One Nigerian woman by the name of Beauty left Nigeria with her husband in the year 2000 to work in Lebanon. They had intended to earn money so that they could in turn 'invest in life' in Nigeria. They lived a rather satisfactory life in Lebanon until the woman lost her job as a domestic worker in 2006. In the same year, the Lebanese war broke out. This combination of factors made the Nigerian family to decide to leave Lebanon abruptly and to move to Istanbul which is considered as a place where new opportunities can be found, including the possibility of entering Europe. Other trajectories in which Istanbul emerged as a(n) (in-between) destination involved going through various other countries such as Egypt, Sudan and Morocco.

Thus, while some migrants aim to reach Istanbul right from the beginning in order to (1) build up a life there; (2) stay only temporarily in the city and return to the countries of origin after a certain period of time; or (3) move onwards to the EU, others end up in the city in more or less unexpected ways. It is furthermore important to note that some migrants also move to Istanbul because all conventional channels to the EU are blocked. Hence, their only option is to make use of extralegal means. This indicates that the border-control policies of Europe have a direct and profound impact on the migratory routes of African individuals, turning these into lengthy and risky trajectories. Moreover, it suggests that locational and trajectory thresholds, including high costs of migration and Turkish transit policies, are not only factors that inform the decision-making process *before* a migrant's departure takes place, as Van der Velde and Van Naerssen (2011, 222) indicate. In fact, these factors may also affect migrants' *actual* mobility processes in direct and indirect ways (Schapendonk 2011). In the next section, the focus is on the en route factors of migration processes by presenting migrants' experiences in Istanbul.

Meeting the Border: Settlement, Transit, and Circularity in Istanbul

Migrants' diverse routes to Istanbul lead also to diverse living situations in the city. Some are embedded in clear institutional frameworks, of which African university students is a good example, whereas the life of others is characterized by irregularity, insecurity, frequent police harassment and lack of a stable income. In this respect, it is relevant to note that there are a few NGOs trying to help improve the life situations of irregular migrants and asylum seekers. The diverse churches, some set up and led by migrants, have been important in giving daily support to the migrants (for an elaboration on this, see Suter 2012). Nevertheless, many migrants complained about their precarious living conditions and about police brutality in Istanbul. They talked about 'raids' that have taken place to scare migrants away from certain areas, racial discrimination and verbal insults, as well as about dubious controls whereby money, clothes and mobile telephones were confiscated from migrants. As illustrated by the following quote of a Comorian student, even migrants who are legal and institutionally embedded face similar problem of oppression and discrimination by the authorities:

> The biggest problem for Africans here is the police situation. One day I was walking through Tarlabasi and I looked in the face of a policeman … he looked back … and the trouble started. He asked me to stop and he checked me. He was saying that I was a suspect in drugs trade, I said I was a student and showed him my card … But he didn't look at it. He asked for my passport, and you know what happened? They took my passport, and gave me back a copy, just a paper. This was two months ago and still I don't have my passport. I have written many letters … to the Ministry of Foreign Affairs here … to my home country. But nothing helped. (Interview Amir, Istanbul, May 2008)

The precarious living conditions in Istanbul tend to compound and complicate migrants' transit situations in the city, meaning that there are many obstacles preventing them from living a peaceful and integrated life. These conditions could even result in some migrants deciding to move to Europe despite their initial plan of remaining in Istanbul.

This context has made migrants very creative in finding ways and means to sustain their daily income. Although there are migrants with relatively settled and well-paid jobs, such as being English language teachers, workers in telephone houses and cybercafes, and semi-professional football players (Steen 2012), most of the migrants interviewed are employed in the large informal sector of the city. The researchers came across, among others, migrants selling watches to tourists and working as domestic workers in the homes of rich families. The quick-to-get jobs, also popularly known as *chabuk chabuk*, meaning 'do it quickly' or 'hurry up' in the Turkish language, have become important in this respect. This informal labour system works as follows: Certain recruiters, also known as 'connection men' by the migrants, go to migrants-concentrated neighbourhoods to hire labourers for short-term work which are often heavy, manual jobs such as carriers at construction sites. These middlemen would shout 'chabuk chabuk' through the streets to attract migrants to join them (see also Steen 2012; Brewer and Yükseker 2006; Suter 2012).

While many migrants do not achieve upward socio-economic mobility through these informal jobs, nevertheless, quite some successful careers in the booming transnational trade between Turkey and African countries were encountered (Steen 2012; see also Özkan and Akgün [2010] for more information on African-Turkish trade relations). While many actors in this particular form of trade are transient and mobile, moving between China, Europe, the Middle East and Africa (Bredeloup 2012b), there are specific roles reserved for migrants who have stayed for a longer period of time in Istanbul. Some of them act as guides and translators and assist the African traders in the markets, others play the role of managers to import/export goods in the name of businesses located in Africa. From starting out as managers, some migrants manage to set up their own businesses later on and become traders in their own right. From then on, they are often involved in circular mobilities between Turkey and their countries of origin. As a result, other forms of trade-related services, such as accommodations and restaurants, were also set up by migrants to support these traders and their trade (Steen 2012; Schapendonk 2013).

What is interesting about this contrast between precarious living conditions and emerging opportunities is the fact that migration aspirations become highly changeable and migration decision-making processes shift in terms of geographical orientations. There are different dynamics at play here. As stated earlier, there are migrants who live a relatively stable life in Istanbul only to 'fall into transit' as a consequence of the 'inconvenient' or less conducive sociopolitical climate or because they think that they can improve their lives by moving to Europe. As one Burundian respondent explained:

> Students who fail [the exams], fall into transit, they are waiting to go to Europe! Like a Congolese friend, he is in France now. Actually, he didn't even fail [his exams]. Many people come here as students but feel that they can have a better life in Europe, so they go! So this also makes you think, what if I fail? I don't know, but I probably would not go back to my country; I probably would go to Europe. (Interview Isac, Istanbul, April 2008)

For others, the opposite is true as their transitory status becomes one of settlement over time. This is illustrated by the case of Caroline, a Nigerian woman who had already lived in Istanbul for seven years when she was met in 2008. In the first years of her stay, she had tried to enter Greece several times but all her attempts had failed. She reflected on her situation:

> Life continues, when you tried something without any success you have to try your luck somewhere else ... I decided to stay here in Istanbul. I only want to rest now, build up my life, raise my children and live the best life I can. I tried to go to Greece several times. But I failed. But I think God helped me with that, God has let me fail to let me see that life is not only about changing places. Life is about satisfaction, about rest. Now I know that life in Greece is not much better than here in Turkey; there is no paradise on earth, not in Europe, not in Canada; you have to create your own paradise. (Interview Caroline, Istanbul, April 2008)

Then there are migrants who are involved in circular mobilities, such as the aforementioned traders. However, some migrants also move between their countries of origin and Turkey in the hope of finally reaching Europe. The author has met a Senegalese migrant who entered Istanbul for the third time on a temporary visa. With his trips, he had tried to move to Europe. Similarly, a Nigerian man who had been in Turkey twice explained he would try his luck to reach Europe when his visa was about to expire. In the event of failure, he would repeat this strategy and renew his visa in his home country, return to Istanbul and try to go to Europe again. Thus, while transit statuses are often associated with a sense of waiting and immobility, some 'transit migrants' are actually mobile, moving across borders (Schapendonk 2012).

Finally, some migrants have predominantly negative experiences of Istanbul as the place they do not want to live, yet they cannot leave either. This was for example the case for four friends from Ivory Coast who entered Istanbul with the help of a football agent. This agent had arranged for these young men to play in a football club on probation. However, he abandoned them and kept their passports and, worse, the club did not hire them after the trial period. As a result, they became undocumented migrants after their visas had expired and they found themselves stuck in Istanbul, unable to move back to their countries of origin. Neither was it easy for them to move onwards to Europe as they could not afford to pay for the cost needed to smuggle them to the continent (see also Bazuin 2011).

Over the Threshold: Passages to Europe

By means of telephone conversations and internet, contact with six respondents was kept up for a period of approximately two years. Of these six migrants, three have managed to reach the

EU. In addition, four respondents who were encountered inside the EU (three in Greece, one in the Netherlands) transited from Istanbul. Thus in total the research includes seven migrants for whom Istanbul was the actual springboard to Europe. Six out of these seven migrants passed the Turkish-European border by means of an irregular boat trip to Greece. They moved from Istanbul to the harbour city of Izmir from where the sea passage was arranged by middlemen. Most of them reached the Greek island of Samos. One respondent, however, had a less turbulent trajectory as he married a Dutch woman during his stay in Istanbul. This marriage was his gateway to Europe; hence he did not have to enter as an irregular migrant. This at least suggests that irregular border crossing is not the only way for Africans to reach Europe from Istanbul. In this respect, it is also important to note that many migrants who claimed asylum in Istanbul were hoping for a resettlement process in other countries in 'the West', such as Canada and the United States.

Although six migrants among the respondents for this research had entered the EU in a similar way (by undertaking a risky journey over sea), their trajectories have been very different. Three migration stories are highlighted below.

Destiny

Destiny is a Nigerian man of 31 years old (in 2008). He arrived at Samos in August 2008 in a rubber boat with approximately thirty other migrants. Once he arrived there, he pretended not to speak English to hide his identity and subsequently the Greek authorities took him to the reception centre on the island. He stayed eight days in 'the camp' and was brought to Athens afterwards. There he applied for asylum, and he was given a 'pink card' by the authorities; this document proves that his application for asylum was underway. From then on, he lived in Athens for several months but he was unhappy there. He found Athens to be a very violent and insecure place with a lot of competition among migrants and thus considered to leave for another place. In the meantime, he started a relationship with a Greek student living in Heraklion (the capital of Crete) so Destiny decided to go to Crete to live with her. During that period he had two contrasting strategies. One was to work on his relationship with his girlfriend in order to marry her so he could regularize his stay in Greece. The second strategy was to save money and leave Greece in order to move to Austria where one of his Nigerian friends lived. To achieve this, the money was saved without the knowledge of his girlfriend. When the relationship did not develop in the way he had expected, Destiny decided to leave Greece as soon as he had saved enough money. He went to Patras, the harbour city of Greece that is heavily controlled against irregular migration from Greece to Italy, and from there he took a ship to Italy. He finally reached Italy in the summer of 2011. He stayed there for two months and then moved on to Switzerland – instead of to Austria, as he had heard that the asylum procedures in Switzerland are less strict. In August 2011, he applied for asylum there and was sent to an asylum centre in a small village not far from Zurich. The Swiss authorities did not grant him asylum and instead started the repatriation procedure. So Destiny had to move again. This time, he returned to Italy. From there, he tried to navigate his way through the EU – hoping to find a good place to live.

Joseph

Joseph is a young Nigerian man who lived together with Destiny in Istanbul. At that time, they were very good friends and saw each other almost every day. Joseph also reached Samos in August 2008 and was sent to Athens after spending twenty-two days in the reception camp. Their trajectories began to separate as Joseph went to the island of Corfu soon after he reached Athens, in the hope of getting a job in the informal tourism industry there. From then on, he started to commute between Corfu and Athens as Corfu was, according to him, 'good for business, but not for living'. In contrast to Destiny, Joseph was able to improve his social and economic situation quite quickly and was beginning to enjoy life in Greece. Hence, his aspiration of reaching Western Europe, in particular the UK or the Netherlands, began to fade away. In the summer of 2012, however, this all changed again. The desperate political and economic crisis in Greece affected Joseph badly. He was especially worried about the recent increase of xenophobic attacks against migrants. During a telephone conversation he assured that he planned to leave Greece within a year because he did not want to stay 'in this troubled country'.

Said

Said, a 42-year-old man from the Comoros Islands, has had a long and complex migration history before he entered the city of Istanbul. He dreamt of reaching France, the country where his brother was living. After staying for seven months in Istanbul, he reached Greece in 2008 in the same way as Destiny and Joseph: by rubber boat. However, his situation in Greece differed from the situations of the Nigerian migrants as he pretended to come from Somalia. Based on this, he managed to obtain refugee status from the Greek authorities. This refugee status, as Said explained, has made it possible for him to travel legally to another European country. He went to France in 2009 and joined his brother in Marseille. However, after the first weeks, Said became frustrated because of the lack of (financial) assistance from his brother. Subsequently, he moved in with another Comorian family in Marseille. According to Said, there is widespread solidarity among Comorians in Marseille; his stay with this family was a good example. In the meantime, he threw away the Greek asylum papers, as this would, according to Said, complicate the process of becoming a French citizen. He was hoping to obtain French residency by marrying Cherolle, a woman with a Comorian background and a French passport. Nevertheless, during a visit in Marseille, he emphasized that his future plans were still quite open:

> I live here for several months but without papers. That makes the situation bad. That is why I say I go to the Netherlands one day, because my situation here is not certain. Like this house, I will stay in this house of course; the people help me and understand me. But I would move out of the house if I hear that there is a small job somewhere! (Interview Said, Marseille, December 2009).

A Synthesis: Turbulence and Thresholds

Even though the outcomes of the above-mentioned stories are different, they have several commonalities too. First, the step-by-step and fragmented character of these processes is self-evident. These migrants have not only changed places, their living situations had also transformed because of changing social connections (in the case of Destiny and Said), changing sociopolitical climates (in the case of Joseph) and changing legal statuses (in the case of Said). In these fragmented and turbulent processes, flexibility and insecurity appear to be the two sides of the same coin. Migrants can easily adapt their plans to changing circumstances because they hardly have any formal obligations in terms of housing, social connections and employment that tie them to the place. This makes them flexible and mobile actors, as argued by Jordan and Düvell (2002). At the same time, the lack of legal status, the lengthy asylum procedures and the limited opportunities to upgrade their socio-economic statuses can make it all the more difficult for these migrants to reach their desired standard of living and/or their aspired destinations. Having passed one threshold, or one physical border, the migrants in question are more often than not confronted with new thresholds and different borders in their trajectories.

In terms of the threshold approach as presented in this book (see also Van der Velde and Van Naerssen 2011), in none of the above-mentioned cases it is very clear where locational factors end and where trajectories factors begin. A refugee status is easily considered a 'keep factor' as it makes staying in a particular country more secure and comfortable. This, however, does not count for Said as he had used this status only as an enabling factor to make his onward journey to France. Destiny's relationship with his Greek girlfriend shows a similar mechanism. Initially, this relationship seemed a factor that would facilitate his stay in Greece but it rather became an enabling factor for his onward movement because it made it easier for him to save money since most of his daily expenses were covered by his girlfriend. Furthermore, when we compare the two Nigerian cases of Destiny and Joseph, we notice that similar factors can have very different outcomes. After a certain period of time, both Joseph and Destiny's economic situation had improved. For Joseph, his improved situation was a reason to stay in Greece but Destiny preferred to use his savings to try his luck elsewhere in Europe and thus moved on to Italy. This suggests that we should investigate migrants' individual mobility processes carefully as mobility and immobility can mean different things to different people in different situations (Adey 2006; see also Schapendonk and Steel 2014). The way individuals value their mobility/immobility is an important factor that influences their migration decision-making processes.

Conclusion: Migrants' (Im)Mobility and Migration Trajectories

Istanbul is an important migration hub for African migrants who are heading for Europe. The growth of the economy and employment in the informal sector of the city has provided opportunities for documented, as well as undocumented migrants in terms of their upward socio-economic mobility. One of the most important opportunities for migrants is to get into the transnational trade sector (Steen 2012; Schapendonk 2013). As a consequence, Istanbul is not only a point of passage for Africans, it is also a destination and a place of circular mobilities in its own right. Istanbul is also regarded as a place where migrants may constantly adjust their future plans, where new dreams emerge and where migratory directions change. It is a threshold place where precarious living conditions and new opportunities are contrasting realities that can coexist.

This contrasting reality is important in understanding the fragmented migration trajectories from Africa in the direction of Turkey, and eventually, Europe. Both the temporal and spatial dimensions of trajectories are important in this respect. Whilst not equating migrant trajectories as migrant life paths, it is, nevertheless, important to realize that similar events have different values for different migrants when these occur in different periods of their lives. As shown in the case of Caroline, the strength of the migrant's motivation to move matters, despite repeated failures in crossing into Europe. Similarly, without falling into geographical determinism, it is fair to say that the evolution of a spatial trajectory also influences migration aspirations. As we have seen from the examples highlighted above, being in Istanbul may intensify the longing to reach Europe in the case of migrants who are confronted with precarious living conditions. However, being in this specific location may also result in new contacts with new information that can profoundly influence future plans. As a result, we can hardly understand the outcomes of these fragmented trajectories by only focusing on migrants' (rational) decision-making before their departures. These trajectories are not the outcome of clear-cut and instant 'go/no-go situations' in which the migrants decide to move or to stay (see also Grillo 2007). As Van der Velde and Van Naerssen (2011) rightly argue, trajectories may involve sequences of decisions, and a series of thresholds may appear to migrants in no particular order. Migrants' trajectories may be affected by multiple plans, unexpected events, (bad) luck, changing political climates and encounters with new people. Trajectories are process-like phenomena, and migrants' decisions must be considered as embedded in these processes.

How then do we make sense of the trajectory as a relevant unit of analysis in migration research? The proposal put forward here is to understand trajectories in the same way as relational geographers understand places (see for instance Massey 2005; Amin 2002). Like places, trajectories are always in the making, always evolving and always affected by 'a set of thrown-together factors' (Massey 2005). This relational starting point, however, does not imply that migrants are perceived as being endlessly and restlessly on the move, never able to find a place to settle down. On the contrary, in the course of this research, many migrants were met who were living happy and satisfied lives in their destinations, also in Istanbul. A relational perspective on migration trajectories does mean, however, that we should not perceive migrants' settlement as something permanent – as something that lasts forever (see also Skeldon 1997). Moreover, it means that we should not equate migrants' settlement with a form of spatial fixity and assimilation. In fact, as underscored in the transnational debate (see for instance Portes 1997; Baas 2009), it is the integrated and institutionally-settled migrant who is in the favourable position to become a transnational actor, moving from one place to another and feeling at home wherever s/he is. Settlement, thus, goes hand in hand with new mobilities which may in turn affect individual migration trajectories.

References

Adepoju, A. 2000. "Issues and Recent Trends in International Migration in Sub-Saharan Africa." *International Social Science Journal* 52 (165): 383–94.

_____ . 2008. "Migration in Sub-Saharan Africa." *Current African Issues* 37. Uppsala: Nordiska Afrikainstitutet.

Adepoju, A., and A. van der Wiel. 2010. *Seeking Greener Pastures Abroad: A Migration Profile of Nigeria.* Ibidan: Safari Books.

Adey, P. 2006. "If Mobility Is Everything Then It Is Nothing: Towards a Relational Politics of (Im) Mobilities." *Mobilities* 1 (1): 75–94.

Akinbingöl, Ö.F. 2003. *Knooppunt Istanbul: Mensensmokkel via Turkije.* Amsterdam: Meulenhoff.

Amin, A. 2002. "Spatialities of Globalisation." *Environment and Planning A* 34 (3): 385–99.

Baas, M. 2009. "Imagined Mobility: Migration and Transnationalism among Indian Students in Australia." PhD diss. Amsterdam: University of Amsterdam.

Baird, T. 2011. "The Missing Migration Component of Turkey-Africa Relations." *Open Democracy,* 1 December. Accessed 15 September 2014. http://www.opendemocracy.net/theodore-baird/missing-migration-component-of-turkey-africa-relations.

Bakewell, O., and G. Jonsson. 2011. "Migration, Mobility and the African City." IMI Working Papers Series 50. Oxford: University of Oxford.

Bazuin, S. 2011. "West African 'Transit' Migrants in Istanbul: A Bridge to Europe?" MSc. Thesis International Development Studies. Utrecht: Utrecht University.

Biehl, K. 2009. "Migration 'Securization' and Its Everyday Implications: An Examination of Turkish Asylum Policy and Practice." CARIM Summer School 2008 – Best Participant Essays Series 2009/01. San Domenico die Fiesole: European University Institute.

Brachet, J. 2005. "Constructions of Territoriality in the Sahara: The Transformation of Spaces of Transit." *Stichproben: Wiener Zeitschrift für kritische Afrikastudien* 5 (8): 237–53.

Bredeloup, S. 2012a. "Sahara Transit: Times, Spaces, People." *Population, Space and Place* 18 (4): 457–67.

_____ . 2012b. "African Trading Post in Guangzhou: Emergent or Recurrent Commercial Form?" *African Diaspora* 5 (1): 27–50.

Brewer, K.T., and D. Yükseker. 2006. "A Survey on African Migrants and Asylum Seekers in Istanbul." MiReKoC Research Projects 2005–2006. Istanbul: Koç University.

Crush, J., and D.A. McDonald. 2000. "Transnationalism, African Immigration, and New Migrant Spaces in South Africa: An Introduction." *Canadian Journal of African Studies* 34 (1): 1–19.

Didem Daniş, A. 2006. "'Integration in Limbo': Iraqi, Afghan, Maghrebi and Iranian Migrants in Istanbul." MireKoc Research Projects 2005–2006. Istanbul: Koç University.

De Bruijn, M., R. van Dijk, and D. Foeken, eds. 2001. *Mobile Africa: Changing Patterns of Movement in Africa and Beyond.* Leiden: Brill.

De Clerck, H.M. 2013. "Sub-Saharan African Migrants in Turkey: A Case Study on Senegalese Migrants in Istanbul." *SBF Dergesi* 68 (1): 39–58.

De Haas, H. 2008. "The Myth of Invasion: The Inconvenient Realities of African Migration to Europe." *Third World Quarterly* 29 (7): 1305–22.

Ellis, S. 2011. *Seasons of Rain: Africa in the World.* London: Hurst and Co.

Falola, T., and O.O.J. Okpeh. 2008. "Introduction." In *Population Movements, Conflicts, and Displacement in Nigeria*, edited by T. Falola and O.O.J. Okpeh, 1–17. Trenton: Africa World Press.

Frambach, N. 2011. "Refugees in Istanbul: Lost between Policy and Practice." MSc. Thesis International Development Studies. Utrecht: Utrecht University.

Furst-Nichols, R., and K. Jacobsen. 2011. "African Refugees in Israel." *Forced Migration Review*, February. Accessed 12 August 2012. http://www.fmreview.org/non-state/Furst-NicholsJacobsen.html.

Gnisci, D. 2008. *West African Mobility and Migration Policies of OECD Countries.* Paris: Organisation for Economic Co-Operation and Development.

Grillo, R. 2007. "Betwixt and Between: Trajectories and Projects of Transmigration."*Journal of Ethnic and Migration Studies* 33 (2): 199–217.

Hahn, H.P., and G. Klute, eds. 2007. *Cultures of Migration: African Perspectives.* Berlin: Lit Verlag.

Hamood, S. 2006. *African Transit Migration through Libya to Europe: The Human Cost.* Cairo: American University.

Içduygu, A., and S. Toktas. 2002. "How Do Smuggling and Trafficking Operate via Irregular Border Crossings in the Middle East? Evidence from Fieldwork in Turkey." *International Migration* 40 (6): 25–54.

Içduygu, A., and D. Yükseker. 2008. "Rethinking the Transit Migration in Turkey: Reality and Re-Presentation in the Creation of a Migratory Phenomenon." Paper prepared for the IMISCOE Conference on "(Irregular) Transit Migration in the European Space: Theory, Politics and Research Methodology," 18–20 April. Istanbul: International Migration, Integration and Social Cohesion in Europe.

Jordan, B., and F. Düvell. 2002. *Irregular Migration: The Dilemmas of Transnational Mobility.* Cheltenham: Edward Elgar.

King, R., R. Black, M. Collyer, A. Fielding, and R. Skeldon. 2010. *People on the Move: An Atlas of Migration.* Berkeley: University of California Press.

Kraler, A. 2005. "The State and Population Mobility in the Great Lakes: What Is Different about Post-Colonial Migrations?" Sussex Migration Working Paper 24. Sussex: University of Sussex.

Martino, F. 2010. "Migrations: Evros, Last Door to Europe." *Observatorio Balcani e Caucaso*, 9 September. Accessed 13 February 2014. http://www.balcanicaucaso.org/eng/Regions-and-countries/Greece/Migrations-Evros-last-door-to-Europe-79696.

Massey, D. 2005. *For Space*. London: Sage.

Nyamnjoh, F.B. 2006. *Insiders and Outsiders: Citizenship and Xenophobia in Contemporary Southern Africa*. Dakar: Codesria Books.

Ozdil, K. 2008. "'To Get a Paper, to Get a Job': The Quiet Struggles of African Foreigners in Istanbul, Turkey." MSc Thesis Sociology and Social Anthropology. Budapest: Central European University.

Özkan, M., and B. Akgün. 2010. "Turkey's Opening to Africa." *Journal of Modern African Studies* 48 (4): 525–46.

Papadopoulos, A.G., K. Charalambos, C. Chalkias, and L.M. Fratsea. 2013. "New Migrant Flows and Their Social Integration in Greece: The Case of Sub-Saharan Africans in a Crisis-Stricken Country." Paper presented at the 4th Norface Migration Network Conference on "Migration: Global Development, New Frontiers," 10–13 April. London: University College London.

Peil, M. 1995. "Ghanaians Abroad." *African Affairs* 94 (376): 345–68.

Pelican, M., and P. Tatah. 2009. "Migration to the Gulf States and China: Local Perspectives from Cameroon." *African Diaspora* 2 (2): 229–44.

Politzer, M. 2008. "China and Africa: Stronger Economic Ties Mean More Migration." *Migration Information Source*, 6 August. Accessed 28 February 2011. http://www.migrationpolicy.org/article/china-and-africa-stronger-economic-ties-mean-more-migration/.

Portes, A. 1997. "Immigration Theory for a New Century: Some Problems and Opportunities." *International Migration Review* 32 (4): 799–825.

Schapendonk, J. 2011. "Turbulent Trajectories: Sub-Saharan African Migrants Heading North." PhD diss. Nijmegen: Radboud University.

———. 2012. "Migrants' Im/Mobilities on Their Way to the EU: Lost in Transit?" *Tijdschrift voor economische en sociale geografie* 103 (5): 577–83.

———. 2013. "From Transit Migrants to Trading Migrants: Development Opportunities for Nigerians in the Transnational Trade Sector of Istanbul." *Sustainability* 5 (7): 2856–73.

Schapendonk, J., and G. Steel. 2014. "Following Migrant Trajectories: The Im/Mobility of Sub-Saharan Africans en Route to the European Union." *Annals of Association of American Geographers* 104 (2): 262–70.

Skeldon, R. 1997. *Migration and Development: A Global Perspective*. Harlow: Longman.

Spaan, E., and D. van Moppes. 2006. "African Exodus? Trends and Patterns of International Migration in Sub-Saharan Africa." Working Papers Migration and Development Series 4. Nijmegen: Radboud University.

Steen, F. 2012. "'Without Trust There Is No Trade': Nigerian Migrants Engaged in Transnational Informal Trade in Istanbul." MSc. Thesis. Utrecht: Utrecht University.

Suter, B. 2012. "Tales of Transit: Sub-Saharan African Migrants' Experiences in Istanbul." PhD diss. Linköping: Linköping University.

Triandafyllidou, A., and T. Maroukis. 2008. "The Case of the Greek Islands: The Challenge of Migration at the EU's Southeastern Sea Borders." In *Immigration Flows and the Management of the EU's Southern Maritime Borders*, edited by G. Pinyol, 63–82. Barcelona: Barcelona Centre for International Affairs.

Van der Velde, M., and T. van Naerssen. 2011. "People, Borders, Trajectories: An Approach to Cross-Border Mobility and Immobility in and to the European Union." *Area* 43 (2): 218–24.

Van Liempt, I. 2007. "Navigating Borders: Inside Perspectives on the Process of Human Smuggling into the Netherlands." PhD diss. Amsterdam: Amsterdam University Press.

Wabgou, M. 2012. "Aiming at Latin America: African Immigration to Mexico." In *Migration in the Service of African Development: Essays in Honour of Professor Aderanti Adepoju*, edited by J.O. Oucho, 113–56. Ibadan: Safari Books.

Watson, I., and J. Bailey-Hoover. 2011. "Football Unites Turkey's 'Marginalized' African Immigrants." *CNN*, 25 October. Accessed 13 August 2012. http://edition.cnn.com/2011/10/25/sport/football/turkey-africa-cup/index.html.

Zoomers, A., G. Rivera-Salgado, M.M.B. Asis, N. Piper, P. Raghuram, M. Awumbila, T. Manuh, and J. Schapendonk. 2009. "Migration in a Globalizing World: Knowledge, Migration and Development." In *Knowledge on the Move: Emerging Agendas for Development-Oriented Research*, edited by H. Molenaar, L. Box, and R. Engelhard, 89–122. Leiden: International Development Publications.

Chapter 17

Immobilized between Two EU Thresholds: Suspended Trajectories of Sub-Saharan Migrants in the Limboscape of Ceuta

Xavier Ferrer-Gallardo and Keina R. Espiñeira

When Spain joined the Schengen Agreement in 1991, the North African city of Ceuta[1] started to gradually turn into a key hub of irregular sub-Saharan migration to the European Union[2] (see Alscher 2005; Berriane and Aderghal 2009; Carling 2007; Driessen 1996; Fekete 2004; Ferrer-Gallardo 2008, 2011; Gold 1999; Mutlu and Leite 2012; Planet 1998; Saddiki 2010; Soddu 2002, 2006). Since then, both the increasing securitization of its border and the fluctuant – though persistent – arrival of migrants have transformed the socio-spatial nature of this territory. These transformations have subsequently influenced migrants' perceptions vis-à-vis the destinations and trajectories they take into consideration during their decision-making process. They have had an impact on spatial behaviour in key migration routes towards the European Union.

In this light, the present chapter addresses how the two different borders of Ceuta, the land border with Morocco and the sea border with the Iberian Peninsula, influence the opportunities of migrants en route toward the EU. In so doing, it builds on previous contributions in the field of transit migration to the EU, which have focused on the north of Africa (to name but a few: Alioua 2008; Barros et al. 2002; Bredeloup 2012; Collyer 2007; Collyer and De Haas 2012; Collyer et al. 2012; De Haas 2008; Düvell 2005, 2012; Khachani 2006; Lahlou 2005; Schapendonk 2012). In concrete terms, it puts the lens on the situation of those sub-Saharan migrants who, having managed to irregularly cross the EU-North African border fence, find themselves stranded in Ceuta. The authors argue that, under these circumstances, the city becomes what they define as a limbo-like landscape[3] (a *limboscape*): a transitional zone, a threshold or midway territory between two different borders, between the hell of repatriation/expulsion and the heaven of regularization,

---

1   Ceuta has a surface area of 19.48 square km and a total perimeter of 30 km long, of which 8 km constitute its land border with Morocco. The city is inhabited by 84,018 people (http://www.ine.es, accessed 1 January 2013).

2   The same applies for the city of Melilla. The circumstances of both cities vis-à-vis irregular migration to the EU are very similar. Nevertheless, in this contribution we will focus on the specific case of Ceuta.

3   Limbo:

    1) *Roman Catholic Theology*: a region on the border of hell or heaven, serving as the abode after death of unbaptized infants (limbo of infants) and of the righteous who died before the coming of Christ (limbo of the fathers or limbo of the patriarchs)

    2) A place or state of oblivion to which persons or things are regarded as being relegated when cast aside, forgotten, past, or out of date

    3) An intermediate, transitional, or midway state or place

    4) Place or state of imprisonment or confinement. See http://www.dictionary.reference.com, accessed 6 November 2014.

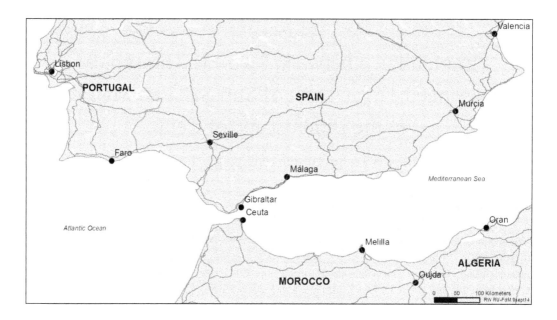

Figure 17.1   Ceuta and Melilla

where the migrants' trajectories towards 'European-EU' are spatially and temporally suspended (see Ferrer-Gallardo and Albet-Mas 2013).

This chapter is structured in three main sections which, to an important extent, coincide with the three basic components of the approach proposed by Van der Velde and Van Naerssen (2011) in order to capture the individual spatial migratory behaviour. The components are borders that are constructed as barriers or deconstructed to facilitate mobility, trajectories as the routes people use to cross borders, and people as agents who decide to migrate crossing borders and following certain routes. The first section (*borders*) explores the double-border dimension of Ceuta and its impact on the mobility of migrants. It describes the origins and changing characteristics of the city's border securitization process as well as the political geographical specificities of this EU-African territory. The second section (*trajectories*) addresses how the strengthening of border controls in the southern EU perimeter has affected the trajectories of sub-Saharan migrants heading for the EU. And, in concrete, how the alteration of the routes has been translated into fluctuating degrees of migratory pressure vis-à-vis the city of Ceuta. The third section (*people*) scrutinizes the circumstances of those sub-Saharan migrants who find themselves stranded in Ceuta and are consequently forced to face a complex period of waiting before the (EU) law (see Van Houtum 2010b).

Finally, it is argued that, over the last years, both the increasing securitization of its borders and the variable though persistent arrival of sub-Saharan immigrants to Ceuta has profoundly reconfigured the socio-spatial disposition of the city. The fact that those migrants who manage to cross the border fence are forced to spend long periods of time waiting in the city has provided Ceuta with a new territorial idiosyncrasy. We suggest that the notion of 'limboscape' (see also Ferrer-Gallardo and Albet-Mas 2013) constitutes a useful tool in order to capture this new spatial idiosyncrasy of Ceuta and that, in turn, it helps to illustrate (and conceptualize) the present proliferation of confinement/encampment practices across the EU, as well as its related forced migrant immobility dynamics.

Borders

The Spanish–Moroccan border became an external EU border in 1986. It then started to serve as a key instrument in EU migration policy, but it also became an outcome of the discussion on the essence of European (Union) identity and its territorial demarcation (see Van der Velde and Van Naerssen 2011, 220). When Spain joined the Schengen Agreement in 1991, its visa regime adjusted to the new situation. The range of legal modifications associated to the 'Schengenization' of the Spanish–Moroccan border, and the arrival of newly perceived threats, notably in the form of irregular immigration from sub-Saharan Africa, came together with the implementation of new securitization techniques and the physical reshaping of the border. In order to prevent the entrance of undocumented immigrants, security controls were reinforced all along the maritime and land border between Spain and Morocco with the financial assistance of EU institutions (see Alscher 2005; Ferrer-Gallardo 2008; Mutlu and Leite 2012; Saddiki 2010).

The tough but to a certain extent ineffective attempts to completely obstruct the movement of some 'undesired' non-EU citizens across the specific border of Ceuta contrast with the elasticity that EU legislation shows when it comes to facilitate the free cross-border flow of 'desired' non-EU citizens (see Ferrer-Gallardo 2011). The economic sustainability of Ceuta largely depends on the interaction with its hinterland. For this reason, Spain exempts visa requirement to citizens of the neighbouring Moroccan province of Tetouan. This exception was incorporated into the Protocol of Accession of Spain to the Schengen Agreement in 1991 with the commitment to maintain tight identity controls to those wanting to travel to the rest of Spanish territory (Planet 2002). The fact that document controls are conducted both at the land crossing point between Morocco and Ceuta and at the maritime crossing point between Ceuta and the Iberian Peninsula implies that the entire territory of Ceuta functions as a threshold between two EU borders, as an intermediate space, as a 19.5 square km border zone.

After Spain's EU entrance, the reconfiguration of the Spanish border regime ran parallel to the reshaping of migratory dynamics in the north of Africa. In the mid-1990s, the growing flows of sub-Saharan migrants heading for the EU implied that Libya, Algeria, Tunisia and Morocco started consolidating as key transit countries, but also as destination countries. Consequently, migratory dynamics in Ceuta's hinterland experienced huge transformations. The border of Ceuta emerged as a new, but relatively low threshold to be crossed within the trajectories of sub-Saharan migrants to the EU. Increasingly, the Spanish African city would be perceived as a less dangerous irregular gate to the EU. Entering Ceuta would therefore become a safer and attractive alternative to the clandestine crossing of the Mediterranean.

In 1995, the irregular access of sub-Saharan citizens to Ceuta (that were not repatriated to their countries of origin nor allowed to cross the maritime border toward the Iberian Peninsula) had already become a frequent phenomenon. This was the source of tensions in the city, where racist attacks and migrant riots claiming for their rights proliferated (see Gold 1999, 2000; Planet 1998). In this context, Spanish and EU authorities decided to undertake the fortification of the land perimeter of the city. In order to halt the increasing flows of irregular immigration, a double metal fence, which height would later on reach six metres, was erected between the city and Morocco. This is how the Ceuta border scenario started to become a paradigmatic example of the EU's sealing-off its outer perimeter. As a consequence, the city would be globally known as an icon of so-called 'Fortress Europe'.

The aforementioned iconic dimension of Ceuta was reinforced by the migration crisis that took place during the autumn of 2005. Prior to the crisis, in 2004, there was a 37 per cent fall in the arrival of immigrants to the shores of the Iberian Peninsula (EC 2005). This occurred due to the implementation of the System of Integrated External Surveillance on Spain's southern coasts, and also due to greater coordination between Spanish and Moroccan border control practices. Following the blockage of the maritime route through the western Mediterranean (across the Strait of Gibraltar), the migratory pressure on the land border of Ceuta increased remarkably.

Particularly after the year 2000, an increasing number of migrants had converged in the surroundings of Ceuta, waiting for an opportunity to be able to enter Ceuta. Migrants started to gather in informal camps near the border fence, in the forests of Belyounech, Morocco (see Soddu 2006). In this context, during the months of September and October of 2005, illegal entries to the city grew substantially. Border guards, both in Morocco and in Spain, harshly repressed the attempts of entry (Blanchard and Wender 2007). Hundreds of migrants crossed the fence and made it into Ceuta. According to the Spanish Ministry of Home Affairs, 5,566 immigrants irregularly entered Ceuta and Melilla in 2005. But there were many more that could not make it. Eleven immigrants were killed and many more were wounded during the events. The crisis reached dramatic proportions and placed the cities under the global focus of media interest.

On the 6th of February 2014 global media attention was focused again on the Ceuta border after 15 migrants drowned while they attempted to reach North African EU soil by swimming from Morocco (*El País,* 15 February 2014). The role of the Spanish Guardia Civil was rather controversial since rubber bullets were used in order to stop the migrants. This opened an intense debate in Spain about the proportionality of securitization measures. The episode also shed light on the irregular push-back of migrants (refoulements) to Morocco that, according to several NGOs, had taken place in the Ceuta border scenario on several occasions (*El País,* 19 February 2014). A new reinforcement of the border fence was announced by the Spanish Minister of Home Affairs, Fernandez Díaz, during a visit to Ceuta on 5 March 2014.

Nearly a decade before that, the events of 2005 had already led to a significant transformation of border securitization practices. The crisis represented an inflection point within migratory and border management dynamics in the EU. The immediate response was the physical reinforcement of the fences and the strengthening of border patrols on both sides of the perimeter. Spanish and Moroccan army units were sent to the border and remained there for a short time. The militarization of the border was temporarily literal. Ceuta and Melilla would thus become the two most heavily securitized border posts of the European Union. Also, surprisingly enough, Morocco agreed to contribute to the monitoring of the borders that, in the Moroccan official discourse, are considered colonial cities, which are still, illegitimately, in Spanish hands. Since then, as Casas-Cortes et al. (2013, 52) argue, Moroccan border and police authorities have increasingly cooperated in managing migrant flows toward the Spanish/EU border – despite the fact that Morocco claims the sovereignty of Ceuta (and hence does not recognize its land border as an official border). Anyhow, efforts have gone beyond the EU land borders in Africa, and have been extended throughout the Moroccan territory by means of Moroccan readmission agreements with various EU member states, the establishment of temporary circular labour migration centres to facilitate temporary legal migration, and other forms of Moroccan cooperation with EU border management imperatives. As denounced by several NGO's and scholarly critical approaches to EU bordering practices, the externalization of EU border controls to third countries has given rise to migration management practices which have not always fully respected the fundamental rights of migrants (see Wender 2004; APDHA 2012; GADEM 2012; Migreurop 2009; MSF-E 2005; Soddu, 2006).

Trajectories

The reinforcement of border controls after 2005 made approaching and jumping the Ceuta border fence increasingly difficult for migrants. Nevertheless, irregular migrants kept on entering Ceuta, using more complex crossing practices, such as hiding in vehicles or swimming around the border fence (Collyer and De Haas 2012). Fortification techniques continued to be developed, but so did ways of subverting them. Despite the gradual strengthening of the Ceuta border, the number of migrants who have attempted and succeeded to irregularly enter the city has not been steady. Quantitative highs and downs vis-à-vis irregular border crossings have run parallel to the fluctuation of migration routes toward the EU. This fluctuation has in its turn been a reaction to the gradual deployment of stricter border securitization measures along the southern EU perimeter. But it has also been caused by the refinement/development of the EU strategy of externalization of border controls (see Pinyol 2012; Zaragoza 2012). As Casas-Cortes et al. (2013, 41) note, particularly after the crisis of Ceuta and Melilla in 2005, and the so-called 'Cayucos crisis' in the Canary Islands in 2006, the security goals of externalization and the creation of policing and geoeconomic regimes beyond the borders of the EU and the Schengen space were given added impetus. To an important extent, this 'added impetus' explains the decrease of irregular entrances in Ceuta and Melilla after 2005 (see Figure 17.2).

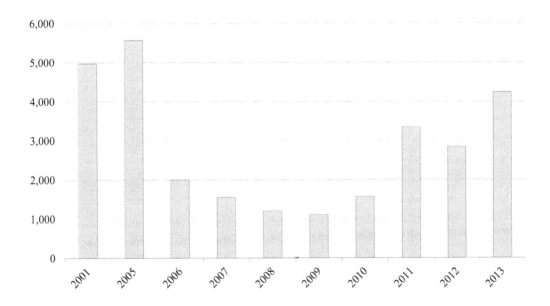

Figure 17.2   Number of irregular entrances in Ceuta and Melilla, 2001–13

Source: compiled by the authors based on data provided by the Spanish Ministry of Home Affairs, 2001–13.

Needless to say, the Ceuta border scenario continues to be a paradigmatic example of how the EU tries to seal off its outer perimeter to stop the arrival of migratory flows. However, in the course of recent years, the landscape of EU border fortification has expanded. Ceuta and Melilla are not the only worldwide known icons of Fortress Europe. Spaces like the Canary Islands, Lampedusa, Malta, the Greek island of Lesbos or the Evros river on the Greek-Turkish border have joined the list (see Cuttitta 2012; Godenau and Zapata-Hernández 2008; Kitagawa 2011; López-Sala and Esteban-Sánchez 2007; Triandafyllidou 2010; Triandafyllidou and Maroukis 2008). Entry points of irregular immigration have multiplied and diversified along the southern border of the EU. Parallel to that, policies and practices designed to cut irregular migration flows to the EU have also intensified.

After the above-mentioned crisis of 2005, the trajectories of sub-Saharan immigrants en route toward Europe detoured. The perception of the Ceuta border as a more difficultly permeable border certainly influenced mobility patterns. Migrants started to follow alternative routes – first towards the Canary Islands and later across the central (Italy, Malta) and eastern (Greece) Mediterranean routes. The increase of arrivals to the Canary Islands followed the implementation of 'Plan Africa' by the Spanish Government and the development of different Frontex[4] operations (Hera, Minerva and Indalo[5]) on the Atlantic coast of Africa (see Pinyol 2012). As a result, migration decreased in the Atlantic route, moving toward Italian, Maltese and Greek coasts. Immediately afterwards, especially due to the deployment of Hermes and Poseidon operations by Frontex, flows shifted to the land border between Greece and Turkey. In this context, in October 2010 Greece asked the other member states to support the Rapid Border Intervention Teams (RABIT) of Frontex. It was the first time that an EU state requested such intervention.

The total number of detections of illegal border crossings outside the EU was similar in 2009 (104,599) and in 2010 (104,049), but increased in 2011 (140,980) (Frontex 2012, 14). Data certified the rechannelling of migratory flows, as well as the fluctuation in the weight of each of the migratory routes to the EU across the Mediterranean (see Figure 17.3). Irregular flows via the

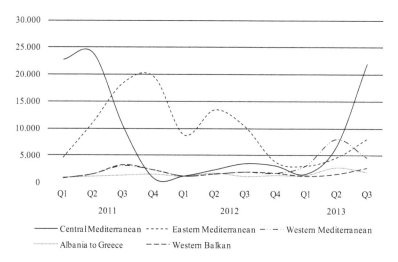

Figure 17.3   Detections of illegal border crossings between border crossing points by main migration routes, 2011–13

Source: Adapted from Frontex 2014.

---

4   Frontex (European Agency for the Management of Operational Cooperation at the External Borders of the Member States of the European Union) is the European Union agency for external border security (see http://www.frontex.europa.eu, accessed 28 October 2014).

5   For the scope of Frontex operations, see http://www.frontexit.org/en/docs/13-map-111-operations-frontex/file (accessed 27 September 2014).

central Mediterranean route increased notably due to the influence that the Arab Spring had in border management bilateral agreements between the EU and North African countries like Libya and Tunisia (see Bialasiewicz 2011; Zapata-Barrero and Ferrer-Gallardo 2012). The number of detected illegal border crossings in the sea border of Greece has fallen considerably over the last years: 28,848 in 2009, 6,175 in 2010 and 1.467 in 2011. Parallel to that, illegal crossings at the land Greek-Turkish border followed the opposite direction and increased from 11,127 in 2009 to 49,513 in 2010 and to 55,589 in 2011 (Frontex 2012). The Greek-Turkish land border became the new Achilles heel of the EU border securitization apparatus. In December 2012 a new border fence was erected there (in Orestiada), in a very similar fashion as the Ceuta border fence was erected almost two decades before.

People

Regarding the specific situation at the borders of Ceuta and Melilla, a glance on the data shows how, despite the fortification of the border, the two cities are still paramount entrance sites for a number of irregular migrants en route towards the European Union (see Figure 17.3). In 2005 the Spanish Ministry of Home Affairs registered 5,566 irregular entrances. In the following years there was a decrease as a result of the strengthening of the border and the changing trajectories of migrants. In 2009 the Spanish authorities registered 1,108 entrances, which marked the lowest number of irregular crossings to Ceuta and Melilla in the last ten years. But since 2010 the number of irregular entrances has increased again. There were 4,235 irregular entrances in 2013. Although the Spanish and EU authorities (with the cooperation of the Moroccan authorities) have been intensively working on and investing in the fortification of the Ceuta border since 2005, it actually remains almost as permeable as it was a decade ago.

In June 2013 Morocco and the EU signed a Mobility Partnership. The Mobility Partnership with Morocco is the fifth of its kind, following those signed with Moldova and Cape Verde in 2008, with Georgia in 2009 and with Armenia in 2011. The partnership includes 'negotiations between the EU and Morocco on an agreement for facilitating the issuing of visas for certain groups of people, particularly students, researchers and business professionals. Negotiations will also continue on an agreement for the return of irregular migrants' (EC 2013).

Morocco already readmits its own nationals based on bilateral readmission agreements signed with Spain, France, Italy and Germany, but, so far, refuses to sign a readmission agreement with the EU. The reluctance of Morocco to sign a readmission agreement with the EU is mostly linked with an unwillingness to readmit irregular non-Moroccans that have passed through the country before entering the European Union (*Statewatch News Online*, 7 March 2013).

At the moment, given the geographical specificity of Ceuta (separated from the rest of the African continent by the border fence and from the European landmass by the waters of the Straits of Gibraltar), migrants who manage to cross the land perimeter of the city find themselves trapped between two different borders. They are still in Africa but already in EU territory, and they cannot cross to the European side of the EU (the aimed destination for most of them). They have crossed the tough threshold that the highly fortified external African borders of the European Union represent. Nevertheless, Ceuta is not the desired destination for these migrants and they must still face an interval of involuntary immobility and the hard task of crossing a second threshold.

Periods of forced waiting in-between for migrants in Ceuta are variable. The decision on asylum applications, as well as the processes for handling the deportation of irregular immigrants and their readmission by countries of origin, can take several years. During this time, most of the immigrants are sheltered in the Centre of Temporary Stay for Immigrants (CETI). The CETI of Ceuta was inaugurated in April 2000, it is managed by the Spanish government, and it is intended to provide basic social services and assistance to irregular migrants and asylum seekers on their arrival to Ceuta. The CETI has room for 512 immigrants. Since its inauguration, more than 23,000 people have been sheltered there. The perimeter of the centre is surrounded by a 3-metre-high fence with video-surveillance cameras on top. It might not look like it but the CETI is actually an open centre. Hence it is not exactly a detention centre. There is a single entrance monitored by private security. Migrants can move in an out of the CETI freely showing an identification card which barcode is contrasted with the reading of the fingerprint (migrants obtain this card after having been identified and registered by Spanish police). But migrants cannot move in and out of Ceuta. The CETI is open to them. But the city is not. Migrants are not confined in the CETI, but they are confined in the city. This makes an important difference between the CETIs and the CIEs (*Centros de Internamiento de Extranjeros* [Internment Centres for Foreigners]) to which many migrants will be transferred after their stay in the CETI. Migrants are detained in the CIE, but are sheltered in the CETI.

Asylum application must be processed in one month. After that date, if the application is not successful, an expelling procedure is opened. The opening of an expelling procedure is usually accompanied by the confinement in a CIE that, as said, acts as a migrant detention centre. This means that those migrants in Ceuta who are going to be expelled must be first transferred to a CIE in the Iberian Peninsula, before the expelling is conducted. It is important to note that, since the adoption of the European Directive on common standards and procedures for returning illegally staying third-country nationals,[6] the maximum period of confinement in a migrant detention centre within the Schengen area is six months. In the case of Spain, the implementation of the Return Directive extends the period of detention up to 60 days. Beyond that period, if migrants have not been expelled, they are released. They are free but in an irregular situation. They are not deported but their legal status is not regularized. They remain in Spain, but without valid documents to legally stay and work there. They remain in a sort of juridical limbo. For them, the CETI becomes a waiting space that precedes a subsequent waiting space (a CIE in the Iberian Peninsula).

This waiting in-between is an uncertain period of time. The output of this waiting time can imply regularization or the obtaining of refugee status, but can also imply expulsion or repatriation to their countries of origin. Some of those migrants try to find alternative ways to illegally cross to the Iberian Peninsula. Particularly since late 2010, hiding in trucks that transfer trash from Ceuta to the Iberian Peninsula has become a way to exit Ceuta and reach the European continent. Hundreds of migrants have been transferred to the northern shore of the Mediterranean blended with trash (see *El País*, 9 January 2003; *El Mundo*, 11 June 2013). Some of them have died in their attempt to do so. In order to stop this, the boundaries of Ceuta's Waste Transfer Plant are now being patrolled and securitized. This entails a desperate attempt to fight against the uncertainty of waiting and a way to escape the context of involuntary immobility where these migrants are.

Close

In the light of the role it plays vis-à-vis the flows of sub-Saharan immigration to the EU, this chapter suggests that the city of Ceuta can be symbolically depicted as a 'threshold territory': as a space of transit towards the EU which is paradoxically situated already in the EU; as a limbo

---

6   Directive 2008/115/EC, available at http://eur-lex.europa.eu/legal-content/EN/TXT/?uri=CELEX: 32008L0115 (accessed 28 September 2014).

space (a limboscape) between two different EU borders; a waiting area that, despite the increasing danger and strengthening of border controls, still seems to be worth entering for many sub-Saharan migrants. Today, as it occurred during the last decade, when Ceuta's iconic twin metal fencing provided powerful visuality to the Fortress Europe/Gated Community spatial paradigm (Ferrer-Gallardo 2008, 2011; Van Houtum and Pijpers 2007), this non-European EU territory is still (or even more) central in our understanding of how the EU project – as well as (im)mobility dynamics across its external borders – are socio-spatially constructed.

By depicting it as a limboscape, we suggest that Ceuta's spatial allegorical strength helps us deepen our conceptualization of the current landscape of EU external socio-spatial b/ordering. In concrete terms, we suggest that the limboscape profile of this EU territory embodies now an icon of the process of scattered and proliferating spatialization through which the human-blacklisting practices (Van Houtum 2010a) of EU border and migration management operate. It sheds light on the diffuse territorial scenery of what we might define as the process of migrant purgation (the waiting between the 'hell' of repatriation and the 'heaven' of regularization) that the geopolitics of EU b/ordering fabricates.

The CETI of Ceuta is sometimes described as a sweet prison by those who live there. As it has been argued, it is in fact an open centre. It is not exactly a detention camp/centre. It is not a prison, just almost – a sugared version of it, one might say. CETI residents can get in and out of the centre as they wish. And the city functions as the CETI's backyard. Immigrants can move freely within the city. But they cannot freely leave it. They are trapped, immobilized in Ceuta. Hence, Ceuta as a whole works as an intermediate territory, as a transitional space. It is a threshold, a waiting area of 19.5 square km where the legal status of immigrants (the granting or denial of the right of access to the EU) is to be resolved, and where the interpretation of Schengen's legal apparatus will determine the extent to which they deserve to have right to free mobility. It entails, therefore, a space/time of provisional oblivion between the heaven of regularization and the hell of repatriation/deportation. It is the region of the diffuse margins between EU space and non-EU space that serves as the temporary abode for those who have managed to irregularly cross the EU external border but have not yet received the baptism or the conviction of the Schengen law. It is a limboscape.

The 'purgatory geopolitics' deployed by the EU b/ordering regime on this particular African territory – often adopting much stricter fashions elsewhere – reverberates simultaneously all over the dispersing geographies that currently conform the outer border of the European Union. The Ceuta case, therefore, illustrates the propagation of a deterritorializing archipelago-like EU landscape of limbos (See Ferrer-Gallardo and Albet-Mas 2013). All together, they incarnate a limbo(land)scape encompassing a constellation of EU and non-EU territories where the selection of access to the EU is performed. Transitional spaces where migrants must undergo the uncertain process of waiting before the EU law. That is where the human-blacklisting bureaucratic machinery of the Schengen regime operates (Van Houtum 2010); the grey zones where an essential part of the EU project is socio-spatially fabricated.

As it did during the last decade when the militarization of its land border shed light to the physical reinforcement of the EU external border, today Ceuta also illuminates the logic of sprinkling of territories of exception (of socio-spatial almostness) the EU is currently engaged in. It invites to keep on digging on the debate about the dissemination of ubiquitous waiting areas where EU practices of socio-spatial b/ordering develop.

So far, much scholarly research has focused on mobility dynamics across the borders of Ceuta. In contrast, forced immobility dynamics within Ceuta remain conceptually underexplored and so does the impact it has on the perceptions and preferences of immigrants whose planned or aspirational trajectories include the crossing the borders of the EU-African territory Ceuta. This contribution has attempted to draw some attention on this regard.

The Mobility Partnership signed between the EU and Morocco in June 2013, which might eventually pave the way to the signing of a Readmission Agreement by Morocco, could transform this border dynamic in the near future. Whether this Readmission Agreement is signed or not will certainly condition the here described limboscape profile of the North African EU city of Ceuta. If it is indeed signed, it might also eventually transform the very essence of the Ceuta border as an attractive threshold to be crossed for irregular migrants in their trajectory to the EU.

References

Alioua, M. 2008. "La migration transnationale – logique individuelle dans l'espace national: L'exemple des transmigrants subsahariens à l'épreuve de l'externalisation de la gestion des flux migratoires au Maroc." *Social Science Information* 47 (4): 697–713.

Alscher, S. 2005. "Knocking at the Doors of 'Fortress Europe': Migration and Border Control in Southern Spain and Eastern Poland." CCIS Working Paper 126. San Diego: Center for Comparative Immigration Studies.

APDHA (Asociación Pro Derechos Humanos de Andalucía). 2012. *Derechos humanos en la frontera sur.* Sevilla: APDHA.

Barros, L., M. Lahlou, C. Escoffier, P. Pumares, and P. Ruspini. 2002. "L'immigration irrégulière subsaharienne à travers et vers le Maroc." Cahier des migraciones internationales 54F. Geneva: International Labour Office.

Berriane, M., and M. Aderghal. 2009. *Etat de la recherche sur les migrations internationales à partir, vers et à travers le Maroc.* Rabat: Equipe de Recherche sur la Région et la Régionalisation.

Bialasiewicz, L. 2011. "Borders, Above All?" *Political Geography* 30 (6): 299–300.

Blanchard, E., and A.S. Wender, eds. 2007. *Guerre aux migrants: Le livre noir de Ceuta et Melilla.* Paris: Syllepse.

Bredeloup, S. 2012. "Sahara Transit: Times, Spaces, People." *Population, Space and Place* 18 (4): 457–67.

Carling, J. 2007. "Migration Control and Migrant Fatalities at the Spanish-African Borders." *International Migration Review* 41 (2): 316–43.

Casas-Cortes, M., S. Cobarrubias, and J. Pickles. 2013. "Re-Bordering the Neighbourhood: Europe's Emerging Geographies of Non-Accession Integration." *European Urban and Regional Studies* 20 (1): 37–58.

Collyer, M. 2007. "In-Between Places: Trans-Saharan Transit Migrants in Morocco and the Fragmented Journey to Europe." *Antipode* 39 (4): 668–90.

Collyer, M., and H. de Haas. 2012. "Developing Dynamic Categorisations of Transit Migration." *Population, Space and Place* 18 (4): 468–81.

Collyer, M., F. Düvell, and H. de Haas. 2012. "Critical Approaches to Transit Migration." *Population, Space and Place* 18 (4): 407–14.

Cuttitta, P. 2012. *Lo spettacolo del confine: Lampedusa tra produzione e messa in scena della frontiera.* Milan: Mimesis.

De Haas, H. 2008. "Irregular Migration from West Africa to the Maghreb and the European Union: An Overview of Recent Trends." IOM Migration Research Series 32. Geneva: International Organization of Migration.

Driessen, H. 1996. "The 'New Immigration' and the Transformation of the European-African Frontier." In *Border Identities: Nation and State at International Frontiers*, edited by T.M. Wilson and H. Donnan, 96–116. Cambridge, MA: Cambridge University Press.

Düvell, F., ed. 2005. *Illegal Immigration in Europe: Beyond Control*. Houndmills: Palgrave Macmillan.

_____. 2012. "Transit Migration: A Blurred and Politicised Concept." *Population, Space and Place* 18 (4): 415–27.

EC (European Commission). 2005. "Visit to Ceuta and Melilla – Mission Report: Technical Mission to Morocco on Illegal immigration, 7th October–11th October 2005." *Europa Press Releases Database*, 19 October. Accessed 27 September 2014. http://europa.eu/rapid/press-release_MEMO-05-380_en.htm.

_____. 2013. "Migration and Mobility Partnership Signed between the EU and Morocco." *Europa Press Releases Database*, 7 June. Accessed 27 September 2014. http://europa.eu/rapid/press-release_IP-13-513_en.htm.

Fekete, L. 2004. "Deaths at Europe's Borders." *Race and Class* 45 (4): 75–89.

Ferrer-Gallardo, X. 2008. "The Spanish–Moroccan Border Complex: Processes of Geopolitical, Functional and Symbolic Rebordering." *Political Geography* 27: 301–21.

_____. 2011. "Territorial (Dis)continuity Dynamics between Ceuta and Morocco: Conflictual Fortification vis-à-vis Cooperative Interaction at the EU Border in Africa." *Tijdschrift voor economische en sociale geografie* 102 (1): 24–38.

Ferrer-Gallardo, X., and A. Albet-Mas. 2013. 'EU-Limboscapes: Ceuta and the Proliferation of Migrant Detention Spaces across the European Union." *European Urban and Regional Studies*. Accessed 27 September 2014. doi: 10.1177/0969776413508766.

Ferrero-Turrión, R., and A. López-Sala. 2012. "Fronteras y seguridad en el Mediterráneo." In *Fronteras en movimiento: Migraciones hacia la Unión Europea en el contexto Mediterráneo*, edited by R. Zapata-Barrero and X. Ferrer-Gallardo, 229–54. Barcelona: Bellaterra.

Frontex. 2012. *FRAN Quarterly: Quarter 4 October–December 2012*. Warsaw: Frontex.

_____. 2014. *FRAN Quarterly: Quarter 3 July–September 2013*. Warsaw: Frontex.

GADEM. 2012. "Recrudescence de la repression envers les migrants au Maroc: Une violence qu'on croyait révolue." *GADEM*, 23 October. Accessed 26 January 2013. http://www.gadem-asso.org/Recrudescence-de-la-repression,147.

Godenau, D., and V.M. Zapata-Hernández. 2008. "The Case of the Canary Islands (Spain): A Region of Transit between Africa and Europe." In *Immigration Flows and the Management of the EU's Southern Maritime Borders*, edited by G. Pinyol, 13–44. Barcelona: Barcelona Centre for International Affairs.

Gold, P. 1999. "Immigration into the European Union via the Spanish Enclaves of Ceuta and Melilla: A Reflection of Regional Economic Disparities." *Mediterranean Politics* 4 (3): 23–36.

_____. 2000. *Europe or Africa? A Contemporary Study of the Spanish North African Enclaves of Ceuta and Melilla*. Liverpool: Liverpool University Press.

Khachani, M. 2006. *La emigración subsahariana: Marruecos como espacio de tránsito*. Barcelona: Barcelona Centre for International Affairs.

Kitagawa, S. 2011. "Geographies of Migration across and beyond Europe: The Camp and the Road of Movements." In *Europe in the World: EU Geopolitics and the Making of European Space*, edited by L. Bialasiewicz, 201–22. Farnham: Ashgate.

Lahlou, M. 2005. "Les migrations irrégulières entre le Maghreb et l'Union européenne: évolutions récentes." CARIM Research Report 2005/03. Florence: European University Institute, Robert Schuman Centre for Advanced Studies.

López-Sala, A., and V. Esteban-Sánchez. 2007. "The Farthest Southern Border of Europe: Immigration and Politics in the Canary Islands?" *Revista migraciones internacionales* 4 (1): 87–110.

———. 2010. "La nueva arquitectura política del control migratorio en la frontera marítima del suroeste de Europa: Los casos de España y Malta." In *Migraciones y fronteras: Nuevos contornos para la movilidad internacional*, edited by M.E. Anguiano and A.M. López-Sala, 75–102. Barcelona: Icaria.

Migreurop. 2009. *Atlas des migrants en Europe: Géographie critique des politiques migratoires.* Paris: Armand Colin.

MSF-E (Médecins Sans Frontières Espagne). 2005. "Violence et immigration: Rapport sur l'immigration d'origine subsaharienne (ISS) en situation irrégulière au Maroc." *Médecins Sans Frontières*, 29 September. Accessed 18 January 2013. http://www.msf.fr/sites/www.msf.fr/files/2005-09-29-MSFE.pdf.

Mutlu, C.E., and C.C. Leite. 2012. "Dark Side of the Rock: Borders, Exceptionalism, and the Precarious Case of Ceuta and Melilla." *Eurasia Border Review* 3 (2): 21–39.

Pinyol, G. 2012. "¿Una oportunidad perdida? La construcción de un escenario euroafricano de migraciones y su impacto en las fronteras exteriores de la Unión Europea." In *Fronteras en movimiento: Migraciones hacia la Unión Europea en el contexto Mediterráneo*, edited by R. Zapata-Barrero and X. Ferrer-Gallardo, 255–80. Barcelona: Bellaterra.

Planet Contreras, A.I. 1998. *Melilla y Ceuta: Espacios Frontera Hispano-Marroquíes.* Melilla: Universidad Nacional de Educación a Distancia.

———. 2002. "La frontière comme ressource: Le cas de Ceuta et Melilla." In *La Méditerranée des réseaux: Marchands, entrepreneurs et migrants entre l'Europe et le Maghreb*, edited by J. Césari, 267–81. Paris: Maisonneuve et Larose.

Saddiki, S. 2010. "Ceuta and Melilla Fences: A EU Multidimensional Border." Paper presented at the ISA Annual Convention "Theory vs. Policy? Connecting Scholars and Practitioners", 17–20 February. New Orleans: International Studies Association.

Schapendonk, J. 2012. "Migrants' Im/Mobilities on Their Way to the EU: Lost in Transit?" *Tijdschrift voor economische en sociale geografie* 103 (5): 577–83.

Soddu, P. 2002. *Inmigración extra-comunitaria en Europa: El caso de Ceuta y Melilla.* Ceuta: Archivo Central.

Soddu, P. 2006. "Ceuta and Melilla: Security, Human Rights and Frontier Control." In *The Mediterranean Yearbook Med. 2006*, 212–14. Barcelona: Institut Europeu de la Mediterrània.

Triandafyllidou, A. 2010. "Control de la inmigración en el sur de Europa (1ª parte): Estrategias de 'cerco' (fencing)." *ARI* 7.

Triandafyllidou, A., and T. Maraukis. 2008. "The Case of the Greek Islands: The Challenge of Migration at the EU's Southeastern Sea Borders." In *Immigration Flows and the Management of the EU's Southern Maritime Borders*, edited by G. Pinyol, 63–82. Barcelona: Barcelona Centre for International Affairs.

Van der Velde, M., and T. van Naerssen. 2011. "People, Borders, Trajectories: An Approach to Cross-Border Mobility and Immobility in and to the European Union." *Area* 43 (2): 218–24.

Van Houtum, H. 2010a. "Human Blacklisting: The Global Apartheid of the EU's External Border Regime." *Environment and Planning D: Society and Space* 28 (6): 957–76.

———. 2010b. "Waiting before the Law: Kafka on the Border." *Social and Legal Studies* 19 (3): 285–97.

Van Houtum, H., and R. Pijpers. 2007. "The European Union as a Gated Community: The Two-Faced Border and Immigration Regime of the EU." *Antipode* 39 (2): 291–309.

Wender, S. 2004. *La situation alarmante des migrants subsahariens au Maroc et les consequences des politiques de l'union Europeene*. Paris: Cimade.

Zapata-Barrero, R., and X. Ferrer-Gallardo, eds. 2012. *Fronteras en movimiento: Migraciones hacia la Unión Europea en el contexto Mediterráneo*. Barcelona: Bellaterra.

Zaragoza, J. 2012. "Justicia global y externalización de políticas migratorias: El caso español." In *Fronteras en movimiento: Migraciones hacia la Unión Europea en el contexto Mediterráneo*, edited by R. Zapata-Barrero and X. Ferrer-Gallardo, 143–70. Barcelona: Bellaterra.

EPILOGUE

Chapter 18

The Threshold Approach Revisited

Martin van der Velde and Ton van Naerssen

During the past three decades, cross-border human mobility has substantially increased. For instance, the migrant stock in the more developed regions of the world rose from 7.2 per cent in 1990 to 10.8 per cent in 2013 (UN 2013). And between 1995 and 2013 international tourism doubled to more than a billion trips (UNWTO 2014). This increasing mobility is partly facilitated by the development of communication and transport technologies. New technologies are accompanied by a new mental order and both are interrelated. In a world of 'global life styles' and transnational diasporas, locations at great geographical distance from one's place of residence are becoming familiar. For instance, worldwide middle-class markers and icons like Starbucks and Nike or well-known tourist attractions like Disneyland, the Eiffel Tower or Mount Kilimanjaro are known to people living thousands of miles away.

Human mobility today is characterized by globalization and the substantial rise in the number of short touristic trips and international meetings and conferences. Yet, people are remarkably immobile when the duration of stay abroad becomes longer, as is the case with labour migration. The idea is that even when circumstances are conducive for people to move, for instance, when they live in one of the EU countries near the border, they are usually not inclined to cross the border except for short visits. To reiterate our arguments in the introductory chapter: In 2008, among the EU-15, only 1.7 per cent of the workers in border regions commuted across the border. When the 12 new member states are added, this share is still only 2.1 per cent (MKW Wirtschaftsforschung 2009). In general, the other side of the border is of great mental distance and, as we explained in Chapter 1, we consider this perception and giving meaning as basic in decision-making processes regarding mobility and migration (see also Salazar 2011).

The threshold approach, with the three key components of people who move, borders delineating (national) territories and routes to intended destinations, is a tool to analyse decision-making in mobility and migration in general. However, we first introduced the threshold approach in the context of labour migration across national borders in the EU. We distinguish three thresholds – the *mental*, the *locational* and the *trajectory* threshold – that, whoever intends to cross borders, have to be overcome (Van der Velde and Van Naerssen 2011). The contributions in this book try to apply the threshold approach beyond the EU context to other regions across the globe. More importantly, they all aim to broaden the concept beyond labour migration to other contemporary forms of human mobility.

Our call to researchers interested in border issues and international migration has attracted submission of a wide variety of papers. Many contributors have even introduced new idioms to clarify the 'thresholds' concept and to enrich our approach. Just to name a few, concerning the weighing of pros and cons when crossing the thresholds: dynamics of immobility and mobility (Chapter 3), strategic and subjective rationality (Chapter 4) and 'borderwork' (Chapter 14). Concerning borders: thin and thick borders (Chapter 8), border barriers (Chapter 11) and internalized borders (Chapter 12). Concerning trajectories and places of transfer: contexts of exit and reception (Chapter 15), forced migrant immobility dynamics and limboscape (Chapter 17).

The contributors to this book come from different backgrounds and have varied experiences in mobility and migration research that encompass a diversity of social groups and regions. Thus, it is not surprising that they differ in the use and appreciation of the threshold approach. While some seem less at ease with the approach, others incorporate the approach without much difficulty. Some authors have limited themselves to one threshold, while others have considered all three with emphasis on a particular one. Although diverse as a whole, the contributions constitute a useful elaboration and refinement of the approach. We have distinguished four major comments, namely, the non-linearity of the decision-making process, the concept of borders and 'bordering', the mobility of different social groups, and the changing context of the decision-making process. We will discuss these consecutively in the following section.

The Non-Linearity of the Decision-Making Process

Some contributors pointed out that the threshold approach seems to imply that the mobility and migration decision-making process is a linear one, following a rigid sequence as suggested by Figure 1.1 in Chapter 1, involving first the mental threshold, then the locational one and finally the trajectory threshold. Since we did not explicitly mention flexibility in the decision-making process when we first came up with this approach, the graph evoked a critical response. Therefore, several contributions in this book intend to clarify the flexible interrelationship between the three thresholds. For instance, it could well be that the crossing of the locational threshold is the first step in decision-making. Such is the case in family reunions when the foreign partners and children emigrate to join their spouses or parents, or when someone living abroad finds employment for a family member or friend. In these cases, the destinations are given and the mental threshold follows the locational one. It could also be that the migrant actually begins the journey without a precise destination. Moreover, even after the migrant has decided to move, he or she may need to rely on an intermediary or agent to facilitate the journey and the latter will ultimately decide on the location and route as well (the spatial trajectory might determine the final destination of the migrant).

The three thresholds are relatively dependent on one another and are not always distinguishable. Smith (Chapter 3) states that '[t]he locational threshold ... conceives of migration as essentially multidimensional. It takes as its point of departure that the perception of the end location and insight into the trajectory that needs to be followed need to be considered together and not in isolation to understand how a decision to set off is made' (p. 32 in this volume).

Figure 1.1 might also suggest that it is a 'once and for all' decision to cross a threshold. This could indeed be the case when a tourist orders a packaged tour to a country or when a labour migrant leaves for the place where he or she has been granted a working permit. However, several authors in this book have argued the contrary, namely that the decision to move is a continuous process of perceiving and weighing risks (costs) and opportunities (benefits) before and during the journey to the destination. Fàbrega and Lim (Chapter 8) state that factors that impact on the mental threshold by respectively lowering difference and indifference on both sides of the border are simultaneously present and are constantly being renegotiated on the ground. Sandberg and Pijpers in their study of Polish circular migrants (Chapter 13) have similarly highlighted that 'the rhythm of circular migration is not determined by notions of closure and/or loopholes only; it depends on contingently opening and closing windows of opportunity as well' (p. 196 in this volume).

Leung and Schapendonk broaden this idea to the other thresholds. For Leung (Chapter 5), the route is always in the making and there is no definite destination to speak of. She argues that 'the strength of the trajectory threshold concept lies in its recognition and emphasis on the dynamic, often unpredictable and extremely contextualized nature of the mobility process' (p. 63 in this volume). She henceforth pleas for more effort in integrating multiple spatialities and temporalities into the trajectory threshold thinking. Schapendonk (Chapter 16), when discussing the trajectory threshold of sub-Saharan migrants and Istanbul as a place of transit, proposes that we should understand trajectories in a similar way as relational geographers (amongst others Massey 2005) understand places, always to be 'in the making, always evolving and always being affected by "a set of thrown-together factors"' (p. 246 in this volume). We can partially operationalize the rather abstract statements by distinguishing between the deliberation phase in the decision-making process (in our mind) and its implementation phase (the actual process of stay and movement). During the latter phase, there may be circumstances and reasons to start the deliberation process again and this cycle could go on and on.

The contextualized nature of the mobility process is indeed proven by the contributions to this book. As with the inherent dynamic nature of the process, however, this is debatable since there can be as much continuity as dynamism involved. Also, contrary to the view that a continuous flow of decisions exists with regards to the thresholds, a migrant may not always be in the position to decide. In many cases, migrants do not have a choice but to rely on intermediaries who will preselect the route for him/her. Hugo and Napitupulu (Chapter 15) clearly suggest that the trajectory threshold does not play a significant role in the decision-making of the majority of the seashore asylum seekers in Australia because the real decision-making power lies in the broker and/or the migration industrial chain these asylum seekers have chosen, trusted and paid for. Ferrer Gallardo and Espiñeira (Chapter 17), referring to the new 'territorial idiosyncrasy' of Ceuta, illuminate the proliferation of confinement and encampment practices across the EU, with the enforced immobility dynamics on the migrants as a result. In the case of repatriated returnees or deportees, there is no mental space to choose between options at all especially since the locational and trajectory thresholds tend to be determined by others (Nyberg Sørensen in Chapter 11). We would like to add that this is also the case with the 16.7 million 'external' refugees at the end of 2013 (UNHCR 2014).

Between the two extremes of constantly making choices and not being able to make any because others hold the decision-making power (both concern a considerable number of people worldwide), lays the largest group of mobile people who can plan their destinations and routes autonomously. Moreover, decision-making in mobility and migration is not a completely rational process but it is also not a totally spontaneous or impromptu one.

'Bordering' and 'Borderwork'

Personal, subjective realities provide the real driving force in considering mobility and migration. Trajectories are meant to bridge the distance between places (localities, regions, countries) of origin and places of destination but the distance should not be measured in terms of physical space only. It is also a mental construct since it equally involves motives to stay or go, such as potential migrants' perception of the labour market that can act as a strong or weak pull factor. Borders are both real and imagined. Border management rules, regulations and physical measures prevent people's mobility but so can people's perception of an impenetrable border.

In examining EU border policies, Baggio (Chapter 12) distinguishes three kinds of borders: national borders, internalized, mental borders and externalized, territorial borders. In particular, the internalized borders as a result of negative perceptions or outright xenophobia of the native population against migrants and foreigners, can act as a repellent that prevent potential migrants from crossing the locational threshold. We term this the strengthening of the unfamiliar.

Sandberg and Pijpers (Chapter 13) propose to include the concept of 'borderwork' by Rumford (2006, 2008) in mobility decision-making. His basic idea is that 'bordering' is not an exclusive activity of nation-states because anyone or any social group can create borders. In this way, borders can become spatially displaced and dispersed within societies, for example, by creating gated communities. More explicitly Sandberg and Pijpers argue that processes of 'bordering' are negotiated within the realm of everyday life practices. This is illustrated by their case studies of circular labour migrants. This matches our view of 'bordering' practices as mental processes that involve factors such as 'difference' and 'unfamiliarity' and Baggio's concept of internalized borders.

Social Groups as Mobile Actors

The various chapters in this book have unanimously highlighted the fact that the meaning of thresholds differs according to the social groups an individual belongs to. In Chapter 4, Jinnah explains how thresholds and 'bordering' differ among different social groups. While individual motivations and decisions constitute the driving force behind cross-border mobility, individual and public resources (economic, social, mental and so on) create thresholds for it. She explains that the locational and trajectory thresholds tend to be more relevant to some people than to others because the amount of financial and other resources that can be mobilized during the migration project determines, to some extent, the strength of each threshold.

Her observation confirms the importance of class, ethnicity and gender in cross-border mobility, which is also our position as we wrote that we 'are inspired by the approaches that pay attention to macro structures, such as the political economy of the world market and international relations, and to the micro structures as well, such as social networks and differences between class, ethnicity and gender' (Van der Velde and Van Naerssen 2011, 220). In a survey assessing the level of interest in a regular passenger train connection between two cities on both sides of the Finnish–Russian border, Izotov and Soinninen (Chapter 7) found distinctly different mobility patterns between the social groups surveyed. The differences in age, occupation, gender and income explain the differences in thresholds and 'bordering' among the various social groups.

In the global network, the spaces of flows and places tend to favour a globalized elite who possesses the financial resources to travel and enjoy the consumption of the symbols of their wealth and power worldwide (Castells 2000, 440ff.; Favell 2008). If mobility has become a class-stratified phenomenon (Bauman 1998, 12), the reverse is also true. Or, in stronger terms, 'Mobility has become the most powerful and most coveted stratifying factor' (90). Leung illustrates this in Chapter 5 with the mobility of academic university employees who attend workshops, conferences and summer schools in other countries. And indeed, 'Even foreign universities are not foreign. After one delivers a lecture, one can expect the same questions in Singapore, Tokyo, Paris or Manchester' (Rundell 2011, 205).

Van Naerssen and Asis (Chapter 6) centre their contribution on the relation between gender and thresholds and argue that, since we all bear different identities and whilst gender matters in border crossings, the intersection of gender with ethnicity and class should not be overlooked.

If the approach is to be policy and politically relevant, it has to be refined and incorporate the diversity of gendered migration. Generally speaking, people play different roles and therefore, possess various identities that influence decisions on why, when and how we cross borders. And vice versa, borders can define our role and identity as locals or foreigners, as welcome or unwelcome guests and/or as intruders or trouble-makers – as in the case of 'foreign hooligans' in football games.

The Changing Thresholds: The Micro and Macro Context

Reflecting on the mental thresholds in the German–Polish border region, Szytniewski (Chapter 2) shows that images of the 'otherness' and 'familiarity' are subject to continuous and changing interpretations, as a result of new and changing border-crossing experiences such as new mobility practices, changing travel restrictions and border policies, new migratory experiences and so on. Hence, she concludes that 'the present form of the mental border threshold seems to be regarded as rather static, whereas images of otherness are subject to continuous interpretation processes as a result of different perspectives and practices, direct and indirect experiences and changes in obtained and assessed knowledge, which are always open to reinterpretation' (p. 19 in this volume). Her argument could be applied to the other thresholds as well.

Perception and subjective meanings of the cross-border mobility of an individual can be decisive in the way thresholds are defined. However, since these can change, so can the height (that is, the level of difficulties) of thresholds. In Chapter 3, Smith argues that the narratives of the young Ghanaians he interviewed, highlight their motivations in crossing borders but also the vagueness in their perception of their destinations. He concludes that bounded rationality not only pertains to what is actually known, but also to what is considered relevant from moment to moment. This relevance is determined by personal circumstances and since these are open to changes, by consequence, the 'migration horizon continually shifts' (p. 39 in this volume; see also Chapter 13 in this volume, dealing with Polish circular migrants).

This, in turn, depends on the changing environment and its impact on several elements, including the individuals' choice whether to be mobile, the desired destination and the way to reach it. As Jinnah explains in Chapter 4, on the *micro level*, the environment depends on the immediate social and economic conditions in which the individual lives. On the macro level, it depends on the society in a broader sense, which he or she has less control over. Contrary to the study on Ghana, Jinnah in Chapter 4 pays explicit attention to the transnational environment of the migrant and structural (external) factors that impact on his or her migration decision-making. However, she also points out that personal circumstances and quantifiable factors, such as financial, human and social capital, do play a role. As was mentioned earlier, the locational and trajectory thresholds are of different relevance to different people because of the different availability of financial and other resources during the migration project.

Leung, in her contribution entitled 'Thresholds in academic mobility' (Chapter 5), refers to the importance of households and families in migrant decision-making as explained in the New Economies of Labour Migration (NELM), but she also argues that in the realm of academic mobility, it is imperative to expand beyond the conventional categories employed in NELM to include, for instance, academic collectives. '[O]ne can also observe the need to rescale academic staff mobility as a strategic act performed within a large social unit, namely the mobile academic's research group, department and university' (p. 61 in this volume). Thus the locational threshold is shaped by multiple actors, institutions and macrostructures, as well as coincidence:

'The interplay between the respective influences is highly contextual in *any* specific mobility trajectory' (p. 63 in this volume; emphasis added by the editors).

Changes at the *macro level* concern socio-economic and institutional changes at the national or larger regional level. More in general, thresholds dynamics depend on migration policy frameworks, labour market developments and social structures and processes of the receiving communities. Changes in these contexts impact on the locational threshold and will enable mobility and transnational activities, but may also result in immobility. In this book, the impact on mobility and especially the mental threshold is most clearly demonstrated in the contributions that deal with border regions. Next to the opening of the German–Polish border, to which we already referred to, these concern the relaxation of the Finnish–Russian one (Chapter 7) and the border between Cambodia and Thailand (Chapter 8). In both cases, mobility has increased, although the impact differs substantially. While it is still limited in Chapter 7 it did lead to greater mobility and economic activities in Chapter 8. Due to the events of 9/11 and the war on terror, an opposite change has taken place at the Canadian-US border (Chapter 9). Here, as Konrad argues, 'The mental border threshold has emerged from relative insignificant in the migrant decision-making process, to loom as a primary consideration and substantial initial barrier in short-term crossing' (p. 132 in this volume).

An important change took place at the Israel-Palestine border when Israel built a wall in an effort to prevent attacks. This is the subject of Elnakhala's contribution (Chapter 10). Her insights add to the threshold approach in a number of ways. First, it deals with the new topic of decision-making by militants as border crossers. Second, it explains how hardened and fortified borders can lead to the use of new tactics by militants who wish to cross borders. It therefore sheds light on the effect of the interaction between fortification and militant border crossings. Third, where the threshold approach considers trajectories as geographical routes followed by border crossers, this chapter posits that the physical movements of people are part of the more general and overarching concept of the tactics of border crossings: militant border crossers may choose to switch the type of their attacks in order to achieve their goals. Just like migrants can change their destinations and routes, militants can change border-crossing tactics and their destinations.

How changes at the macro level affect the meaning of the locational border for return migrants is the subject of two other contributions in this book. The first concerns Bolivian migrants in Spain and what Nijenhuis in Chapter 14 calls, 'the changing contexts of exit and reception' (p. 199 in this volume) of the migration corridor. The context of exit concerns the motivations and aspirations of the migrant, the economic conditions in the area of origin and the attitude of the sending country towards migrants. Factors of the context of reception are the societal response, the labour market and the opportunities for integration in the receiving country. For migrants in Spain who lost their jobs, the main option is to cross the border again and return. In other words, the economic crisis has changed the locational threshold for Bolivian migrants. Nyberg Sørensen (Chapter 11) presents another perspective on the change of the locational threshold of (forced) returnees. Being forced to move back from the original destination, the locational threshold of returnees may change from being a poor but secure home community to becoming an even poorer and also insecure location. The familiar could become unfamiliar. From a threshold perspective, we should thus explore the conditions people are returning to and what prevents them from staying safely in their home country. A joint outcome for the individuals, families and communities affected is their increased exposure to violence and insecurity. The difference between the two contributions do not only concern the type of returnees but also the contextual level: macro in the case of Nijenhuis, micro in the case of Nyberg Sørensen.

Refining and Expanding the Threshold Approach

This book aims to consider whether the threshold approach can be applied to other cross-border human mobilities rather than to labour migration only. The answer is positive, provided certain conditions are met. There are reasons to refine and expand – in the sense of further explanation and clarification – the original threshold approach and in this way, the contributions in this book undoubtedly enrich the approach.

Some contributors suggest that the concept of 'border' should be broadened and loosened from its territorial connotation. This is in fact what we do when using ideas, such as 'mental border', 'familiarity' and 'bordering'. Internalized borders, transnational social networks and 'borderwork' by ordinary people and their organizations are integral parts of the approach and they should be studied in relation to the 'keep' and 'repel' factors impacting on cross-border movements. However, while we acknowledge the impact of internalized borders and daily 'bordering practices' on mobility decision-making, the effects of the territorial ones, such as securitization of the borders should not be forgotten.

Another clarification concerns the flexibility of the approach. To explain mobility by using the three thresholds as moments in the decision-making process is feasible and valid, provided their sequence in the use of the approach is not fixed (as might be suggested by Figure 1.1) and concerns an iterative process. That is to say, during the mobility/migration journey, people can change the destination and the trajectory they had in mind, pending individual and contextual changes. In our approach, we emphasize the process of decision-making as a *flow* of a number of decisions that are to be made, which intersect or straddle across at least three thresholds. This stream in time is relatively easy to follow if the decision-making involves a rather straightforward process while in other cases it is more complicated to study for several reasons, such as the need for continuous decision-making when many borders have to be crossed. The iterative process involves both the mental process of passing the three thresholds as well as the actual location and spatial movement of the decision-making individual. In the amended graphical representation of the approach (Figure 18.1) this is expressed first of all by a division of the 'mobility arena' in a lower half that represents the mental process as it concerns the three thresholds, and a higher part representing the actual stay and movements in the material world. Secondly the arrows indicate mutual influences and indirect feedback mechanisms in the amended graphical representation of the approach (see the central rectangle of the mobility arena in Figure 18.1).

A third clarification concerns the mutual relationship between the individual decision-making actor and his/her life environment consisting of the social, economic and physical milieus. It seems that some contributors interpret the approach as based on individual choices only but this is not our intention if only because we are conscious of the fact that no individual acts in a vacuum. Reading the foregoing chapters, however, makes it clear that there is certain consensus among the contributors that the threshold approach should incorporate more explicit contextual factors at the micro and macro level that impact on the perception and meaning attached to the three thresholds by the individual decision-maker.

We may conceive this in the following way and expand the original graph. Besides analysing the *mental flow*, we should also look at the decision-making *field*, taking into account all factors that impact on the decision taken at a given moment in time – for example the economic situation in a region, the strength of the social network, the composition of the household and so on. Since the field in which the decision-making actor acts is an all encompassing, versatile and complex one, this coincides with what Hoerder (2002, 20) calls,

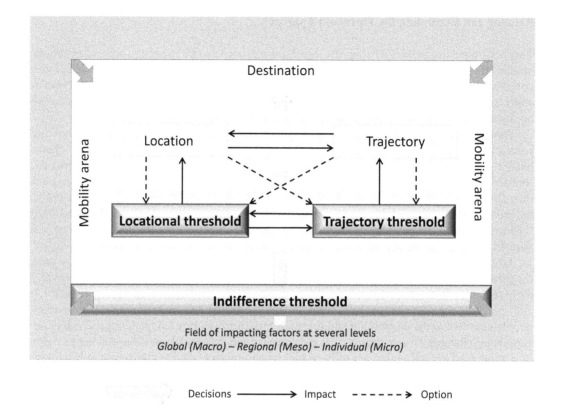

Figure 18.1   Field of factors, thresholds and decision-making in spatial mobility
Source: the authors.

a 'holistic material-emotional approach' to migration. The analysis of macro, meso and micro level (personal qualities and resources) impact factors and their differences, can facilitate the analysis.

Thereby, the greatest challenge will be to incorporate the changes in the environment of the decision-maker and the associated 'dynamics of mobility and immobility' in the decision-making process and in our approach. Or, since the factors in the field change in time and impact on the individual decision-making process, for a full analysis we have to study both the flow and the field in mobility decision-making. In figure 18.1 this is graphically portrayed by a second rectangle surrounding the mobility arena, the displaying the field of factors. The thick arrows at the four corners represent the permanent impact during the journey.

The central position of the mobility arena and the graphic accentuation of the thresholds indicate that the mental flow of decisions remains the very purpose of our approach. This is in our opinion still a neglected theme in mobility and migration decision-making. However, the expanded threshold approach incorporating real movements in space and the field of impacting factors will hopefully open windows to new empirical inquiries. Ultimately, we also think that the threshold approach contributes to an increased understanding of the geography of mobility around the globe, both in the academic sense as well as in a policy context.

References

Bauman, Z. 1998. *Globalization: The Human Consequences.* Cambridge: Polity Press.

Castells, M. 2000. *The Rise of the Network Society: The Information Age; Economy, Society and Culture.* 2nd ed. Vol. 1. Oxford: Blackwell.

Favell, A. 2008. *Eurostars and Eurocities: Free Movement and Mobility in an Integrating Europe.* Malden: Wiley-Blackwell.

Rundell, J., ed. 2011. *Aesthetics and Modernity: Essays by Agnes Heller.* Plymouth: Lexington Books.

Hoerder, D. 2002. *Cultures in Contact: World Migrations in the Second Millennium.* Durham, NC: Duke University Press.

Massey, D. 2005. *For Space.* London: Sage.

MKW Wirtschaftsforschung. 2009. *Scientific Report on the Mobility of Cross-Border Workers within the EU-27/ EEA/EFTA Countries.* Munich: MKW.

Rumford, C. 2006. "Theorizing Borders." *European Journal of Social Theory* 9 (2): 155–69.

———. 2008. "Introduction: Citizens and Borderwork in Europe." *Space and Polity* 12 (1): 1–12.

Salazar, N.B. 2011. "The Power of Imagination in Transnational Mobilities." *Identities: Global Studies in Culture and Power* 18 (6): 576–98.

UN (United Nations). 2013. "Trends in International Migrant Stock: The 2013 Revision." Accessed 8 July 2014. http://esa.un.org/unmigration/TIMSA2013/migrantstocks2013.htm.

UNHCR (United Nations High Commissioner for Refugees). 2014. "Facts and Figures about Refugees." Accessed 25 August 2014. http://www.unhcr.org.uk?about-us/key-facts-and-figures.

UNWTO (United Nations World Tourism Organization). 2014. "UNWTO World Tourism Barometer." Accessed 8 July 2014. http://mkt.unwto.org/en/barometer.

Van der Velde, M., and T. van Naerssen. 2011. "People, Borders, Trajectories: An Approach to Cross-Border Mobility and Immobility in and to the European Union." *Area* 43 (2): 218–24.

Chapter 19

Borders as Resourceful Thresholds

Henk van Houtum

It is a privilege to have been asked to write a short afterword in this inspiring collection of articles. The question was if I could look back at the historical career and further prospect of the central concept of the book, the threshold approach, and embed this concept in the wider debate on border studies. Well then, let me begin by complimenting the editors to have followed, even meticulously, a central conceptual structure, on the basis of which and around which the individual or set of authors have written their contribution. By choosing a central conceptual framework this edited volume has most certainly gained cohesion and academic parallelism in such a way that the contributions interestingly feed into and complement each other.

For the basis of this conceptual framework on borders as thresholds of indifference an extended version of the work is used that Martin van der Velde, one of the two editors of this book, and myself have written in an article in 2004. It is good to explain the background of this article, as it says something about the past track the concept of thresholds has travelled. At the time, we wanted to focus on explaining the labour migration and mobility patterns in the EU. Many labour market studies deal with mobility. But because the dominant mode of practice of about 97 per cent of the workers is cross-border immobility, not mobility, we realized that in order to explain mobility patterns in the EU, we needed to start from the other side, in order to get a much better academic grip on the concept of immobility. Despite the many efforts of the EU to bring about a homogeneous labour market the dominant pattern was and remains to be, labour market *im*mobility. That is, there may be an increasingly barrier-low internal market for capital, information and services, but for labour, the actual moving house for a job in another member state is, excluding the outlying case of Luxembourg, relatively small. On average somewhere around three per cent or less. What most academics has been puzzling is why relatively so few people move across the border, even in cases where it would make perfect sense, in strictly economic terms. We found that most frameworks trying to explain the mobility of labour were based on (adapted versions of neoclassical) rational decision models in which the structural difference between the foreign and domestic labour market was put central. The general idea of most of these models was that if the profitable difference in pull and push factors between foreign and domestic market would be high enough, people would go. Yet, most models were not able to explain the relative persistence of people to not move even in cases of enlarged or enlarging varying welfare differences in the EU. Apparently, and not surprisingly, the mobility of goods and information across borders is a whole different field than the moving of people themselves with their feelings, emotions, identities and behaviours. So, our main aim was to better understand why most people most of the time did not even consider the possibility of moving across the border. To this end, what we in fact particularly did in that article of 2004 was to enrich the concept of the border in these models (Van Houtum and Van der Velde 2004). We found that in most of these rational decision models, borders are dominantly seen as a mere cut-off line, a threshold of a potentially profitable difference between diverging labour markets. The decision-making process in these models is often imaginatively based on an evaluation of the characteristics and opportunities of the present (*home*) and a possible new location (*away*), after which a decision is made to become mobile (*go*) or stay put (*stay*). This Cartesian worldview of human action, which has found its present translation in mainstream economics in the (bounded) rational agent, still motivates EU labour market policymaking. We enriched this meagre and all-too naive understanding of borders with existing debates in border studies at the time, in which national borders should not so much be understood (anymore) as economistic differences alone but as socially meaningful constructions of power to a varying degree internalized by the people living within the territorial outlines of the country. Borders were hence not seen as static lines of differences but rather as continuously reproduced social phenomena, as human all too human (Paasi 1996). We postulated that the bordering of our mental orientation and territorial (id)entity is preventing the existence of a large-scale cross-border or transnational labour market in the European Union. Despite many years of European integration, the national border still produces a difference in the imagination of belonging and as such it produces an attitude of *indifference* towards the market on what is perceived as the 'other side', as an abroad, an out-land (Van Houtum and Van der Velde 2004). Apparently still, this mechanism of distanciation helps to gain control in order to gain a social focal point, a selection of social priorities. The space beyond a state then becomes a space of withdrawal, of mental 'emptiness', often resulting in a conservative tendency towards cross-border activities. That what is beyond the constructed differentiating border of comfort (*difference*) is socially made legitimate to be neglected (*indifference*) (Van Houtum and Van der Velde 2004). Interestingly enough, migrants from outside the EU are perhaps more 'European' than those born inside the EU, in the sense of deliberately moving across borders into the EU, which is something Joris Schapendonk, for instance, has recently taken up (see also his Chapter 16 in this volume). For many EU citizens, the market across the border is apparently something that is still seen as

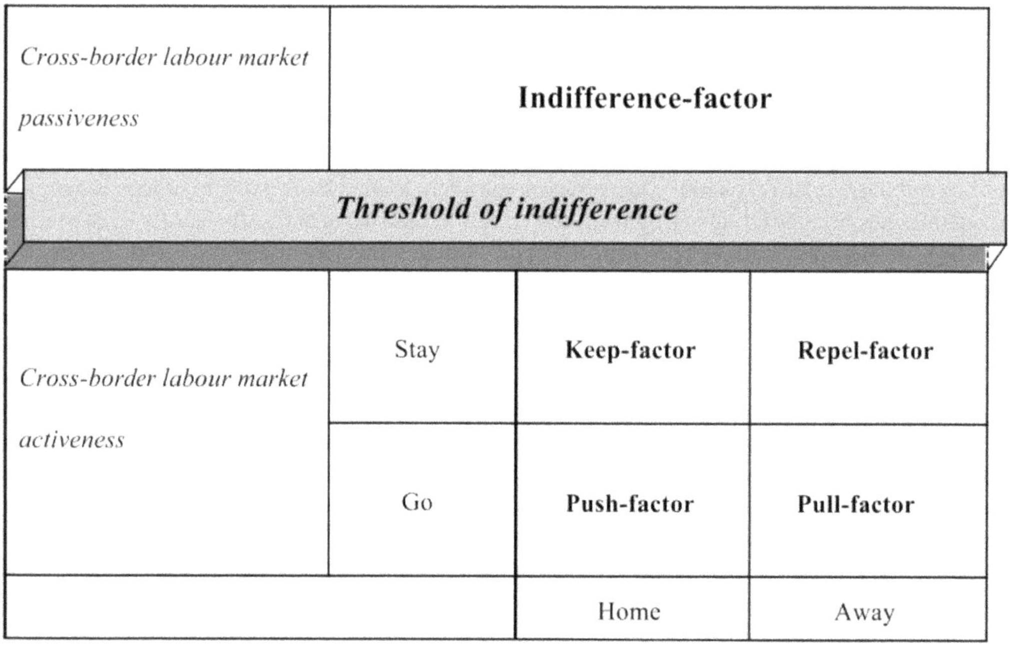

Cross-border labour market passiveness	Indifference-factor		
Threshold of indifference			
Cross-border labour market activeness	Stay	**Keep-factor**	**Repel-factor**
	Go	**Push-factor**	**Pull-factor**
		Home	Away

Figure 19.1   The threshold of indifference
Source: Van Houtum and Van der Velde 2004.

something that can be neglected for the daily social practices in the own country. We suggested that it is the inclusion of the attitude of such a nationally habitualized *indifference* that may help to explain why most workers do not even consider seeking work across the border. The reality is thus that the majority of workers does not surpass the threshold of indifference – only a small group will 'enter' the bottom part of the scheme, the active attitude part in which cross-border mobility is taken into full consideration. This resulted in the model above.

Earlier, Ton van Naerssen and I had reframed the idea of borders as social constructs into the idea that borders should not thus be understood as nouns, as finite, but rather as continuous work in progress, hence as verbs (Van Houtum and Van Naerssen 2002; Van Houtum 2010a, 2010b). To this end, we coined the term *bordering*. And to make clear that with the continuous process of border-making there is a continuous process of ordering a society and antagonizing another, we used the terms 'bordering', 'ordering' and 'othering' (see also Van Houtum 2010b). Newman (2006) picked this up in his article on the distinction between the borders and the bordering approach. Our bordering approach was later used in the edited volume *B/Ordering Space* (Van Houtum et al. 2005). In this volume, the American critical theorist Hooper, at the time working for the Nijmegen Centre for Border Research, introduced the term border-work to describe this social making of borders, something that was later elaborated and extended by Rumfold (2008) in his interesting book on how citizens make and unmake borders in Europe (see also the chapter by Sandberg and Pijpers in this volume).

It is most importantly the achievement of Martin van der Velde, Bas Spierings and Ton van Naerssen to steadily have further expanded, sharpened and tested the principle of the threshold of indifference that was coined in 2004. In their articles of 2008, 2011 and 2013 Martin worked together with Bas to refine and extend the idea of the threshold of indifference into the concept of *unfamiliarity* when it concerned cross-border shopping (Spierings and Van der Velde 2008, 2013; Van der Velde and Spierings 2010). They empirically tested the approach and found out that in order for cross-border shopping to occur, a certain 'comfortable unfamiliarity' is quintessential, as it provokes a certain adventurous curiosity that is commercially beneficial. They stated that in order to promote cross-border attention and interaction – or increase international shopping mobility – and prevent cross-border aversion and avoidance – or increase international shopping immobility – European urban and regional policies should aim to help to produce processes of productive (un) familiarization (Spierings and Van der Velde 2013). Borders as markers of differences between countries are a necessity for people to become mobile and visit 'the other side'. This interesting conclusion resonates with the work that recently has come up in border studies, summed under the terms *borderscapes* or *borderscaping*, a fusion of 'borders' with 'scaping'. The latter comes from the Dutch term *scheppen* and the German word *schaffen*, which means to create or design, in which consciously is sought to use the difference a border as productive seducer makes (Eker and Van Houtum 2013; Van Houtum and Spierings 2012; Brambilla 2014; Sohn 2013; Buoli 2015). This view has important new consequences for policies in borderlands, also in the EU where policies have for decades now largely been aiming at obliterating borders that were only seen as barriers and not resources.

Martin also worked with Ton to apply the threshold approach that was originally applied for migration inside the EU towards migration towards the EU (Van der Velde and Van Naerssen 2011). And that is a relevant extension of the approach, because also outside the EU, where welfare differences with members states in the EU are higher, the dominant pattern is actually sedentarism, non-migration. Where tourism has exponentially expanded over the recent decades, moving across the border to actually stay is still the exception rather than the rule (see also the chapter of Van der Velde and Van Naerssen in this volume). Worldwide the percentage of migrants is, according to the UN (2011), about 3 per cent. There is a persistent idea and even fear of mass migration or even invasion into the EU but this is not backed by the actual figures. Again therefore, as in the case of cross-border mobility and shopping in the EU, the perception of international human mobility in public media and political arenas is often heavily overestimated and overstressed compared with the reality, that is, the dominance of immobility. For the case of international migration outside the EU, Van der Velde and Van Naerssen redefined the threshold of indifference as the *mental border threshold*, which I understand from the perspective of wishing to stimulate mobility, because it emphasizes the border as barrier, a mental distance (see also Van Houtum 1999), but arguably is conceptually fitting less well when one wishes to explain sedentarism from the perspective of those who wish to stay. Those who willingly wish to stay put and do not even consider moving abroad generally arguably would not speak of a mental border, but of indifference. To this mental border threshold they interestingly added the *locational threshold* (whether or not the migrant can find a suitable location) and *trajectory threshold* (whether or not he or she reaches his or her destination because of the difficulties encountered during the journey) (see also Figure 1.1 in this volume).

So, ten years after our coining of the concept of the threshold of indifference that aimed to explain cross-border labour market immobility (Van Houtum and Van der Velde 2004), a whole series of publications have followed contributing to an increased richness of the 'Nijmegen' approach. The concept has travelled and evolved. By persistently expanding the approach to other fields and by inviting other scholars to help build if further, Martin and others have sharpened and refined the debate on borders, (labour) migration and sedentarism with the approach.

This book provides an important new step in this debate. In this book Martin and Ton have invited a range of distinguished scholars to reflect on the approach from their own research expertise and field, and in diverse borders sites around the world. Some have stayed close to the approach as offered in Chapter 1 by Van der Velde and Van Naerssen, others have dealt with it more loosely. This has rendered many new insights and deliberations, resulting in possible new directions of the approach as summed in, the still perhaps rather complex, Figure 18.1. It has been made convincingly clear in the various chapters that two possible enrichments of the approach lie in the continuous changing and shifting of the interpretation of the threshold over time and space, as well as in the inclusion of social groups an individual belongs to as this may lead to different perceptions of the border. And it has made clear that the approach obviously also has limits. In the interesting chapter on the Israel-Palestine border of Doaa' Elnakhala who addresses the border tactics in a rather different manner, namely in terms of hostility and violence, the approach is clearly less well applicable. And as Xavier Ferrer-Gallardo and Keina Espiñeira and Ninna Nyberg Sørensen have argued in their chapters, the approach certainly differs in terms of applicability to various cases of migration. For refugees and especially clearly for forced returnees, the idea of being able to choose, even to some extent, is quite different than for cross-border workers in an inner-EU borderland. It shows that the approach that begun with the explanation of the low percentage of cross-border workers has again travelled and is meeting new limits and new challenges, at it should.

It does not stop here, I am sure. Adding to the comments made by the editors in their concluding chapter, I would argue that an important challenge for the future of the approach is to better combine the idea of a border as a threshold and a border as a seduction or resource. When we for example zoom in on the inner borders of the EU, for a long time in the integration process of the EU the national borders were interpreted as a threshold in terms of a barrier, something that needed to get rid of. This led to a kind of internal market score sheet logic, in which the attention is solely focused on the speed of deleting the borders as barriers. It is only in recent times that spatial developers and planners are really and more willingly coming to terms with the reality that for the near future the internal borders of the EU will not disappear. As has been argued above, the cross-border (labour market) immobility is not an anomaly, but the dominant pattern. In spatial development terms this should not be seen only as pitiful, but as a critical potential to make much better use of. In the book *Borderland* (Eker and Van Houtum 2013) we tried to work with possible scenarios for this, which could possibly further enrich the unfamiliarity approach of Spierings and Van der Velde (see above). We made a design continuum of the border, ranging from the barrier as a seductive and resourceful difference marker on the one hand (the scenario of the border as a resourceful threshold), to the border being totally open, thereby turning the border into a historical relict of a time that once was on the other (the scenario cross-border community). This leads to a range of possible designs for border landscapes. Maybe it is worthwhile to see if such a continuum rather than an either/or could be applied to the case of migration and mobility. Rather than trying to create a communality across borders and breaking down all possible barriers, it might sometimes be much more inspiring and innovative to emphasize the productive difference between borderlands and markets on either side of the border. Seductive differences are much more appealing than forcing markets or people to integrate. The same perhaps holds for the external borders of the EU. Here one sees a complete opposite logic when compared to the inner borders, namely a harsh and discriminative closure for many migrants, leading to an increasing number of unnecessary casualties (Van Houtum 2010a; see also Baggio's chapter in this volume). Where the inner border logic surely is ready for further refinement and more creative redesign, arguably more towards productive and seductive closure, the external border clearly needs more creative openness. In this way, the policy logics towards the internal and external borders which are dominantly and falsely kept separate, and which also in border studies are wrongly kept as separate debates, should and can be connected.

This connects to my final point. The lesson of this book for me is that after ten years of enriching and expanding the 'Nijmegen' approach on 'borders as thresholds', the debate should arguably step out of the academic circle more often and become more policy relevant, something which the editors clearly recognize themselves, as they conclude this in the last chapter. Zooming in on the EU again, too often, as expressed above, in policy regarding borders and borderlands there has been an emphasis on stimulating mobility/migration inside the EU, thereby almost obstinately ignoring the reality on the ground, which is sedentarism. And when it concerns external borders, too long now, has there been an one-sided policy focus on stopping migrants, an inflexibility that is erroneous, even deadly, which needs to be broken.

I am glad that I can conclude that many new challenges lie ahead. That means that the book has done a good job indeed, as a good debate usually leads to new questions. It is for this reason that I compliment the editors and authors with a compelling contribution to the rich and blossoming field of border studies. I look forward to the road ahead.

References

Brambilla, C. 2014. "Exploring the Critical Potential of the Borderscapes Concept." *Geopolitics* 1–21.

Brambilla, C., and H. van Houtum. 2012. "The Art of Being a 'Grenzgänger' in the Borderscapes of Berlin." *Agora* 28 (4): 28–31.

Buoli, A., ed. 2015. "Border/Scapes: Borderlands Studies and Urban Studies in Dialogue." *Territorio: The Journal of Urbanism.*

Eker, M., and H. van Houtum. 2013. *Borderland: Atlas, Essays and Design; History and Future of the Border Landscape.* Blauwdruk: Wageningen.

Newman, D. 2006. "Borders and Bordering: Towards an Interdisciplinary Dialogue." *European Journal of Social Theory* 9 (2): 171–86.

Paasi, A. 1996. *Territories, Boundaries, and Consciousness: The Changing Geographies of the Finnish–Russian Boundary*. Chichester: J. Wiley & Sons.

Rumford, C. 2008. *Citizens and Borderwork in Contemporary Europe*. London: Routledge.

Sohn, C. 2013. "The Border as a Resource in the Global Urban Space: A Contribution to the Cross-Border Metropolis Hypothesis." *International Journal of Urban and Regional Research* 38 (5): 1697–711.

Spierings, B., and M. van der Velde. 2008. "Shopping, Borders and Unfamiliarity: Consumer Mobility in Europe." *Tijdschrift voor economische en sociale geografie* 99 (4): 497–505.

Spierings, B., and M. van der Velde. 2013. "Cross-Border Differences and Unfamiliarity: Shopping Mobility in the Dutch-German Rhine-Waal Euroregion." *European Planning Studies* 21 (1): 5–23.

Van der Velde, M., and B. Spierings. 2010. "Consumer Mobility and the Communication of Difference: Reflecting on Cross-Border Shopping Practices and Experiences in the Dutch-German Borderland." *Journal of Borderland Studies* 25 (3–4): 195–210.

Van der Velde, M., and T. van Naerssen. 2011. "People, Borders, Trajectories: An Approach to Cross-Border Mobility and Immobility in and to the European Union." *Area* 43 (2): 218–24.

Van Houtum, H. 1999. "Internationalisation and Mental Borders." *Tijdschrift voor economische en sociale geografie* 90 (3): 329–35.

——. 2010a. "Human Blacklisting: The Global Apartheid of the EU's External Border Regime." *Environment and Planning D: Society and Space* 28 (6): 957–76.

——. 2010b. "Waiting before the Law: Kafka on the Border." *Social and Legal Studies* 19 (3): 285–97.

Van Houtum, H., O. Kramsch, and W. Zierhofer, eds. 2005. *B/Ordering Space*. Aldershot: Ashgate.

Van Houtum, H., and B. Spierings. 2012. "De grens als begin" [The border as a beginning]. *Agora* 28 (4): 4–5.

Van Houtum, H., and M. van der Velde. 2004. "The Power of Cross-Border Labour Market Immobility." *Tijdschrift voor economische en sociale geografie* 95 (1): 100–107.

Van Houtum, H., and T. van Naerssen. 2002. "Bordering, Ordering and Othering." *Tijdschrift voor economische en sociale geografie* 93 (2): 125–36.

Index

For Product Safety Concerns and Information please contact our EU
representative GPSR@taylorandfrancis.com
Taylor & Francis Verlag GmbH, Kaufingerstraße 24, 80331 München, Germany

www.ingramcontent.com/pod-product-compliance
Ingram Content Group UK Ltd.
Pitfield, Milton Keynes, MK11 3LW, UK
UKHW051945210425
457613UK00034B/1223